AF358839

Physical and Mechanical Properties of Wood- and Bamboo-Based Materials

Physical and Mechanical Properties of Wood- and Bamboo-Based Materials

Editors

Xuehua Wang
Yan Wu

Basel • Beijing • Wuhan • Barcelona • Belgrade • Novi Sad • Cluj • Manchester

Editors
Xuehua Wang
Nanjing Forestry University
Nanjing
China

Yan Wu
Nanjing Forestry University
Nanjing
China

Editorial Office
MDPI AG
Grosspeteranlage 5
4052 Basel, Switzerland

This is a reprint of articles from the Special Issue published online in the open access journal *Forests* (ISSN 1999-4907) (available at: https://www.mdpi.com/journal/forests/special_issues/ZV9DSS4W4A).

For citation purposes, cite each article independently as indicated on the article page online and as indicated below:

Lastname, A.A.; Lastname, B.B. Article Title. *Journal Name* **Year**, *Volume Number*, Page Range.

ISBN 978-3-7258-2569-1 (Hbk)
ISBN 978-3-7258-2570-7 (PDF)
doi.org/10.3390/books978-3-7258-2570-7

Cover image courtesy of Xuehua Wang

Contents

About the Editors . . . . **vii**

Xuehua Wang
Topicalities in the Research and Application of Bamboo and Wood
Reprinted from: *Forests* **2024**, *15*, 1917, doi:10.3390/f15111917 . . . . **1**

Shiyu Cao, Jiagui Ji, Haowei Yin and Xuehua Wang
The Influence of Treatment Methods on Bending Mechanical Properties of Bamboo Strips
Reprinted from: *Forests* **2024**, *15*, 406, doi:10.3390/f15030406 . . . . **5**

Xiaomei Liao, Xuan Fang, Xin Gao, Songlin Yi and Yongdong Zhou
Effect of High-Intensity Microwave Treatment on Structural and Chemical Characteristics of
Chinese Fir
Reprinted from: *Forests* **2024**, *15*, 516, doi:10.3390/f15030516 . . . . **19**

Yifan Ma, Yu Luan, Lin Chen, Bin Huang, Xun Luo, Hu Miao and Changhua Fang
A Novel Bamboo–Wood Composite Utilizing High-Utilization, Easy-to-Manufacture Bamboo
Units: Optimization of Mechanical Properties and Bonding Performance
Reprinted from: *Forests* **2024**, *15*, 716, doi:10.3390/f15040716 . . . . **34**

Wenjia Liu, Ling Zhu, Anca Maria Varodi, Xinyou Liu and Jiufang Lv
The Effect of Wet and Dry Cycles on the Strength and the Surface Characteristics of Coromandel
Lacquer Coatings
Reprinted from: *Forests* **2024**, *15*, 770, doi:10.3390/f15050770 . . . . **51**

Yongyue Zhang, Jiayao Li, Yun Lu and Jiangtao Shi
Highly Mechanical Strength, Flexible and Stretchable Wood-Based Elastomers without
Chemical Cross-Linking
Reprinted from: *Forests* **2024**, *15*, 836, doi:10.3390/f15050836 . . . . **68**

Weiwei Yang, Wanrong Ma and Xinyou Liu
Evaluation of Deterioration Degree of Archaeological Wood from Luoyang Canal No. 1 Ancient
Ship
Reprinted from: *Forests* **2024**, *15*, 963, doi:10.3390/f15060963 . . . . **81**

Hongbo Li, Qipeng Zhu, Pengchen Lu, Xi Chen and Yu Xian
The Gradient Variation of Location Distribution, Cross-Section Area, and Mechanical Properties
of Moso Bamboo Vascular Bundles along the Radial Direction
Reprinted from: *Forests* **2024**, *15*, 1023, doi:10.3390/f15061023 . . . . **92**

Weiwei Yang, Wanrong Ma, Xinyou Liu and Wei Wang
Consolidation and Dehydration Effects of Mildly Degraded Wood from Luoyang Canal No. 1
Ancient Ship
Reprinted from: *Forests* **2024**, *15*, 1089, doi:10.3390/f15071089 . . . . **106**

Ying Li, Chunmiao Li, Xueyong Ren, Fuming Chen and Linbi Chen
Optimizing the Preparation Process of Bamboo Scrimber with Bamboo Waste Bio-Oil Phenolic
Resin Using Response Surface Methodology
Reprinted from: *Forests* **2024**, *15*, 1173, doi:10.3390/f15071173 . . . . **119**

Siyang Ji, Qunying Mou, Ting Li, Xiazhen Li, Zhiyong Cai and Xianjun Li
The Novel Applications of Bionic Design Based on the Natural Structural Characteristics of
Bamboo
Reprinted from: *Forests* **2024**, *15*, 1205, doi:10.3390/f15071205 . . . . **133**

Jianhua Lyu, Jialei Wang and Ming Chen
Effects of Heat Treatment on the Chemical Composition
and Microstructure of *Cupressus funebris* Endl. Wood
Reprinted from: *Forests* **2024**, *15*, 1370, doi:10.3390/f15081370 . . . . **148**

Youna Hua, Wei Wang, Jingying Gao, Ning Li and Zening Qu
A Study on the Effects of Vacuum, Nitrogen, and Air Heat Treatments on Single-Chain Cellulose
Based on a Molecular Dynamics Simulation
Reprinted from: *Forests* **2024**, *15*, 1613, doi:10.3390/f15091613 . . . . **161**

Xuehua Wang, Siyuan Yu, Shuotong Deng, Ru Xu, Qi Chen and Pingping Xu
Effect of Bamboo Nodes on the Mechanical Properties of *Phyllostachys iridescens*
Reprinted from: *Forests* **2024**, *15*, 1740, doi:10.3390/f15101740 . . . . **175**

About the Editors

Xuehua Wang

Xuehua Wang, Associate Professor at the College of Furnishings and Industrial Design, Nanjing Forestry University, majoring in Wood Science and Technology. Her main research areas are wood and bamboo properties, modification, and application in building and furniture. In this domain, Dr. Wang has led two projects funded by the National Natural Science Foundation of China, one project under the "14th Five-Year Plan" National Key R&D Program, one project within the strategic alliance for wood and bamboo industry technology innovation, and four school–enterprise cooperation initiatives. She has authored over 50 papers in professional journals, secured 17 invention and utility model patents, contributed to the development of six standards in the field, and published a textbook for university classes.

Yan Wu

Yan Wu, Doctoral Supervisor, stands as an exceptional young pillar teacher within the "Blue Project" in Jiangsu Province, China. Her expertise lies in the research and development of green coating technology, functional wood coatings, and innovative wood household materials. She has led and contributed to the successful completion of over 10 national, provincial, and ministerial projects, including the National Natural Science Foundation of China Youth Fund, the National Natural Science Foundation of China, the 948 Project of the State Forestry Administration, the Natural Science Foundation of Jiangsu Province, the China Postdoctoral Special Grant, and the First Class Grant. Wu Yan has been granted 13 national invention patents and has authored more than 100 academic papers, with more than 40 of them being indexed in SCI and EI. Additionally, she has published one monograph and collaborated on the compilation of a foreign language monograph.

Editorial

Topicalities in the Research and Application of Bamboo and Wood

Xuehua Wang

College of Furnishings and Industrial Design, Nanjing Forestry University, Nanjing 210037, China; wangxuehua@njfu.edu.cn

Citation: Wang, X. Topicalities in the Research and Application of Bamboo and Wood. *Forests* **2024**, *15*, 1917. https://doi.org/10.3390/f15111917

Received: 28 October 2024
Accepted: 30 October 2024
Published: 30 October 2024

Wood and bamboo are significant biomass materials with a broad historical and practical value in human civilization. In today's society, as the concept of sustainable development gains widespread acceptance, these natural resources are increasingly attracting attention. Due to their environmental sustainability, exceptional strength-to-weight ratio, commendable seismic resilience, esthetic appeal, and ease of processing, bamboo and wood occupy an irreplaceable position in architecture, furnishing, transportation, and other sectors. However, wood- and bamboo-based materials are not devoid of drawbacks, including variable dimensional stability, limited fire resistance, and susceptibility to degradation. These have prompted academic research on bamboo and wood materials to fully explore their potential and maximize resource utilization.

The purpose of this Special Issue is to provide readers with the latest advancements in the processing and application of wood and bamboo. It covers a wide range of topics, from the basic properties of bamboo and wood and the impact of various treatment processes on their performance to the development and application of bamboo and wood materials and the conservation of archeological wood. Leading experts in this field share their research findings, achievements, and future visions, offering new perspectives on innovative and efficient utilization of wood and bamboo. These contributions highlight vibrant and groundbreaking research areas related to wood and bamboo.

The articles in this Special Issue are categorized into four themes: fundamental properties and mechanisms, the influence of treatment processes on performance, material development and biomimetic applications, and the conservation of archeological wood.

1. Fundamental Properties and Mechanisms

Fundamental performance research serves as the cornerstone for understanding material properties. Through systematic experimentation and theoretical analysis, we can gain profound insights into the physical, mechanical, and chemical basic attributes of bamboo and wood. These studies not only provide a scientific basis for the selection of materials but also lay the foundation for subsequent processing and application.

The bamboo node is one of the structures of bamboo, and its presence will significantly affect the processing efficiency and product quality of bamboo. Currently, when studying the influence of bamboo nodes on the mechanical properties of bamboo, bamboo strips are mostly used as the experimental unit, and there is a lack of research on round bamboo. Wang et al. [1] investigated the mechanical properties of round bamboo pipes in three states: internodes, no diaphragm between internodes, and diaphragm between internodes. It demonstrates that the bamboo diaphragm and nodes have a significant positive effect on the bending properties across the transverse grain, radial ring stiffness, and shear properties along the grain. Radial stiffness showed the most significant increase. But they negatively impact tensile properties.

The bamboo structure is viewed as a natural functionally graded composite material, with the vascular bundles' volume density, type, and size spatially varying continuously. The distribution of vascular bundles is closely related to their mechanical properties.

Li et al. [2] investigated the variation in the location and distribution, cross-section area and mechanical properties of single vascular bundles along the longitudinal and radial directions with respect to their location from the base, middle, and top sections of the bamboo culm, respectively. It was found that the strength and Young's modulus of vascular bundles are all exponentially increased from the inner side to the outer side along the radial direction at the three height positions.

2. The Influence of Treatment Processes on Performance

Different treatment methods for bamboo and wood materials, such as heat treatment, chemical treatment, modification, etc., can greatly improve the performance of bamboo and wood materials. In-depth study of these treatment methods will help guide the industrial production of bamboo and wood materials and broaden their application fields.

Heat treatment is an advanced process. It not only optimizes the physical and chemical properties of wood but also significantly increases the durability of wood. However, the choice of heat treatment medium can have a profound impact on the properties of the wood. Hua et al. [3] found that the vacuum heat treatment may enhance the structural stability of the single-chain cellulose more effectively than the nitrogen and air treatments. Therefore, the vacuum heat treatment can better maintain wood stiffness and deformation resistance, thus improving wood utilization.

Heat treatment is also studied in depth in another article. Lyu et al. [4] concluded that the relative lignin content decreased with increasing treatment temperature, which was significant at lower negative pressures. Cellulose crystallinity showed a change rule of first increasing and then decreasing throughout the heat treatment range. The morphology and structure of the cell wall remained stable throughout the heat treatment range, but slight deformation and rupture occur at high temperatures.

Another treatment, high-intensity microwave (HIMW) treatment, is a time-saving and environmentally friendly method widely applied in the wood processing industry [5], but it will cause damage to the timber structure. It is very important to grasp the crack characteristics and damage mechanisms to regulate the treatment effect and expand its application. Liao et al. [5] studied the crack morphology, cell wall damage, cell wall chemical composition, and crystal structure of cellulose after HIMW treatment. It was found that the initial moisture content (MC) and microwave energy density (MWED) significantly influenced the crack characteristics and cell wall structure. However, it slightly influenced the chemical composition and crystalline structure of the cellulose of the Chinese fir cell wall.

Bamboo strips are the most frequently used bamboo material in industrial applications. Thermal treatment and chemical treatment can significantly change the mechanical properties of bamboo strips. A comprehensive understanding of the different treatment methods is of great significance for bamboo processing. Cao et al. [6] revealed that water boiling and 25% NH_3 treatments were better than the 15% NaOH treatment for enhancing bamboo bending properties. Water boiling with a treatment duration of 10 h was useful in optimal flexibility and plasticity in bamboo bending applications.

In the preparation of bamboo composite timber, multiple factors significantly affect its quality and performance. Exploring the interaction of factors and determining the optimal process parameters is crucial for the production of high-quality recombinant bamboo. Li et al. [7] systematically investigated the effects of hot-pressing temperature, time, and bio-oil phenolic resin (BPF) resin solid content on the modulus of rupture (MOR) and modulus of elasticity (MOE). It was found that the effects of the main factors decreased in the order of resin solid content, hot-pressing temperature, and hot-pressing time. The optimal process was determined to be 150 °C, 27.5 min, and 29% resin solids.

3. Material Development and Biomimetic Applications

Exploring new, sustainable, and high-performance materials is the key to promoting technological innovation and industrial upgrading. The following articles focus on cutting-edge developments in materials development, especially their innovative processes and design concepts.

The elastic deformation ability of wood under the action of external forces is limited, which limits its wide application as an elastic material. To solve this problem, Zhang et al. [8] synthesized a wood-based elastomeric material with strong mechanical properties by some pretreatments and multiple freeze–thaw cycles without using any chemical cross-linking agent. Stretchable wood-based elastomers with high mechanical strength and toughness have potential future applications in biomedicine, flexible electronics, and other fields.

Bamboo–wood composites have found extensive applications in the container flooring, furniture, and construction industries. Aiming at the problems of low utilization rate and high adhesive content of bamboo and wood composite materials, Ma et al. [9] successfully developed a novel bamboo–wood composite material that was made into flattened and slotted bamboo units through efficient integrated processing. Response surface analysis was used to optimize the hot-pressing process parameters. The grooving treatment enhances adhesive permeability and bonding. Compared with homogeneous flat bamboo composites, grooved bamboo composites exhibit superior tensile toughness and ductility.

Bamboo's unique composite gradient structure makes it widely regarded as an extremely efficient natural structure and material with high application value. Ji et al. [10] studied the changes in the vertical structure of bamboo and designed a bionic cantilever beam that combines the characteristics of hollow structure, uneven node distribution, and certain taper, which significantly improved its bending performance. At the same time, they imitated the radial distribution of vascular bundles and designed a thin-wall tube with a "dendritic" partial pressure structure, which exhibited excellent lateral compression performance in transverse compression.

4. The Conservation of Archeological Wood

Wood is an important natural material and was one of the earliest substances utilized by humans. Numerous wooden objects discovered during archeological excavations embody the cultural legacy of past generations, possessing substantial scientific, historical, and artistic importance. Therefore, the restoration and protection of archeological timber has become particularly important.

Since the mode of wood degradation exhibits significant variability across different archeological locations, a comprehensive understanding of the extent of deterioration in archeological wood is essential for implementing scientific, accurate, and rigorous restoration and preservation plans. Yang et al. [11] revealed the degradation process and characteristics of the wood through a comprehensive analysis of the physical, mechanical, morphological, and chemical properties of the archeological wood samples from the Luoyang Canal No. 1 site. They claim that the degradation of archeological wood is mainly due to the decomposition of cellulose and hemicellulose, while lignin exhibits relatively stable properties.

To ensure the conservation of waterlogged archeological wood, sustainable, safe, and effective methods must be implemented, with consolidation and dehydration being crucial for long-term preservation. Yang et al. [12] evaluated the dimensional changes, hygroscopicity, and mechanical properties of wood from the Luoyang Canal No. 1 Ancient Ship after 45% methyltrimethoxysilane (MTMS) and 45% trehalose treatment. It was found that trehalose was better at filling pores and reducing shrinkage, while MTMS significantly reduced the hygroscopicity and surface hydrophilicity of wood and had little impact on the appearance of wood, which was more suitable for reinforcing lightly degraded waterlogged archeological wood.

Research on the degradation mechanism of coating materials is also crucial for the preservation of cultural heritage. Liu et al. [13] used alternating wet and dry cycle experiments to evaluate the protective effect of Coromandel coatings on wooden substrates. It was found that in Coromandel coatings, the lacquer layer provides the best protection for the wooden substrate, while the ash coating is the most fragile. The degradation rate of the Coromandel specimens increases with rising temperatures.

5. Conclusions

This Special Issue presents a valuable collection of studies on the research of bamboo and wood. The collected articles not only deepen our understanding of the basic properties of bamboo and wood and its treatment technology but also provide a scientific basis and strategic guidance for the sustainable development and application of bamboo and wood materials, the effective protection of archeological timber, and the long-term healthy development of forest ecosystems. The content of this Special Issue will promote the harmonious development of human material needs while ensuring the sustainable use of the earth's resources so that humans and nature can live in harmony.

Funding: This research received no external funding.

Acknowledgments: The author would like to thank all authors contributing to this Special Issue. The author would also like to thank Xinyu Feng and Zexuan Xia for helping with this Editorial.

Conflicts of Interest: The authors declare no conflicts of interest.

References

1. Wang, X.; Yu, S.; Deng, S.; Xu, R.; Chen, Q.; Xu, P. Effect of Bamboo Nodes on the Mechanical Properties of *Phyllostachys iridescens*. *Forests* **2024**, *15*, 1740. [CrossRef]
2. Li, H.; Zhu, Q.; Lu, P.; Chen, X.; Xian, Y. The Gradient Variation of Location Distribution, Cross-Section Area, and Mechanical Properties of Moso Bamboo Vascular Bundles along the Radial Direction. *Forests* **2024**, *15*, 1023. [CrossRef]
3. Hua, Y.; Wang, W.; Gao, J.; Li, N.; Qu, Z. A Study on the Effects of Vacuum, Nitrogen, and Air Heat Treatments on Single-Chain Cellulose Based on a Molecular Dynamics Simulation. *Forests* **2024**, *15*, 1613. [CrossRef]
4. Lyu, J.; Wang, J.; Chen, M. Effects of Heat Treatment on the Chemical Composition and Microstructure of *Cupressus funebris* Endl. Wood. *Forests* **2024**, *15*, 1370. [CrossRef]
5. Liao, X.; Fang, X.; Gao, X.; Yi, S.; Zhou, Y. Effect of High-Intensity Microwave Treatment on Structural and Chemical Characteristics of Chinese Fir. *Forests* **2024**, *15*, 516. [CrossRef]
6. Cao, S.; Ji, J.; Yin, H.; Wang, X. The Influence of Treatment Methods on Bending Mechanical Properties of Bamboo Strips. *Forests* **2024**, *15*, 406. [CrossRef]
7. Li, Y.; Li, C.; Ren, X.; Chen, F.; Chen, L. Optimizing the Preparation Process of Bamboo Scrimber with Bamboo Waste Bio-Oil Phenolic Resin Using Response Surface Methodology. *Forests* **2024**, *15*, 1173. [CrossRef]
8. Zhang, Y.; Li, J.; Lu, Y.; Shi, J. Highly Mechanical Strength, Flexible and Stretchable Wood-Based Elastomers without Chemical Cross-Linking. *Forests* **2024**, *15*, 836. [CrossRef]
9. Ma, Y.; Luan, Y.; Chen, L.; Huang, B.; Luo, X.; Miao, H.; Fang, H. A Novel Bamboo–Wood Composite Utilizing High-Utilization Easy-to-Manufacture Bamboo Units: Optimization of Mechanical Properties and Bonding Performance. *Forests* **2024**, *15*, 716. [CrossRef]
10. Ji, S.; Mou, Q.; Li, T.; Li, X.; Cai, Z.; Li, X. The Novel Applications of Bionic Design Based on the Natural Structural Characteristics of Bamboo. *Forests* **2024**, *15*, 1205. [CrossRef]
11. Yang, W.; Ma, W.; Liu, X. Evaluation of Deterioration Degree of Archaeological Wood from Luoyang Canal No. 1 Ancient Ship. *Forests* **2024**, *15*, 963. [CrossRef]
12. Yang, W.; Ma, W.; Liu, X.; Wang, W. Consolidation and Dehydration Effects of Mildly Degraded Wood from Luoyang Canal No. 1 Ancient Ship. *Forests* **2024**, *15*, 1089. [CrossRef]
13. Liu, W.; Zhu, L.; Varodi, A.M.; Liu, X.; Lv, J. The Effect of Wet and Dry Cycles on the Strength and the Surface Characteristics of Coromandel Lacquer Coatings. *Forests* **2024**, *15*, 770. [CrossRef]

Article

The Influence of Treatment Methods on Bending Mechanical Properties of Bamboo Strips

Shuyu Cao [1], Jiagui Ji [2], Haowei Yin [1] and Xuehua Wang [1,*]

1 College of Furnishings and Industrial Design, Nanjing Forestry University, Nanjing 210037, China; shuyu@njfu.edu.cn (S.C.)
2 Zhejiang Sanjian Industry & Trade Co., Ltd., Lishui 323800, China
* Correspondence: wangxuehua@njfu.edu.cn

Abstract: This study aimed to obtain a comprehensive understanding on bamboo as a curve-member manufacturing material by comparative analysis of how different treatment methods on bending properties improve the effect on bamboo strips. In order to achieve this purpose, bamboo strips were subjected to water boiling, 15% NaOH, and 25% NH_3 impregnation; the impact of physical, mechanical and chemical properties were explored. The results revealed that: (1) Water boiling significantly affected crystallinity, cellulose, and lignin content, with a treatment duration of 10 h showing the most favorable results for flexibility and plasticity, greatly improving bending performance. (2) An amount of 15% NaOH treatment significantly increased bending MOE and plastic displacement by 73% and 122.7%. However, it led to a noticeable decrease in bending strength (MOR). A treatment above 8 h could cause irreversible damage to bamboo strips. (3) The improvement of 25% NH_3 on bamboo bending ability was lower than water boiling. The effects of chemical composition were obvious in the initial five days and changed little after five days. Generally, water boiling for over 10 h is suitable for applications with significant bending requirements. While for maintaining bamboo color, original strength, and bending performance, 25% NH_3 for five days was recommended, and 15% NaOH was not advised for improving bamboo bending performance and its applications.

Keywords: bamboo bending ability; treatment methods; comparison; toughness enhancement; chemical composition

Citation: Cao, S.; Ji, J.; Yin, H.; Wang, X. The Influence of Treatment Methods on Bending Mechanical Properties of Bamboo Strips. *Forests* **2024**, *15*, 406. https://doi.org/10.3390/f15030406

Academic Editor: Alain Cloutier

Received: 20 January 2024
Revised: 12 February 2024
Accepted: 17 February 2024
Published: 21 February 2024

1. Introduction

China produces bamboo resources which are widely distributed. Bamboo has short growth cycles, and possesses high strength and bending ability. Bamboo has extensive applications in various fields such as construction, furniture, and handicraft production. Despite the extensive research dedicated to the bending applications of wood [1], the utilization of bamboo in bending modification and application are relatively constrained [2], necessitating further exploration and investigation.

The bending performance of bamboo strips is influenced by treatment methods, such as heat treatment and chemical treatment, which significantly alter the mechanical properties of bamboo strips [3]. In recent years, scholars have conducted preliminary studies on the chemical composition changes of bamboo strips under different treatment temperatures and conditions [4]. Fang et al. found that alkali water boiling not only removes extractives but also affects the color of bamboo, thereby influencing its utilization [5]. Chu et al. investigated the structural changes of bamboo under different heat treatment temperatures and acid-base or glycerol treatment media. An increase in crystallinity was found at a hydrothermal temperature of 135 °C [6]. An et al. proposed that high crystallinity and oriented structure are crucial factors leading to poor bamboo bending performance, suggesting a reduction in crystallinity obtained through alkali treatment [7]. Xu suggested that bamboo color darkens, MOE and flexural strength decreases, and weight loss increases after heat treatment [8]. Zhang et al. discovered a decrease in cellulose and hemicellulose content in bamboo after heat treatment, accompanied by an increase in lignin and extractives content [9]. The aforementioned studies mainly focused on the influences of single variations by infrared spectroscopy, XRD changes, surface properties, or changes in chemical composition, and were mostly about raw bamboo or bamboo fiber; therefore there is a lack of multidimensional comparison about bamboo strips [10].

Bamboo strips are the most widely used bamboo material in industrial applications. A comprehensive understanding of the different treatment methods holds significant importance for bamboo processing, offering insights that can be translated into practical industrial considerations. In order to meet the demand for high flexibility in bending applications, emphasizing the universality and versatility of bamboo, bamboo strips of *Phyllostachys edulis* (Carrière) J. Houz., which is known as moso bamboo, were obtained as the research object. Three different treatment methods, that is, water boiling, 15% NaOH impregnation, and 25% NH_3 impregnation, were used to enhance bending performance and the processing advantages of bamboo in this study. The bamboo strips underwent treatments of varying durations, and their physical and mechanical properties were tested. The changes in cellulose, hemicellulose, and lignin content before and after treatment were determined. Furthermore, X-ray diffraction (XRD) and Fourier-transform infrared spectroscopy (FTIR) were employed to confirm the influence of different treatment methods on the chemical composition and mechanical properties of bamboo [11]. The aim of this study was to provide a theoretical foundation for toughness enhancement in bamboo bending processing.

2. Materials and Methods

2.1. Materials

Moso bamboo strips were obtained from Zhejiang Sanjian Industrial & Trade Co., Ltd., Zhejiang province, Lishui, China. The selected bamboo strips were free from mold, discoloration, and were of uniform greenish-yellow color. Three-year-old bamboo tubes were obtained through a systematic sampling approach. In order to guarantee uniformity of the specimens, the green and yellow layers of the bamboo tube were eliminated, and the middle of the bamboo was selected and processed into bamboo strips (Figure 1). The bamboo strips were prepared according to the Chinese standard of GB/T 15780-1995 [12] testing methods for physical and mechanical properties of bamboo, with dimensions of 160 mm × 20 mm × 5 mm (longitudinal × tangential × radial).

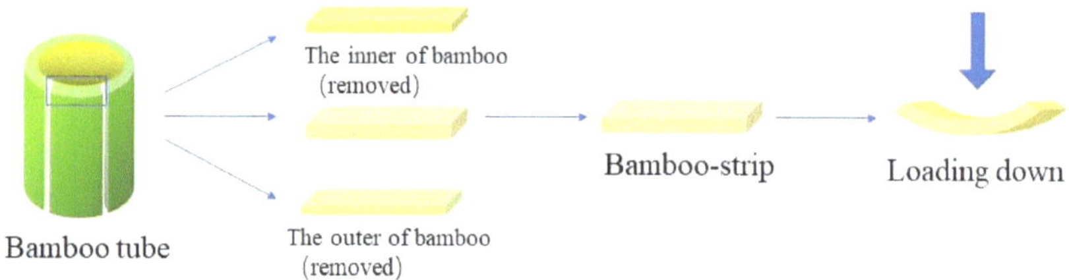

Figure 1. Diagram of the sampling location.

2.2. Sample Treatment

Three different treatment methods were applied to the bamboo strips and different treatment durations were used for each method (Table 1). The three treatment methods were hydrothermal boiling, 15% NaOH (mass content), and 25% NH_3 (mass content) solution impregnation. Following the water boiling treatment, the bamboo strips were immediately prepared for mechanical property testing. While after 15% NaOH and 25% NH_3 solution impregnation, the samples' surfaces were cleaned using water before mechanical performance testing.

Table 1. Treatment methods for bamboo strips.

Treatment Method	Treatment Duration	Other Information
Control	-	Room temperature
Water boiling	2 h, 4 h, 6 h, 8 h, 10 h	135 °C
15% NaOH	2 h, 4 h, 6 h, 8 h, 10 h	Room temperature
25% NH$_3$	5 days(d), 7 d, 9 d, 11 d, 13 d	Room temperature

2.3. Testing Methods

The bamboo strips were first subjected to bending mechanical property testing, and the destruction status after testing was recorded. Subsequently, their density was determined after drying. The dried bamboo was then ground into powder, and subjected to chemical composition testing (cellulose, hemicellulose, lignin), Fourier-transform infrared spectroscopy (FTIR), and X-ray diffraction (XRD) analysis.

A mechanical testing machine, MMW-50, Jinan Nair Testing Machine Co., Ltd., Shandong province, Jinan, China, was utilized for mechanical property testing. The mechanical property, modulus of rupture (MOR), and modulus of elasticity (MOE), were tested using the three bending method, according to Chinese standard GB/T 15780-1995 [12], and loading was at a rate of 10 mm/min. The elastic displacement (De) and plastic displacement (Dp) were obtained from the load-displacement curve by the equivalent elasto-plastic energy method (Figure 2). Each set of experiments was repeated 5 times, and the standard deviation was set as error.

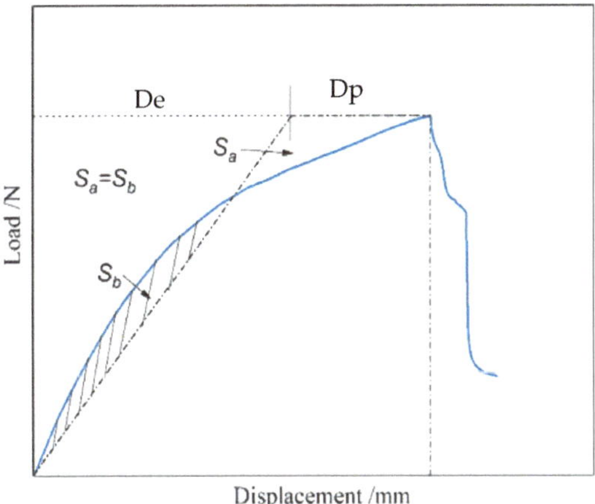

Figure 2. Testing method for De and Dp.

Chemical composition determination followed the methods of the National Renewable Energy Laboratory (NREL) in the United States. The treated bamboo strips were of the same specifications, thoroughly dried in a dry oven (DHG-9070A, Shanghai Jinghong Experiment Co., Ltd., Shanghai, China), and the content of cellulose, acid-insoluble lignin (AIL), and hemicellulose was determined, with each sample undergoing six parallel tests.

Infrared spectroscopy analysis utilized a Fourier-transform infrared spectrometer (VERTEX 80V, Bruker, Luken, Germany) and adopted the KBr pellet method. The powder, sieved through a 200-mesh sieve, was dried and made into KBr pellets (the ratio of bamboo power to KBr was 1:100). Infrared spectroscopy scanning was performed in transmission mode, ranging from 400 cm^{-1} to 4000 cm^{-1}, at a resolution of 4 cm^{-1} with 32 scans.

X-ray diffraction analysis utilized an X-ray diffractometer (AXIS UltraDLD, Shimadzu, Milton Keynes, UK). The powder, sieved through a 200-mesh sieve, was dried, placed on a test slide, and the crystallinity of the sample was analyzed using X-ray diffraction. The

X-ray source was CuKa, with a voltage of 40 kV, and a current of 40 mA. The scanning rate was 5°/min, and the diffraction angle ranged from 5° to 80°.

3. Results

3.1. Physical and Mechanical Properties

As water boiling time increased, the bamboo strips gradually changed from light yellow to reddish-brown (Figure 3a). As treatment time increased, the color deepened. Additionally, after drying, the volume and mass of the bamboo strips decreased. Mechanical testing revealed a small amount of bamboo fiber fracture on the surface, with the main body of bamboo strips remaining intact and maintaining relatively high strength.

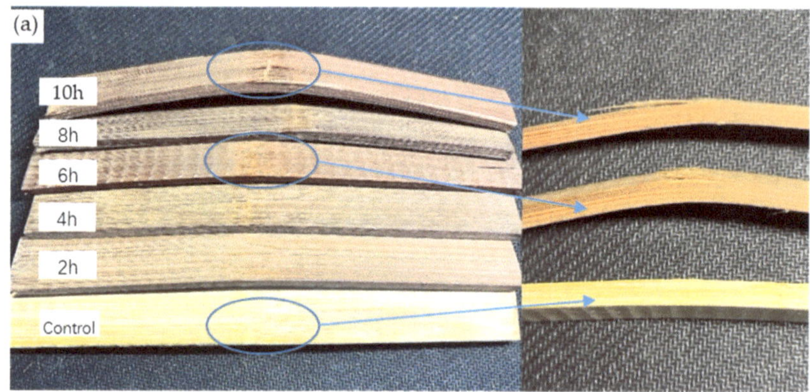

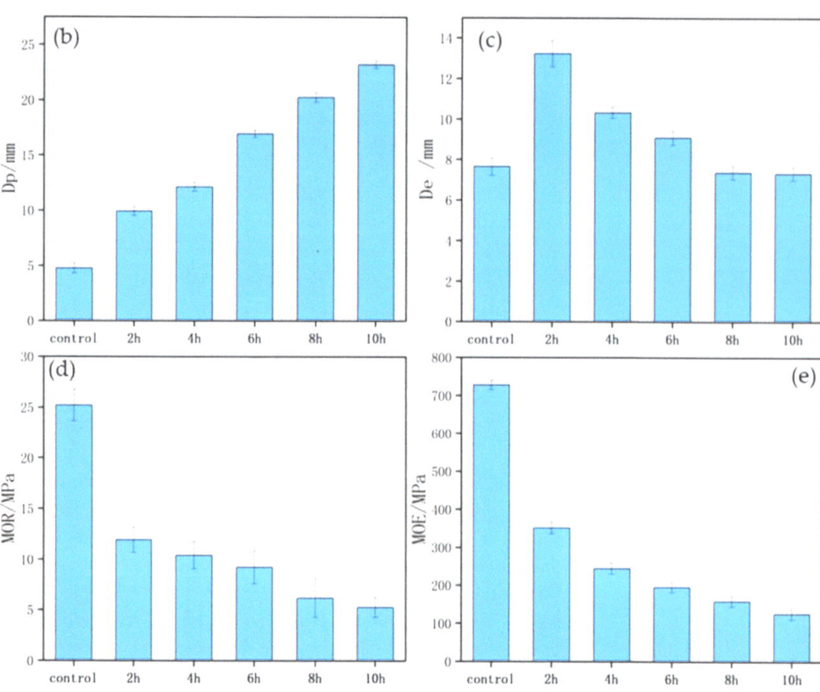

Figure 3. Mechanical and physical properties of bamboo strips using water boiling: (**a**) color and failure status, (**b**) plastic displacement (Dp), (**c**) elastic displacement (De), (**d**) bending strength (MOR) and (**e**) modulus of elasticity (MOE).

As water boiling time increased, both the modulus of rupture (MOR) (Figure 3d) and modulus of elasticity (MOE) (Figure 3e) significantly decreased. After 10 h, the MOR decreased by approximately 50% compared to 2 h, representing a reduction of 79.2% compared to the control, while the bending MOE also decreased significantly, showing an 82.5% reduction compared to the control. The change in elastic displacement was evident. At 10 h, elastic displacement increased by 387.9%, compared to the control (Figure 3b). The density decreased as water boiling time increased (Figure 4a). Compared to the control, the density decreased from 0.75 g/cm^3 to 0.56 g/cm^3 after 2 h of treatment. For 6 h and 8 h, there was a further significant decrease in density, MOR, and MOE. The density and mechanical performance reached minimum values after 8 h of treatment, with a 48% reduction in density compared to the control.

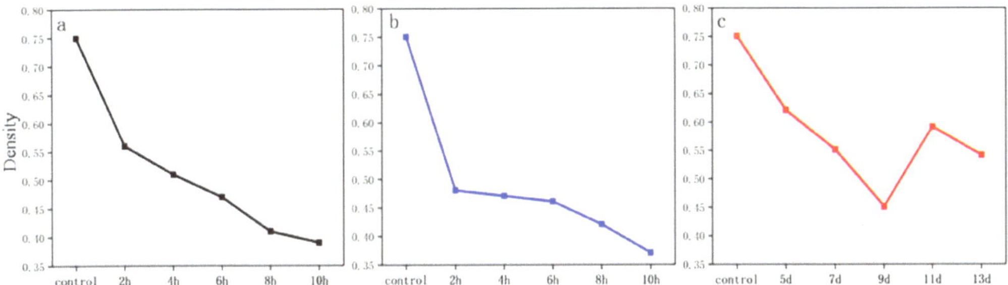

Figure 4. Density changes of bamboo strips for different treatments: (**a**) water boiling, (**b**) 15% NaOH, and (**c**) 25% NH$_3$.

The results aligned with previous studies about the impact of different treatments on bamboo properties. Furuta et al. [13–15] found that the MOE of bamboo slices decreased with increasing water temperature. When the temperature was 160 °C and time was 6 min, the softening effect on bamboo was better. As the temperature and duration increased, the hemicellulose structure changed obviously, which facilitated the softening of the bamboo. Similar to previous research, this study showed that a temperature of 135 °C and heating time adjusted from 2 to 10 h would be a suitable method for bamboo strips. The water boiling process provided reliable guidance for the batch processing and production of thick bamboo strips.

With an increased duration of 15% NaOH treatment, the bamboo strips gradually changed from light yellow to yellow-brown, which was darker than for the water boiling treatment (Figure 5a). The sample surface exhibited a large area of alkali attachment and residue. Moreover, the bamboo fiber damage was significant. As treatment time increased, the yellow-brown color deepened, the area of alkali attachment increased, and bamboo fiber damage intensified. After the mechanical test, bamboo fibers showed a brittle fracture at the force application site. The number and extent of fractures increased with the prolonged treatment duration. However, the volume and mass change after drying were relatively small.

The MOR and MOE of bamboo strips decreased after 15% NaOH treatment and exhibited an initial decrease followed by an increase between 2 and 8 h (Figure 5c–e). At 4 h, the MOR and MOE showed little change compared with 2 h. However, the MOR significantly decreased after 6 h, and the MOE experienced a precipitous drop after 8 h. Compared to the control, the MOR decreased by 57.3% and MOE decreased by 73% at 10 h. Elastic and plastic displacements (Figure 5b) decreased with the increasing treatment duration, ultimately reducing by 52.2% and 122.7%, respectively. Density exhibited a noticeable decline after treatment (Figure 4b). All mechanical properties reached their minimum values at 10 h.

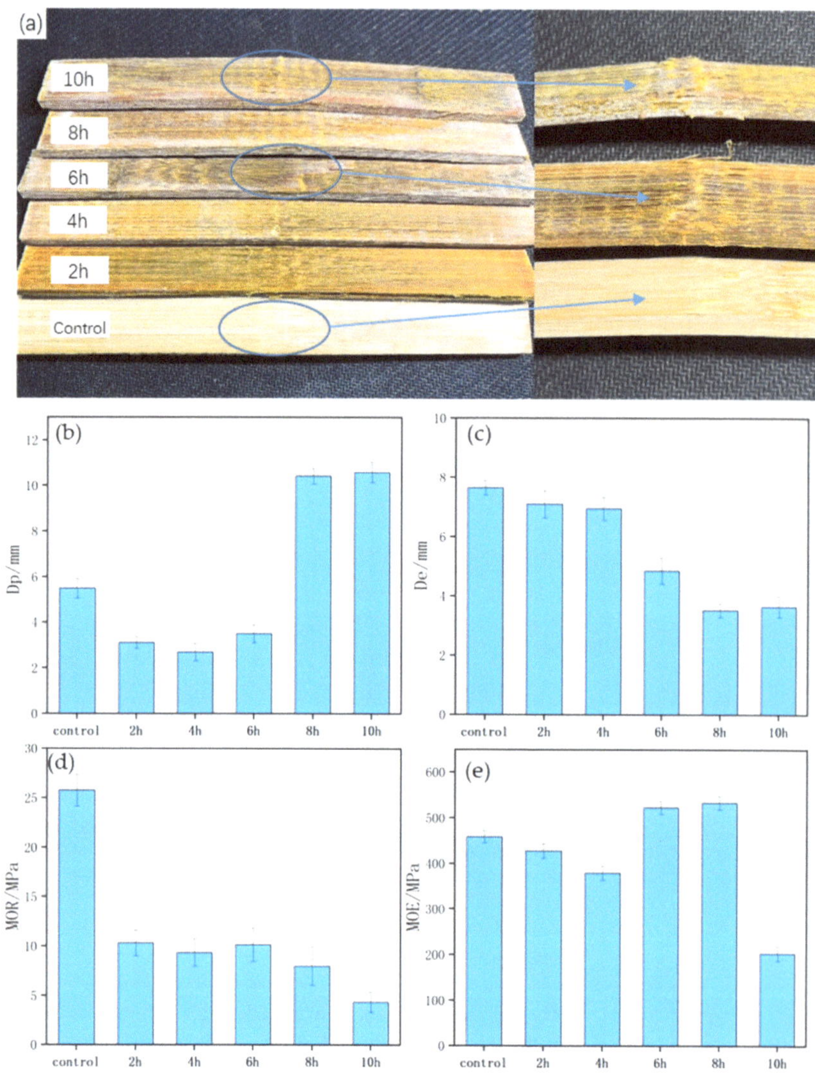

Figure 5. Mechanical and physical properties of bamboo strips using 15% NaOH treatment: (**a**) color and failure status, (**b**) Dp, (**c**) De, (**d**) MOR, and (**e**) MOE.

As 25% NH_3 treatment time increases, bamboo strips gradually changed from light yellow to reddish-brown, and the deepening of the reddish-brown color was relatively small (Figure 6a). After the mechanical test, bamboo fibers on the surface at 9 days(d) showed little fractures. A small amount of NH_3 solution residue was easily wiped clean. At 13 d, there were noticeable fiber fractures on the sample surface, while the change in volume and mass after drying was relatively small.

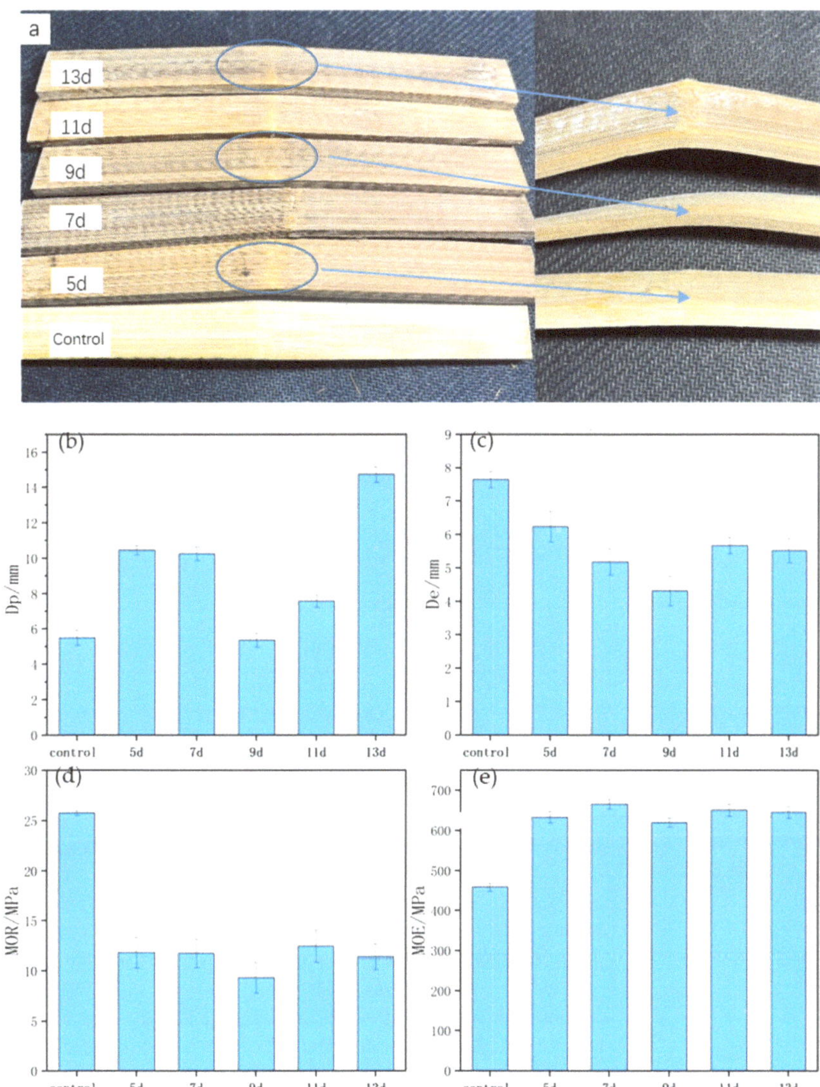

Figure 6. Mechanical and physical properties of bamboo strips using 25% NH3 treatment: (**a**) color and failure status, (**b**) Dp, (**c**) De, (**d**) MOR, and (**e**) MOE.

MOR of bamboo strips decreased, while the MOE increased after 25% NH_3 treatment. Between 5 d and 11 d, it exhibited an initial decrease then increase, and there was a slight decrease at 13 d compared to 11 d. The mechanical property was lowest at 9 d, elastic displacement, plastic displacement, density (Figure 6c), MOR, and MOE decreased by 43.4%, 12.6%, 63.8%, 40%, 66.8%, and 49.2%, respectively, compared to the control (Figure 6b–e).

Li et al. [15] found that the concentration of ammonia solution was 25%, while the sodium hydroxide solution between 10% and 15% was more suitable for the softening treatment of bamboo strips. Based on the previous research, the processing time was optimized and refined in this study, and the results of the alkali treatment were roughly the same as previous studies in terms of ammonia treatment. However, in the sodium hydroxide treatment, this research found that it was more destructive to bamboo, which might be due to the extended concentration and soaking time.

3.2. X-ray Diffraction (XRD)

The 2θ positions of the XRD diffraction peaks for bamboo were observed at 15.8° and 22°, representing characteristic peaks of the cellulose structure [16]. Throughout the treatment process, there was no significant change in the crystalline structure of cellulose [17], while the crystal plane angles and crystallinity changed with treatment duration (Figure 7, Table 2).

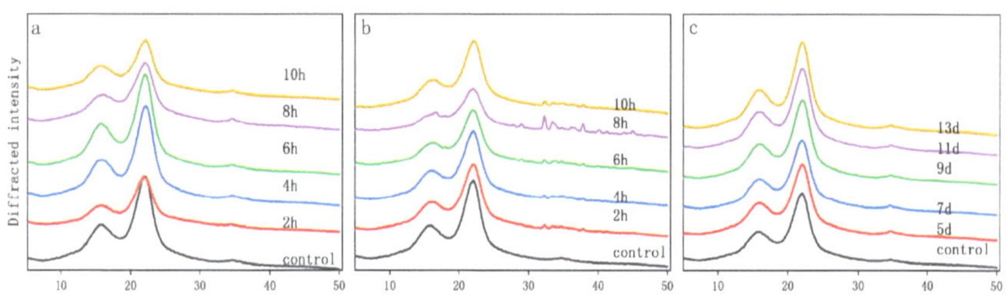

Figure 7. X-ray diffraction of bamboo strips for different treatment methods: (**a**) water boiling, (**b**) 15% NaOH, and (**c**) 25% NH_3 treatment.

Table 2. Crystallization characteristics of bamboo strips using different treatment methods.

Treatment Method		2θ/°	RC/%	Treatment Method		2θ/°	RC/%	Treatment Method		2θ/°	RC/%
Control	-	21.65	51.45		-	-	-		-	-	-
	2 h	22.15	59.66		2 h	22.02	61.32		5d	21.85	53.32
	4 h	22.42	61.89	15%	4 h	21.96	59.45	25%	7d	21.86	52.45
Water	6 h	22.56	62.22	NaOH	6 h	21.65	58.62	NH_3	9d	22.02	52.43
boiling	8 h	22.64	63.34		8 h	21.45	57.78		11d	22.04	52.23
	10 h	22.78	64.75		10 h	21.19	54.45		13d	22.03	51.89

Note: 2θ—002 crystal plane angle, RC—Relative crystallinity.

The diffraction peaks of the crystal plane were concentrated at a range of 21° to 22.7° which indicates that the 101 crystal plane angle (2θ) of bamboo strips did not show significant changes after processing [18]. The treatment could only reach the noncrystalline region of cellulose and could not penetrate the crystalline region. It did not alter the results in the crystalline region [19], and there was no change in the crystalline layer distance [20]. This result was consistent with the research findings of Sun.

As to the crystallinity, the crystallinity gradually increased as water boiling treatment time increased. The crystallinity noticeably decreased with increasing treatment time in 15% NaOH immersion, and the decrease in crystallinity was not significant, showing a relatively small change in 25% NH_3 immersion. The decrease in crystallinity after alkali treatment indicated an irregular molecular arrangement inside the material, weakening its stability, which was related to the degradation of cellulose in the crystalline region [21]. Additionally, alkali treatment changed the microcrystal width of the bamboo microfibers, leading to decreased stability. After water boiling treatment, the crystallinity of the bamboo strips increased, causing changes in thermal stability [22]. Under the influence of heat, hydroxy groups between cellulose molecular chains in the amorphous region undergo condensation reactions, causing a rearrangement of cellulose molecular chains and crystallization in the quasi-crystalline region, thereby increasing the relative crystallinity of the bamboo strips [23].

3.3. Infrared Spectroscopy

In the infrared spectroscopy region of wavenumbers between 4000 cm^{-1} and 2000 cm^{-1}, the spectra mainly reflect stretching vibrations of hydrogen-containing groups and the presence of triple bonds and cumulative double bonds. Strong absorption peaks at 3450 cm^{-1} and 2945 cm^{-1} represented the stretching vibrations of –CH and –OH groups, respectively. This region underwent some shifts during water boiling, but the shift effects were relatively minor in 15% NaOH and 25% NH3 treatments (Figure 8). Under water boiling, the hydroxyl group became active, and the water in the bamboo provided space for the hydroxyl group to move [24].

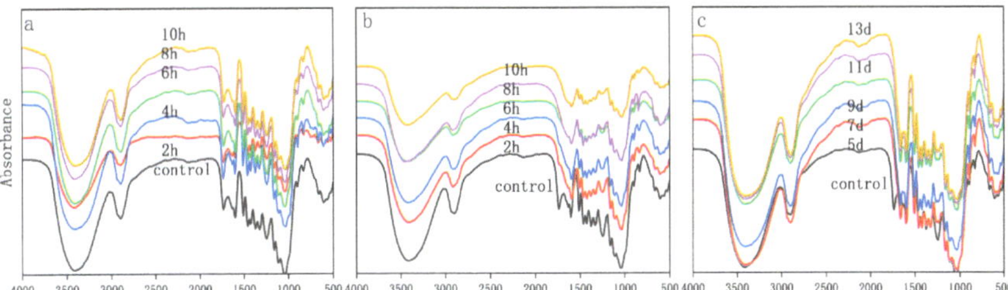

Figure 8. Infrared spectra diffraction of bamboo strips for different treatment methods: (**a**) water boiling, (**b**) 15% NaOH, and (**c**) 25% NH$_3$ treatment.

In the infrared spectroscopy region of wavenumbers between 2000 cm^{-1} and 800 cm^{-1}, more information about the changes in chemical composition was provided. In the water boiling and 15% NaOH treatments, the intensity of the absorption peaks gradually weakened with increasing treatment time, while there was almost no change in the 25% NH$_3$ treatment.

Therefore, we primarily discussed the first two treatment methods. The highest peak at wavenumber 1750 cm^{-1} was caused by ester bonds of acetyl, ester, or carboxyl groups [25]. The peak at 1632 cm^{-1} represented the stretching vibration of the conjugated carbonyl group, and its weakening indicated the damage to hydroxyl and conjugated carbonyl groups [26]. The absorption peaks at 1600^{-1} and 1500 cm^{-1} showed no significant changes, suggesting the relative stability of the benzene ring framework without apparent alterations. The weakening of the absorption peak at wavenumber 1244 cm^{-1} was due to the stretching vibration between the benzene ring and oxygen bonds. Changes in the intensity of the absorption peak at wavenumber 1376 cm^{-1} was caused by C–H stretching vibrations on phenolic hydroxyl groups. The weakening of the absorption peak at wavenumber 1043 cm^{-1} indicated a reduction in the quantity of C–O bonds.

3.4. Chemical Composition

All three treatment methods affect the chemical constituent content. The relative content of cellulose and hemicellulose all decreased. Compared with the control group, after the three treatments, the relative contents of cellulose and hemicellulose decreased significantly, while lignin increased slightly. In water boiling, the relative content of cellulose and hemicellulose gradually increased, while lignin increased lightly as treatment time increased. In the 15% NaOH treatment, the relative content of cellulose and hemicellulose initially decreased and then increased, with a gradual increase in lignin's relative content as treatment time increased. In the 25% NH$_3$ treatment, the relative content of cellulose and hemicellulose changed slightly in a pattern; initially, it decreased and then increased, with a slight decrease in lignin's relative content as the duration of the treatment increased (Figure 9).

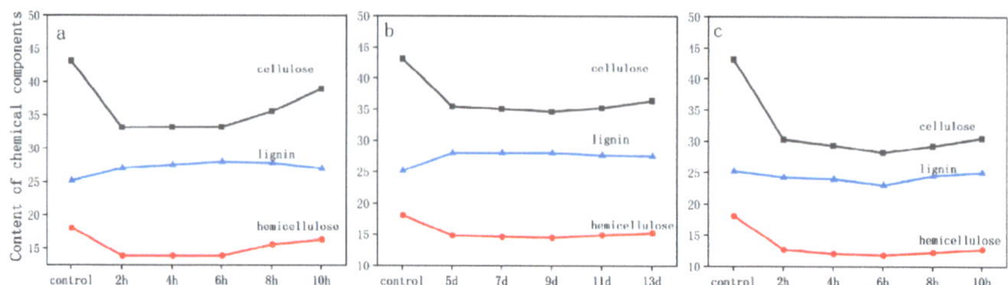

Figure 9. Infrared spectra diffraction of bamboo strips for different treatment methods: (**a**) water boiling, (**b**) 15% NaOH, and (**c**) 25% NH$_3$ treatment.

The chemical composition concentration changes of bamboo strip after water boiling were the main reason for alterations in bamboo color, mechanical properties, and density [27]. The result of the cellulose and hemicellulose relative content that increased with treatment time was consistent with the findings of Meng et al., who suggested that the chemical constituent content increase was due to the difficulty of macromolecules to degrade and other small molecule substances were lost because of thermal reactions or dissolution in the alkaline solution [28]. Due to the relatively good thermal stability of lignin, its content was less affected but increased compared to the control. This increase was not a substantial increase in lignin content, but rather a result of the degradation of cellulose and hemicellulose, leading to an increase in lignin's relative percentage content in the bamboo matrix [29]. Additionally, the degradation of cellulose and hemicellulose, being similar to lignin, would be mistakenly calculated as acid-insoluble lignin in the current detection method, further increasing the relative content of lignin [30].

In 15% NaOH and 25% NH$_3$ treatments, both were alkaline, and they had a certain decomposition effect on cellulose and hemicellulose. Therefore, the relative content of cellulose and hemicellulose initially decreased, reaching the lowest point at 6 h (15% NaOH and 5 d (25% NH$_3$). With increasing treatment time, the alkaline solution disrupted the cell wall structure of bamboo, causing more small molecule substances to dissolve [31], and the relative content of cellulose and hemicellulose to increase again [32]. Regarding lignin, the 15% NaOH had a stronger decomposition ability compared to the 25% NH$_3$, resulting in much lower lignin content than the control and the other two treatment methods. Bamboo strips treated with 15% NaOH also exhibited noticeable discoloration and structural damage (Figure 2).

3.5. Correlation and Comprehensive Analysis

Based on the above data, a correlation was shown in Figure 10. The density of the bamboo strips has a strong positive correlation with MOR, which was 0.92 and 0.93 respectively. It has a weak positive correlation with the elastic displacement of 0.27, and a weak negative correlation with the plastic displacement with a correlation of −0.35. There was a moderate positive correlation between cellulose, hemicellulose, and MOR with a correlation of 0.59 and 0.52, respectively. Cellulose and hemicellulose both had a weak positive correlation with plastic displacement, with a correlation of 0.36 and 0.40 respectively. Lignin had a moderate positive correlation of 0.57 with plastic displacement but a weak positive correlation of 0.27 with elastic displacement. Lignin had more of an effect on plastic displacement than on elasticity displacement. The correlation between elastic displacement and all other parameters was low. Density, cellulose, and hemicellulose of bamboo strips had a great impact on MOR, but had a relatively small impact on plastic displacement. The modulus and elastic displacement were affected by multiple factors.

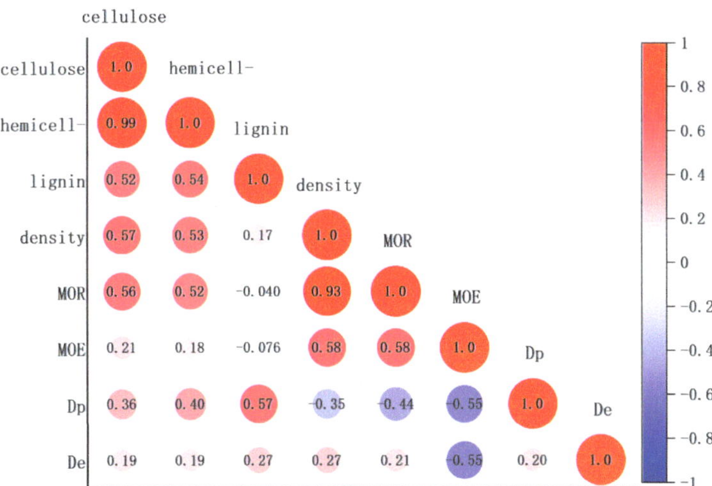

Figure 10. Correlation diagram of mechanical and physical properties.

After water boiling treatment, the MOE (the ability of the bamboo strips to resist bending stress) showed a clear downward trend from untreated to 10 h. The plastic displacement at 2 h was the highest, which had an obvious downward trend and reached a minimum value after 8 h and remained unchanged (the higher the total displacements, the better the deformation performance). This change was consistent with the decreasing trend of crystallinity [33], and the relevant groups decreasing in the infrared spectrum. It showed that water boiling changed the content and arrangement of cellulose and hemicellulose in bamboo strips, while the crystallinity and the related groups decreased. The increase in cellulose and lignin was the main reason for the increase in the deformation effect. The significant reduction in MOE after 8 h also greatly enhanced flexibility and plasticity of bamboo strips. Water boiling has a significant impact on the crystallinity, cellulose content and arrangement [34,35], and lignin content of bamboo strips. A 10-h treatment was most beneficial to its flexibility and plasticity.

MOR and plastic displacement of the bamboo strips treated with 15% NaOH all showed a downward trend from untreated to 10 h. Compared with the control, MOR showed a cliff-like decrease. This change was consistent with XRD. All the crystallinity, cellulose, and lignin contents had a downward trend, indicating that 15% NaOH treatment greatly changed the content and arrangement of cellulose and hemicellulose in bamboo strips and reduced the crystallinity [36,37]. The MOR and MOE first decreased and then increased in a small range at 6 h. The change dropped cliff-like at 10 h compared with control. The overall change was consistent with the significant change in the XRD diffraction intensity, indicating the damage caused by long-term alkali treatment. The effect of short-term treatment on crystallinity, cellulose, and lignin was obvious [38] (Table 2, Figure 7b). While the long treatment effect was mainly concentrated on crystallinity, it would have a certain impact on cellulose and lignin in the short term [39]. If the treatment duration exceeds 8 h, it would cause irreversible damage to the bamboo strips.

The mechanical properties of the bamboo strips treated with 25% NH$_3$ first decreased and then increased. Compared with the control group, MOR decreased precipitously. This change was related to decreasing cellulose and lignin content [40]. As the treatment time exceeded more than five days, the cellulose content changed significantly and was related to mechanical properties. Meanwhile, hemicellulose and lignin had no obvious changes in the more than five day samples. The 25% NH$_3$ treatment affected the content of chemical components in bamboo [41]. It had a certain impact on hemicellulose and lignin in a short time. After five days, it mainly affected cellulose, and the best effect was nine days.

Comparing these three treatment methods against MOE, 25% NH_3 and water boiling had similar effects, both were better than 15% NaOH treatment, while the change in elastic displacement after water boiling was the smallest, and the strength properties of bamboo strips after water boiling was the optimal. In MOE, the changes in 15% NaOH and water boiling were similar, and the plastic displacement increased by water boiling was three times that of 15% NaOH, and toughness and bending properties were significantly better than the 15% NaOH treatment. The density change after these three treatment methods decreased by 40%–50%, and the difference was small. As to the surface status changes, the fiber breakage after water boiling and 25% NH_3 treatment was significantly less than that of the 15% NaOH treatment, and the color and texture retention was in a better condition than 15% NaOH. Based on various data analysis and appearance changes, the treated bamboo strips had excellent performance by water boiling in terms of toughness, bending performance, strength and color and texture. The ranking for these three treatments was in order, water boiling > 25% NH_3 > 15% NaOH in processing.

4. Conclusions

In terms of toughness, bending performance, strength, color, and texture, the ranking of treatment methods was water boiling > 25% NH_3 > 15% NaOH. Water boiling has a significant impact on bamboo crystallinity, cellulose content and arrangement, and lignin content, with a treatment duration of 10 h being the most favorable for flexibility and plasticity, greatly enhancing its bending performance.

The 15% NaOH treatment effected bamboo destructively. Although MOE and plastic displacement increased by 73% and 122.7%, respectively, compared to the control, 15% NaOH treatment directly damaged the fiber structure, resulting in a significant decrease in bamboo strength. A short treatment time had a certain impact on cellulose and lignin, with the impact mainly concentrated on crystallinity. A prolonged treatment time exceeding 8 h brought irreversible damage to bamboo.

The improvement of bamboo bending performance by 25% NH_3 was lower than water boiling, with the impact on various mechanical properties being approximately 50% of water boiling. Its impact mainly changed the content of chemical components, with a short treatment time having a certain impact on hemicellulose and lignin. The optimal effect occurred after nine days.

This study provides a reference for improving the toughness and bending performance of bamboo. In general, water boiling and 25% NH_3 treatments were better than the 15% NaOH treatment for enhancing bamboo bending properties. Water boiling with a treatment duration of 10 h was useful in optimal flexibility and plasticity in bamboo bending applications. In order to maintain the natural color of bamboo, balancing original strength and bending performance, 25% NH_3 for nine days is recommended. Using 15% NaOH was not recommended as it damaged the bamboo structure obviously. Based on these results, further study could focus on refining process parameters of water boiling and 25% NH_3 treatments, and also using multiple thicknesses of bamboo strips, to explore the optimal treatment process to decrease production cost, improve applicability to bamboo material, thus facilitating industrial production of bamboo strips with different specifications.

Author Contributions: Conceptualization, X.W.; formal analysis, S.C.; data curation, S.C. and H.Y.; writing—original draft preparation, S.C.; writing—review and editing, H.Y.; project administration, J.J.; funding acquisition, X.W. All authors have read and agreed to the published version of the manuscript.

Funding: This research was funded by Natural Science Foundation of China, grant number 31800471.

Data Availability Statement: Data are contained within the article.

Acknowledgments: We thank Zhejiang Sanjian Industry & Trade Co., Ltd. for providing bamboo in this test.

Conflicts of Interest: Author Jiagui Ji is employed by the company Zhejiang Sanjian Industry & Trade Co., Ltd. And the company provided bamboo strips in the research. The remaining authors declare that the research was conducted in the absence of any commercial or financial relationships that could be construed as a potential conflict of interest. The funders had no role in the design of the study; in the collection, analyses, or interpretation of data; in the writing of the manuscript; or in the decision to publish the results.

References

1. Tamang, M.; Nandy, S.; Srinet, R.; Das, A.K.; Padalia, H. Bamboo Mapping Using Earth Observation Data: A Systematic Review. *J. Indian Soc. Remote Sens.* **2022**, *50*, 2055–2072. [CrossRef]
2. Zhao, Y.; Feng, S.; Huang, R. A Review of Researches on Wood Bending Techniques. *World For. Res.* **2010**, *23*, 40–44.
3. Tang, Y.; Li, J.; Shen, Y.; Jin, Y.; Wang, Y.; Li, Y. *Phyllostachys edulis* with high temperature heat treatments. *J. Zhejiang A F Univ.* **2014**, *31*, 167–171.
4. Fei, B.; Su, Q.; Liu, H.; Fang, C.; Ma, X.; Zhang, X.; Sun, F. Research progress of bamboo winding technology. *J. For. Eng.* **2022**, *7*, 25–33.
5. Fang, W.; Niu, S.; Shen, D.; Cai, J.; Zhuang, R.; Li, Y. XRD and FTIR Analysis on Bamboo Culm Treated by High-temperature Hot Water Extraction. *Zhejiang For. Sci. Technol.* **2015**, *35*, 47–50.
6. Chu, J.; Ma, L.; Zhang, J. The Chemical Composition of Bamboo after Heat Pretreatment with Fourier Infrared Spectrum Analysis. *Spectrosc. Spectr. Anal.* **2016**, *36*, 3557–3562.
7. An, X.; Wang, H.; Li, W.; Yu, Y. Tensile mechanical properties of fiber sheaths microdissected from moso bamboo. *J. Nanjing For. Univ. (Nat. Sci. Ed.)* **2014**, *38*, 6–10.
8. Xu, X.; Liu, F.; Jiang, L.; Zhu, J.Y.; Haagenson, D.; Wiesenborn, D.P. Cellulose Nanocrystals vs. Cellulose Nanofibrils: A Comparative Study on Their Microstructures and Effects as Polymer Reinforcing Agents. *ACS Appl. Mater. Interfaces* **2013**, *5*, 2999–3009. [CrossRef]
9. Zhang, Y.; Hosseinaei, O.; Wang, S.; Zhou, Z. Influence of hemicellulose extraction on water uptake behavior of wood strands. *Wood Fiber Sci. J. Soc. Wood Sci. Technol.* **2011**, *43*, 244–250.
10. Xu, Y.; Liu, X.; Liu, X.; Tan, J.; Zhu, H. Influence of HNO_3/H_3PO_4-$NANO_2$ mediated oxidation on the structure and properties of cellulose fibers. *Carbohydr. Polym.* **2014**, *111*, 955–963. [CrossRef]
11. Fang, X.; Xu, J.; Guo, H.; Liu, Y. The Effect of Alkali Treatment on the Crystallinity, Thermal Stability, and Surface Roughness of Bamboo Fibers. *Fibers Polym.* **2023**, *24*, 505–514. [CrossRef]
12. *GB/T 15780-1995*; Testing Methods for Physical and Mechanical Properties of Bamboos. Institute of Chinese Academy of Forestry Timber Industry: Beijing, China, 1996.
13. Furuta, Y.; Nakajima, M.; Nakatani, T.; Kojiro, K.; Ishimaru, Y. Effects of the Lignin on the Thermal-Softening Properties of the Water-Swollen Wood. *J. Soc. Mater. Sci. Japan* **2008**, *57*, 344–349. [CrossRef]
14. Hao, J.; Liu, W.; Sun, D. Effect of heat treatment on color of bamboo strips. *J. Bamboo Res.* **2012**, *31*, 34–38.
15. Li, Q.; Lin, J.; Chen, Z.; Wu, Q.; Li, X. Influences of Visual Properties of *Phyllostachys heterocycla* cv.pubescens Surface by Alkaline Hydrothermal Pretreatment. *J. Northwest For. Univ.* **2017**, *32*, 213–217.
16. Pandey, K.K.; Pitman, A.J. FTIR studies of the changes in wood chemistry following decay by brown-rot and white-rot fungi. *Int. Biodeterior. Biodegrad.* **2003**, *52*, 151–160. [CrossRef]
17. Meng, F.; Yu, Y.; Zhang, Y.; Yu, W.; Gao, J. Surface chemical composition analysis of heat-treated bamboo. *Appl. Surf. Sci.* **2016**, *371*, 383–390. [CrossRef]
18. Yang, Z.; Jiang, Z.; Fei, B.; Liu, J. Application of Near Infrared(NIR) Spectroscopy to Wood Science. *Sci. Silvae Sin.* **2005**, *41*, 177–183.
19. Qin, L. *Effect of Thermo-Treatment on Physical, Mechanical Properties and Durability of Reconstituted Bamboo Lumber*; Chinese Academy of Forestry: Beijing, China, 2010.
20. Sun, R.H.; Li, X.J.; Liu, Y.; Hou, R.G.; Qiao, J.Z. Effects of high temperature heat treatment on FTIR and XRD characteristics of bamboo bundles. *J. Cent. South Univ. For. Technol.* **2013**, *33*, 97–100.
21. Hosseinaei, O.; Wang, S.; Rials, T.G.; Xing, C.; Taylor, A.M.; Kelley, S.S.; Hosseinaei, S.W.O.; He, C.; Yao, X.; Xue, J.; et al. Effect of Hemicellulose Extraction on Physical and Mechanical Properties and Mold Susceptibility of Flakeboard. *Prod. J.* **2001**, *61*, 31–37. [CrossRef]
22. Hosseinaei, O.; Wang, S.; Rials, T.G.; Xing, C.; Zhang, Y. Effects of Decreasing Carbohydrate Content on Properties of Wood Strands. *Cellulose* **2011**, *18*, 841–850. [CrossRef]
23. Overend, R.P.; Chornet, E. Fractionation of Lignocellulosics by Steam-aqueous Pretreatments. *Phys. Eng. Sci.* **1987**, *321*, 523–536.
24. Yang, S.; Jiang, Z.; Ren, H. Determination of Crystallinity of Bamboo Fiber Using X-ray Diffraction. *J. Northeast For. Univ.* **2010**, *38*, 75–77.
25. Li, X.; Liu, Y.; Gao, J. FTIR and XRD Analysis of Wood Treated at High Temperatures. *J. Beijing For. Univ.* **2009**, *31*, 104–107.
26. Chen, F.; He, Y.; Wei, X.; Han, S.; Ji, J.; Wang, G. Advances in strength and toughness of hierarchical bamboo under humidity and heat. *J. For. Eng.* **2023**, *8*, 10–18.
27. Huang, M.; Zhang, X.; Yu, W.; Li, W.; Liu, X.; Zhang, W. Mechanical Properties and Structural Characterization of Bamboo Softened by High-Temperature Steam. *J. For. Eng.* **2016**, *1*, 64–68.
28. Zhu, J.; Yu, B.; Cao, M.; Wang, X. Effect of Microwave Treatment on Round Bamboo Softening. *For. Ind.* **2023**, *60*, 15–19.
29. Zhao, R.; Fu, D.; Sun, T. Influence of Different Softening Treatments on Bamboo Quality. *J. Jiamusi Univ. (Nat. Sci. Ed.)* **2009**, *27*, 637–640.

30. Mo, J.; Zhang, W. Physical and Mechanical Properties of Thermally Treated Bamboo Based on Near-Infrared Spectroscopy Technology. *J. For. Eng.* **2019**, *4*, 32–38.

31. Li, W.; Lin, W.; Yang, W.; Tang, K. Preparation and Structure Characterization of Bamboo Fiber by Alkali-Boiling and NaClO Oxidation. *For. Ind.* **2021**, *58*, 6–10.

32. Ju, K.; Kil, M.; Ri, S.; Kim, T.; Kim, J.; Shi, W.; Zhang, L.; Yan, M.; Zhang, J.; Liu, G. Impacts of dietary supplementation of bamboo vinegar and charcoal powder on growth performance, intestinal morphology, and gut microflora of large-scale loach Paramisgurnus dabryanus. *J. Oceanol. Limnol.* **2023**, *41*, 1187–1196. [CrossRef] [PubMed]

33. Chen, B.; Hu, J.; Jin, Y.; Zheng, A.; Zhuan, S.; Lin, J.; Guan, X. Prediction for Compressive Strength Parallel to Grain of Zenia insignis Plantation Based on Fourier Infrared Spectroscopy. *J. Southwest For. Univ. (Nat. Sci.)* **2022**, *42*, 178–183.

34. Shi, Y.; Liu, J.; Lü, W.; Wang, J.; Ni, L. Preparation and Properties of Wood Modified with Acidic and Alkaline Silica Sols. *Wood Ind.* **2019**, *33*, 21–24+33.

35. Smith, C.A.; Want, E.J.; O'Maille, G.; Abagyan, R.; Siuzdak, G. XCMS: Processing mass spectrometry data for metabolite profiling using nonlinear peak alignment, matching, and identification. *Anal. Chem.* **2006**, *78*, 779–787. [CrossRef] [PubMed]

36. Xia, J.; Sinelnikov, I.; Han, B.; Wishart, D.S. MetaboAnalyst 3.0—Making metabolomics more meaningful. *Nucleic Acids Res.* **2015**, *43*, W251–W257. [CrossRef] [PubMed]

37. Pluskal, T.; Castillo, S.; Villar-Briones, A.; Orešič, M. MZmine 2: Modular framework for processing, visualizing, and analyzing mass spectrometry-based molecular profile data. *BMC Bioinform.* **2010**, *11*, 395. [CrossRef] [PubMed]

38. Lancaster, C.; Espinoza, E. Evaluating agarwood products for 2-(2-phenylethyl) chromones using direct analysis in real time time-of-flight mass spectrometry. *Rapid Commun. Mass Spectrom.* **2012**, *26*, 2649–2656. [CrossRef] [PubMed]

39. Selvius, D.; Armitage, R. Direct identification of dyes in textiles by direct analysis in real time-time of flight mass spectrometry. *Anal. Chem.* **2011**, *83*, 6924–6928. [CrossRef]

40. Li, G.; Zhang, Y.; Zhao, C.; Xue, H.; Yuan, L. Chemical variation in cell wall of sugar beet pulp caused by aqueous ammonia pretreatment influence enzymatic digestibility of cellulose. *Ind. Crops Prod.* **2020**, *155*, 112786. [CrossRef]

41. Cajka, T.; Riddellova, K.; Tomaniova, M.; Hajslova, J. Ambient mass spectrometry employing a DART ion source for metabolomic fingerprinting/profiling: A powerful tool for beer origin recognition. *Metabolomics* **2011**, *7*, 500–508. [CrossRef]

 *forests*

Article

Effect of High-Intensity Microwave Treatment on Structural and Chemical Characteristics of Chinese Fir

Xiaomei Liao [1,2,3], Xuan Fang [1], Xin Gao [1], Songlin Yi [2] and Yongdong Zhou [1,*]

1. Research Institute of Wood Industry, Chinese Academy of Forestry, Beijing 100091, China; liaoxiaomei@njfu.edu.cn (X.L.); fangxuan@caf.ac.cn (X.F.); gaoxin@caf.ac.cn (X.G.)
2. College Materials Science and Technology, Beijing Forestry University, Beijing 100083, China; ysonglin@bjfu.edu.cn
3. College Furnishings and Industrial Design, Nanjing Forestry University, Nanjing 210037, China
* Correspondence: zhouyd@caf.ac.cn

Citation: Liao, X.; Fang, X.; Gao, X.; Yi, S.; Zhou, Y. Effect of High-Intensity Microwave Treatment on Structural and Chemical Characteristics of Chinese Fir. *Forests* 2024, 15, 516.
https://doi.org/10.3390/f15030516

Academic Editor: Chunping Dai

Received: 2 February 2024
Revised: 1 March 2024
Accepted: 5 March 2024
Published: 11 March 2024

Abstract: High-intensity microwave (HIMW) treatment is a time-saving and environmentally friendly method widely applied in the wood processing industry. It enhances wood permeability, making it suitable for drying and impregnation modification. This study aimed to investigate the effects of HIMW on macroscopic and microscopic cracks, tracheid cell wall damage, and the chemical structure of Chinese fir [*Cunninghamia lanceolata* (Lamb.) Hook] wood. Through the use of a camera, optical microscope, scanning electron microscope, transmission electron microscope, Fourier-transform infrared spectroscopy, and X-ray diffraction, the morphology of cracks, cell wall damage, the chemical composition of the cell wall, and the crystalline structure of cellulose treated with HIMW were examined and analyzed. The results revealed that the initial moisture content (MC) and microwave energy density (MWED) significantly influenced the crack characteristics and cell wall structure and slightly influenced the chemical composition and crystalline structure of cellulose of the Chinese fir cell wall. HIMW treatment can produce different characteristics of wood cracks. The size and number of cracks were significantly increased with the increase in MWED, and more cracks were found in low MC. Microcracks caused by HIMW treatment tended to initiate at the ray parenchyma, resulting in the stripping of ray cells along its radial direction. Meanwhile, the cracking of adjacent cell junctions, the rupturing of the pit margo and pit torus, and cell wall parts tearing along the direction of microfibers occurred as a result of the HIMW treatment. The most severe damage to the cell walls occurred at the interface of S1/S2, S1, and ML layers, and the cell walls were torn in the S2 layer. There were no significant changes in the FTIR spectra of the HIMW treatment samples. Hemicellulose degradation occurred first, which increased with the increase in MWED. The recrystallization of cellulose and the lignin content increased because of the change in the aromatic C=O groups. As MWED increased, both the crystallinity index (CI) and cellulose crystal width (D_{200}) of the samples that underwent HIMW treatment increased accordingly, and the number of low-MC samples was greater than that of the high-MC samples. The findings contribute to understanding the crack characteristics and damage mechanism induced by HIMW treatment on wood. This study provides valuable insights into regulating the effects of HIMW treatment and expanding its application in wood processing, such as wood drying and functionalized impregnation, according to the specific end-use requirements.

Keywords: microwave; Chinese fir; wood structure; cell wall

1. Introduction

Chinese fir [*Cunninghamia lanceolata* (Lamb.) Hook] is a widely cultivated and fast-growing wood species in China. However, it has inherent drawbacks, such as low density, weak strength and stiffness, susceptibility to biological decay, poor dimensional stability, poor permeability in heartwood, and a tendency to crack during drying [1]. These limitations make it unsuitable for direct use in furniture, flooring, wooden structures and construction [2]. Consequently, there is a need to enhance the quality of Chinese fir wood for its broader application across the wood industry [3]. One common approach is to improve its permeability, especially in the heartwood, to facilitate drying and impregnation.

Fluid flow within softwood occurs through three primary pathways. Longitudinally, fluid flow moves through the lumina of the tracheid, the apertures of pits, and the micropores of bordered pit membranes [4,5]. In the tangential direction, fluid flow primarily occurs through tracheids, bordered pits, and ray cells serving as the main radial pathways [6,7]. Additionally, a microcapillary system within tracheid cell walls allows fluid penetration in longitudinal, tangential, and radial directions [8]. Enhancing wood permeability involves opening up these primary fluid pathways.

Microwave treatment is an effective and environmentally friendly method for wood drying and pretreatment before modification, with successful industrial applications [9,10]. HIMW treatment, in particular, provides a greater amount of energy, resulting in higher temperature and steam pressure within the wood in a shorter period of time [9]. The use of continuous feeding devices in HIMW treatment increases efficiency and overcomes the limitations of conventional methods, which struggle to achieve desired treatment effects on large specimens [9–11]. The effectiveness of microwave treatment depends not only on specific parameters such as power, energy application mode, pulse or continuous treatment, exposure time, and speed of timber movement but also on factors like wood species, initial MC, and sample dimensions [9–13].

Under high-intensity irradiation, the water present in wet wood absorbs electromagnetic energy and rapidly evaporates, generating high internal vapor pressure within the wood cells. This pressure leads to the formation of cracks and micro-cracks in the thin-walled cells, creating new pathways for fluid flow and increasing the permeability of the wood [9,12,14–16]. Torgovnikov and Vinden [9] categorized the modification of softwood into three degrees based on the extent of cell damage. A low degree of modification included the rupturing of pit membranes of wood cells and the melting and replacement of resin in channels. At a moderate degree, the pit membranes and ray parenchyma cells underwent further rupturing, and resin underwent boiling and replacement. At a high degree, the primary cell wall of the wood ruptured, including the ray parenchyma cells and tracheids, and cavities formed in the wood. The level of damage varied depending on the processing parameters and wood sample condition.

Low-intensity microwave treatment can cause the middle lamellar between ray parenchyma cells and longitudinal tracheids to crack, bordered pit membranes in the tracheids cell wall to become damaged, and micro-fibrils on the pit margo to rupture. Furthermore, micro-cracks were discovered in cross-field pit apertures, and the aspirated pit was slightly opened [14,16,17]. When the microwave intensity or processing time increased, the pit border separated from the tracheid cell wall, both the pit margo and the pit torus were broken thoroughly, and cracks appeared in the tracheid cell wall along the direction of microfibers. Meanwhile, the cross-field pit membranes disappeared, the ray parenchyma cells were seriously damaged, the tracheid walls were fractured, and macro-cracks were generated [14–16]. HIMW treatment can cause changes in the chemical composition of wood, including hemicellulose, lignin, and cellulose. The crystallinity and cellulose crystal width of the wood cell wall fluctuated with microwave treatment [18]. While some studies have investigated the structural damage of wood during microwave treatment, there is limited research on the changes in the cell wall. The cell wall, which serves as the fundamental component and primary structure of wood, is a thin layer formed through the deposition of cellulose, hemicellulose, and lignin according to specific biological mechanisms. It consists of multiple layers, including the primary wall (P) and secondary wall (S1, S2, and S3 layers), with adjacent cell walls connected by a middle lamella (ML) [19,20].

Despite the known potential for microwave treatment to cause damage to the macro and microstructure of wood, most studies have focused on the damage to intercellular and tissue cell types. Limited research has been conducted on cell wall damage location and the chemical structure following microwave treatment. However, it is important to understand that the macroscopic properties of wood are largely influenced by the cell wall structure and properties at the microscopic level. The macroscopic destruction observed in wood results from the cumulative damage to the cell wall. Therefore, investigating the destruction of the cell wall plays a crucial role in understanding the macroscopic damage of wood. It is crucial to study and elucidate the structural damage and property changes that occur in the wood cell wall during HIMW treatment. Therefore, the objectives of this research were to examine the morphology and size of macroscopic cracks in Chinese fir wood treated with HIMW and analyze the changes in crack morphology and chemical structure at the cell wall level.

2. Materials and Methods

2.1. Materials

The wood used in this study was obtained from fast-growing Chinese Fir. These logs were sourced from a forest farm in Congjiang County, Guizhou Province, China. The diameter of the trunk at breast height was over 40 cm, and the rotation of the trees was about 30 years. The logs were sawn into wood planks with dimensions of $2200 \times 120 \times 50$ mm^3 (longitudinal × tangential × radial, L × T × R). The treated samples with dimensions of $1200 \times 120 \times 50$ mm^3 were obtained from these planks, while the remaining planks were used as control samples. To ensure uniformity, the initial MC of the samples was adjusted to two groups: low MC (L) ranging from 30% to 60%, and high MC (H) ranging from 60% to 90%. The average oven-dry density of the wood was measured as 0.35 g/cm^3.

2.2. High-Intensity Microwave (HIMW) Treatment

The microwave treatment was conducted using continuous feeding high-intensity microwave equipment (WXD 200 L; Sanle, Nanjing, China). This equipment operates at a frequency of 915 MHz and offers adjustable power ranging from 1 kW to 200 kW. The feeding speed can be step-less adjusted between 0.6 m/min and 6 m/min. The resonant cavity of the equipment has dimensions of $550 \times 340 \times 280$ mm^3 (length × width × height), and the microwave energy is emitted from four feed ports of the resonant cavity, with a maximum power of 25 kW at the upper and lower feed port and 75 kW at the left and right feed port (Figure 1b).

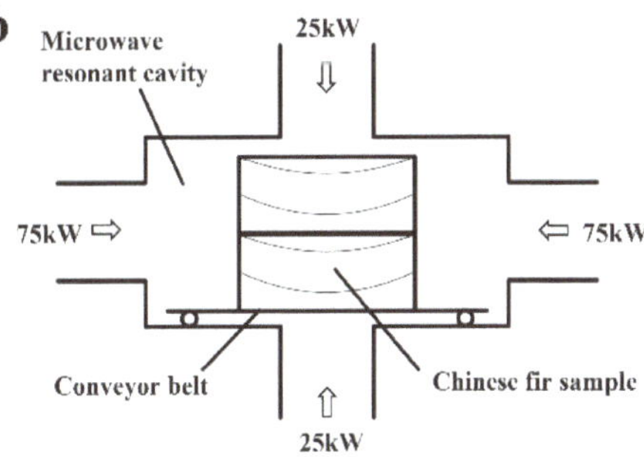

Figure 1. WXD200L microwave equipment. (**a**) Photograph of microwave equipment. (**b**) Four feed port of microwave resonant cavity.

Microwave energy density (MWED) was utilized to quantify the energy absorbed by the samples. MWED represents the energy absorbed per unit volume of the wood sample, and it is influenced by the sample's volume within the microwave radiation, the microwave power, and the duration of exposure. The MWED can be calculated using Equation (1):

$$M = \frac{E}{V} = P \cdot \frac{t}{V} = P \cdot \frac{(l/v)}{V} \tag{1}$$

where M is the microwave energy density in microwave processing (kWh/m^3); E is the microwave energy absorbed by wood (kWh); V is the volume of wood irradiated by microwaves (m^3); P is the microwave power absorbed by wood (kW); t is the processing time in microwave resonance cavity (h); l is the length of the microwave resonant cavity (m) and v is the wood-feeding speed (m/h).

In this study, a microwave input power of 60 kW, 80 kW, and 100 kW, a feeding rate of 1.0 m/min, and a processing duration of 33 s were employed. During microwave processing, it is noted that not all input MWED was absorbed; a portion was reflected back and absorbed by the water load-bearing MW source. To avoid serious cracks caused by HIMW, two planks were stacked feeding during treatment (Figure 1b). The effective energy density acting on the sample was determined by measuring the power absorbed and the reflected power (Table 1).

Table 1. The energy density of Chinese fir during HIMW treatment.

Input Power (kW)	Reflected Power (kW)	Absorbed Power (kW)	Effective MW Energy Density (kWh/m^3)
60	17.9	42.1	58
80	22.28	57.72	80
100	27.81	72.19	100

2.3. Macroscopic Observation

To evaluate the surface checks of Chinese fir samples subjected to HIMW treatment, photographs were captured employing a GR3 camera (Ricoh, Tokyo, Japan). Subsequent analysis and quantification of the surface checks were performed utilizing using Image 1.54d software.

2.4. Optical Microscope Observation

Samples with a dimension of $5 \times 2 \times 2$ mm^3 (L $\times$ T $\times$ R) were prepared from the HIMW treatment samples and the control samples. These samples underwent dehydration using a graded series of acetone and were subsequently embedded in Spurr's low-viscosity resin. Transverse semi-thin sections, approximately 1 µm thickness, were obtained from the embedded samples by a glass cutter and mounted on glass slides. These sections were stained with a 1% w/v solution of toluidine blue in 0.1% boric acid on a hot plate at approximately 60 °C. The stained sections were then observed under an optical microscope (Axio scopeA1; Zeiss, Oberkohen, Germany) to capture the images for analysis.

2.5. Scanning Electron Microscope (SEM) Observation

Samples measuring $5 \times 5 \times 5$ mm^3 (L $\times$ T $\times$ R), extracted from the center position of both the HIMW-treated specimens and the control, were prepared for observation under SEM. Transverse sections (TS), radial longitudinal sections (RLS), and tangential longitudinal sections (TLS) of the samples were prepared by trimming with a microtome (M205C; Leica, Frankfurt, Germany). The trimmed sections were subsequently subjected to adjustment under constant temperature and humidity conditions. Subsequently, they were dried in an oven at 60 °C until a constant weight was achieved. To enhance the conductivity, the sections were coated with gold (Au) using a coating device (Quorum SC7620; East Sussex, UK). The coated sections were then observed using an SEM (ZEISS Gemini 300, Zeiss, Oberkohen, Germany).

2.6. Transmission Electron Microscope (TEM) Observation

The samples used for TEM observation were identical to those for optical microscope. Transverse ultrathin sections, approximately 70 nm thickness, were obtained from the embedded samples using an Ultramicrotome (EM UC7; Leica, Frankfurt, Germany) with histology diamond knives. Following the sectioning process, the ultrathin sections were transformed into copper mesh grids. These grids were then stained with lead citrate and a 50% ethanol-saturated uranyl acetate solution for 5–10 min each. Finally, the stained sections were observed using TEM equipment (Hitachi7650, Ltd., Tokyo, Japan), which ran at an acceleration voltage of 80 kV.

2.7. Fourier-Transform Infrared Spectroscopy (FTIR) Analyzing

The samples were first ground into a powder using a grinder. The resulting material was then screened through a 200 mesh sieve and dried at 60 °C until a constant weight for analysis. Dried powder samples were analyzed using an FTIR spectrometer (Nicolet 6700; Nikola, Boston, MA, USA) in transmission mode to obtain the FTIR spectra. The spectra were collected in the wave number range of 4000–400 cm^{-1}, with 32 scans performed at a resolution of 4 cm^{-1}. To ensure accurate analysis, the FTIR spectra were baseline-corrected using software at specific wave numbers 1800 cm^{-1}, 1548 cm^{-1}, and 840 cm^{-1}.

2.8. X-ray Diffraction (XRD)

The relative crystallinities of the dried powder samples, which were 100–200 mesh size, were determined using an X-ray powder diffractometer (D8 Advance; Bruker, Ettlingen, Germany). CuKα radiation (λ = 1.541 Å) was employed at 40 kV and 40 mA. To conduct the test, the powder samples were placed in a sample box, flattened, and positioned horizontally within the instrument. Intensity measurements were taken in the range of $5° < 2\theta < 40°$, with a scan step size of 0.05° and a scanning speed of 0.1°/s. The crystallinity index (CI) was calculated using Equation (2) [21]:

$$CI = \frac{I_{002} - I_{am}}{I_{002}} \times 100\% \tag{2}$$

where CI expresses the relative degree of crystallinity; I_{002} is the intensity of the crystalline peak at $2\theta = 22.2°$; and I_{am} is the height of the minimum between the 200 and 101 peaks at $2\theta = 18.6°$. Three measurements were performed for each sample group.

The width of the crystallite was determined using the Scherrer equation, which relies on the measurement of the diffraction fringe width of the X-ray reflection in the crystal region. Specifically, the crystalline size D_{200} was calculated based on the diffraction pattern obtained from the 200 lattice planes of cellulose, using Equation (3):

$$D_{200} = \frac{K\lambda}{\beta \cos\theta} \tag{3}$$

where D_{200} is the average size of the crystallites; K is the Scherrer constant (K = 0.9); λ is the X-ray wavelength (λ = 0.1541 nm); β is the peak width of the (200) profile at half maximum (FWHM) in radians; and θ is the half of the (200) Bragg diffraction peak position in radian [22].

3. Results and Discussion

3.1. Macroscopic Surface Cracks Analysis

The wood samples subjected to HIMW treatment exhibit surface cracks, the number and size of which vary depending on different process parameters. To analyze these surface cracks, photographs were captured using a camera, and the original images were subsequently converted into grayscale and processed for edge detection using ImageJ software (refer to Figure 2). The edges of the cracks were identified and quantitatively measured to determine their length and width.

The number of surface cracks and the corresponding length and width of each crack after HIMW treatment are shown in Figure 3. The number of surface checks increases with higher MWED values. When the MWED was 80 kWh/m^3, the number of cracks increased sharply, and the crack size of the high MC group was larger than that of the low MC group. While the MWED was increased to 100 kWh/m^3, samples with lower MC exhibited more surface cracks than those with higher MC. As presented in Table 2, when the MWED ranges from 58 kWh/m^3 to 100 kWh/m^3, and the MC of the sample is lower, the average length o cracks increases from 28.69 mm to 49.27 mm, a rise of 71.7%. Similarly, the mean width o the widest crack nearly doubles from 0.123 mm to 0.240 mm. Conversely, when the MC o the sample is higher, the average length of cracks increases from 20.28 mm to 44.30 mm representing a substantial increase of 118.4%. Additionally, the average width grows from 0.15 mm to 0.199 mm, a rise of 32.7%.

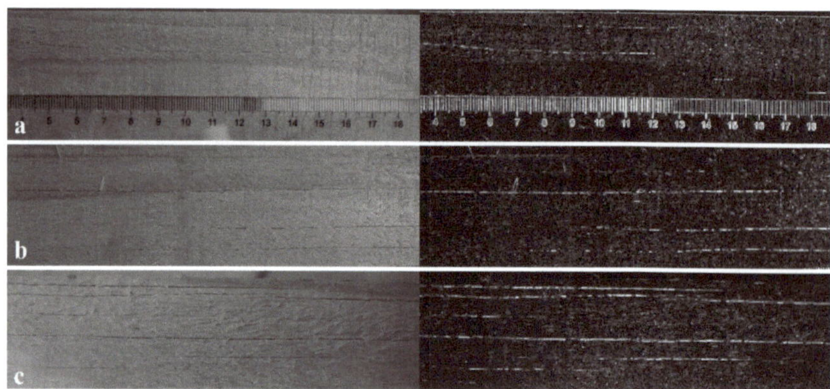

Figure 2. Surface cracks caused by HIMW treatment. (**a**) MWED is 58 kWh/m^3, and the initial MC is 53% (58 L). (**b**) MWED is 80 kWh/m^3, and the initial MC is 48% (80 L). (**c**) MWED is 100 kWh/m^2 and the initial MC is 47% (100 L).

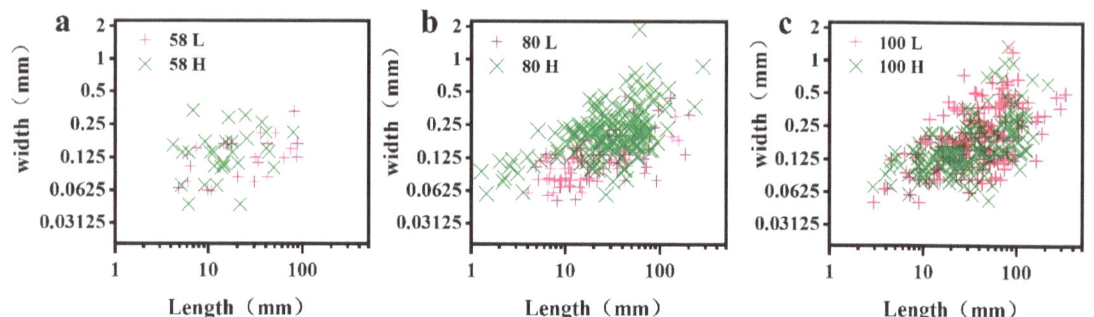

Figure 3. Characteristic of surface cracks caused by HIMW treatment. (**a**) The MWED is 58 kWh/m^2 (**b**) The MWED is 80 kWh/m^3. (**c**) The MWED is 100 kWh/m^3.

Table 2. Surface cracks characteristic of HIMW treatment.

Samples	Width/mm	Length/mm	Length Proportion of Totality/%				Width Proportion of Totality/%			
			Below 10 mm	10–40 mm	40–80 mm	Above 80 mm	Below 0.1 mm	0.1–0.2 mm	0.2–0.4 mm	Above 0.4 mm
58 L	0.123 ± 0.056	28.69 ± 23.09	28.9	44.4	20.0	6.7	42.2	55.6	2.2	0
58 H	0.150 ± 0.073	20.28 ± 16.84	27.6	57.2	17.2	0	17.2	62.1	20.7	0
80 L	0.130 ± 0.067	31.82 ± 34.43	24.7	49.5	17.9	7.9	34.7	52.6	11.6	1.1
80 H	0.245 ± 0.191	40.55 ± 35.85	11.5	46.5	45.8	7.7	7.7	36.8	44.5	11
100 L	0.240 ± 0.194	49.27 ± 46.37	9.3	43.5	32.3	14.9	15.5	38.8	32	13.7
100 H	0.199 ± 0.171	44.30 ± 37.18	12.8	47.7	25.6	13.9	15.1	52.9	24.4	7.6

The length and width of the cracks observed were categorized into four grades, and the proportion of each grade to the total number of cracks was determined. The length of crack was categorized into below 10 mm, 10–40 mm, 40–80 mm, and above 80 mm, and the width was categorized into below 0.1 mm, 0.1–0.2 mm, 0.2–0.4 mm, and above 0.4 mm (Table 2). The number of cracks in each grade was counted as a percentage of the total number of cracks. In the case of MWED ranging from 58 kWh/m^3 to 100 kWh/m^3, particularly evident in the 80 H (MWED is 80 kWh/m^3 and high MC) group, there was a significant decrease in the proportion of cracks with lengths below 10 mm. Conversely, the proportion of cracks with lengths above 40 mm was substantially increased. The proportion of the grades in 10–40 mm was not significant. Furthermore, significant differences were observed in the proportion of cracks with a width below 0.2 mm, especially widths below 0.1 mm, in the 58 L (MWED is 58 kWh/m^3 and low MC) group and 80 L group compared to other groups. With an increase in MWED, the proportion of cracks with widths above 0.2 mm substantially increased, indicating an increase in the number of larger cracks in the groups above 80 H group.

The occurrence of wood cracks is affected by the initial MC of the sample, where samples with a lower MC demonstrate an increased likelihood of crack formation. This phenomenon can be attributed to the rapid heating rate within the wood, which accelerates the evaporation process, resulting in a significant pressure differential between the internal and external environments of the wood. Such a differential facilitates the easy formation of cracks within the wood structure. In contrast, the heating rate of wood with a higher MC was slow because the microwave energy is absorbed by the excess water; meanwhile, it exhibits a more uniform internal heating temperature [14,23]. When comparing samples with the same MC, increased energy absorption correlates with a reduced duration required for water evaporation, resulting in higher steam pressure and a greater destructive power force upon the wood's structure. Consequently, internal structural damage extends to the surface, giving rise to cracks [9].

3.2. Damage in Cell Wall Analysis

3.2.1. Optical Microscope Analysis

To analyze the damage patterns characterized by cell wall cracking, samples treated with HIMW were subjected to detailed observation through optical microscopy and TEM techniques. These observations were conducted on semi-thin and ultra-thin sections to facilitate a comprehensive analysis of the cell wall structural alterations induced by the treatment. As shown in Figure 4, it can be observed that when Chinese fir is treated with HIMW, the initial damage location often occurs near the wood ray. Cracks form along the radial wood rays, and there appears to be separation and peeling between the ray cells and tracheids (Figure 4a). When the MWED reaches 80 kWh/m^3, the cracks in the radial wood rays widen further, and the tracheids near the ray cells separate and extend along the tangential direction (Figure 4b). This is mainly due to the fact that the ray cells are parenchyma cells, which absorb microwave energy and rapidly heat up, generating vapor pressure that impacts and destroys the intercellular layer of the ray cells with lower intensity [9,14]. With an increase in MWED to 100 kWh/m^3, the separation between tracheids is further expanded, and cracks typically start at one wood ray and end at another ray (Figure 4c). This is primarily caused by initial fracture cracks occurring in the tracheid of the cross-field region, resulting in transverse wall breakage. The cross-field region is prone to rupture because the microfibril of wood rays and tracheids cell walls are arranged in different directions [16].

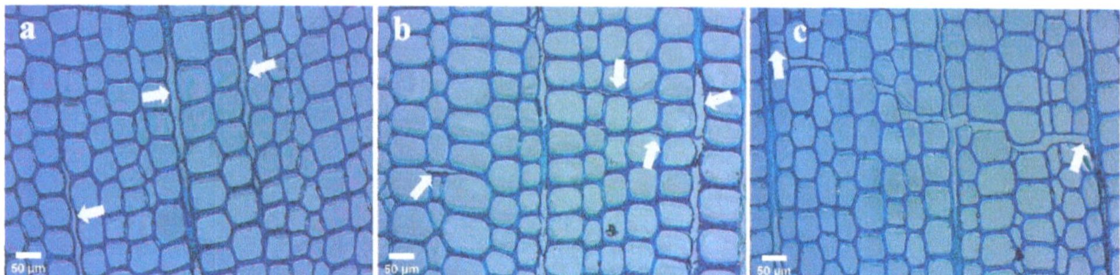

Figure 4. Semi-thin section images of HIMW-treated specimen. (**a**) TS plane of HIMW (60 L) specimen. (**b**) TS plane of HIMW (80 L) specimen. (**c**) TS plane of HIMW (100 L) specimen. The arrows in the figures indicate the location of the damage.

3.2.2. SEM Analysis

Figure 5 presents SEM observations of the sections (TS, RLS, TLS) of the control and HIMW treatment specimens. In the TS plane of the control specimens (Figure 5a), the integrity between the tracheids and wood rays is clearly observed. However, distortion is evident in the HIMW treatment specimens (Figure 5b), and there is a gap of approximately 30 µm in the wood ray position. Notably, the cell wall separation between the tracheids often occurs near the wood rays (Figure 5c). The ray parenchyma cells, being structurally weaker, are initially damaged during the HIMW treatment process [18].

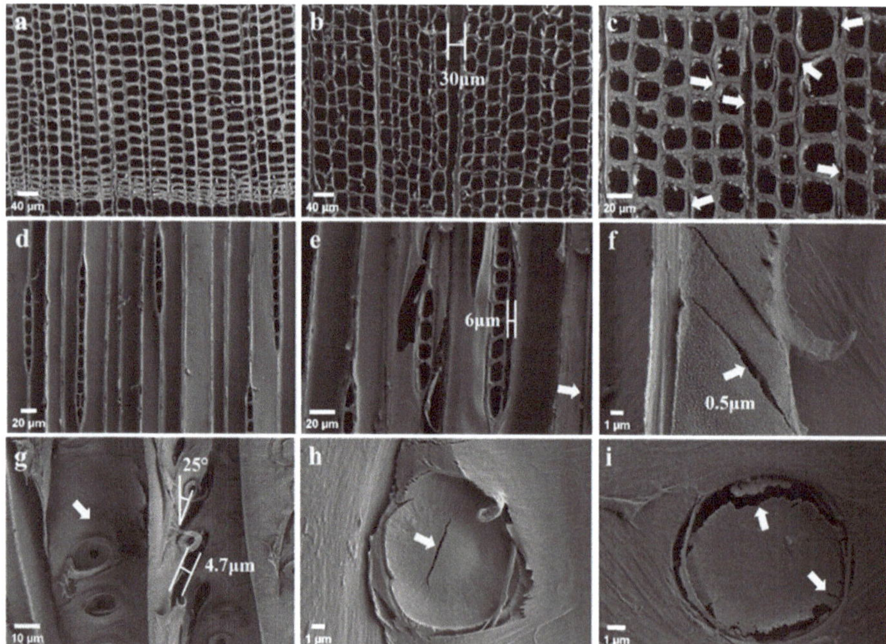

Figure 5. SEM images of control and HIMW-treated specimen. (**a**) TS of control specimen. (**b**) TS of HIMW (80 L) specimen. (**c**) TS of HIMW (80 L) specimen. (**d**) TLS of control specimen. (**e**) TLS of HIMW (80 L) specimen. (**f–i**) RLS e of HIMW (100 L) specimen. The arrows in the figures indicate the location of the damage.

These observations were also noted in the TLS plane, where the MWED was 80 kWh/m. It was observed that the intermediate lamellae between the ray parenchyma cells and the longitudinal tracheids were cracked (Figure 5e); meanwhile, in the control specimen, integrity was maintained (Figure 5d). Furthermore, the compound middle lamella (CML between the tracheids exhibited tears and micro-cracks (Figure 5e). These phenomena can be attributed to the vapor pressure generated by the rapid evaporation of water inside the wood [17,24]. Upon further increasing the MWED to 100 kWh/m^3, a portion of the weakened tracheid walls were torn, and the cracks propagated along the direction of microfibers, which are parallel to the microfiber angle of the S2 layer of the wood cell wall. The crack tilt is approximately 25° (the microfiber angle of the S2 layer of Chinese fir is typically 25–30° away from the fiber axis [25]) (Figure 5f). These cracks also occurred in the bordered pit (Figure 5g), leading to a rupture of the bordered pit into a 4.7 μm width crack along the microfibers of the S2. This finding may be attributed to the physical and chemical properties of the S2 layer, which are believed to dominate the cell wall [14,24]. In wood fiber materials, compared to those in the parallel direction, the strength and stiffness values perpendicular to the wood fiber direction are lower [26]. Consequently, it is more likely to experience vertical breakage under steam pressure. The adjacent microfibers separate, resulting in the formation of cracks that propagate along the orientation direction of the microfibers.

In the heartwood of Chinese fir, the bordered pit membranes are typically covered with amorphous materials, resulting in the pit membrane being completely encrusted and blocking the micropores on the pit margo. This significantly hinders permeability [27]. Additionally, the pit torus (Figure 5h) and pit margo (Figure 5i) were also ruptured and cracked due to the HIMW treatment. The HIMW treatment can also dissolve the amorphous materials on the pit margo, expose the numerous interwoven microfibrils, and increase the number and diameter of the pores. The high-pressure vapor formed within the lumen of the tracheid cell had difficulty escaping through the closed migration paths (pit aspiration). Consequently, the vapor pressure may lead to cracks in the weak region formed by closely aligned bordered pits [14,17].

3.2.3. TEM Analysis

To further examine the damage to the cell wall layer, TEM was utilized to observe the cell wall damage (Figure 6). The TEM image reveals detailed cracks on the cross-section of the fractured tracheid. Comparing it with the control specimen (Figure 6a), it is evident that the interface debonding occurs at the S1/S2 interface, splitting at the S1 layer and CML (Figure 6b). This may be attributed to the microfibril angles being completely different in the S1 layer and S2 layer and the irregular arrangement angle of the microfibers in the S1 layer. Additionally, cracks in the S2 layer of the cell wall are roughly torn (Figure 6c). This indicates that the cracks observed under SEM are mainly due to the tearing of the S2 layer in the tracheid cell wall (refer to Figure 5f,g). It is speculated that the fracture process of the tracheid wall initiates from the fracture of the S1/S2 interface and S1, and then the crack extends to the S2 layer (Figure 6c).

The bordered pit of Chinese fir has aspirated pits (Figure 6d), and the impermeable pit torus blocks the path between tracheid cells, significantly affecting breathability. The pit margo, a reticular structure composed of numerous interwoven microfibrils, facilitates fluid flow between adjacent tracheids through its micropore pore structure [28]. The vapor pressure generated by HIMW causes the pit margo to break and dissolve, leading to some pit torus resilience. Additionally, HIMW treatment can result in the tearing of the pit border (Figure 6e). The direction of the arrangement of the microfibers near the bordered pit was different. The microfibers are arranged to detour the pit aperture, and the cracks extend along the edge of the pit aperture.

HIMW treatment of wood aims to enhance permeability, which can be effectively achieved by introducing appropriate cracks. Analyzing and measuring the morphology of these cracks can serve as an evaluation criterion for assessing the effectiveness of microwave treatment. By carefully examining the cracks, the appropriate energy density can be chosen for processing based on the intended use of the wood.

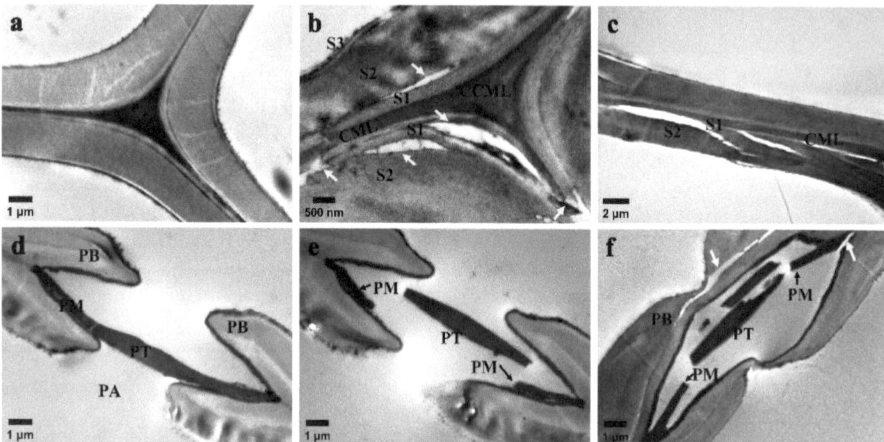

Figure 6. TEM images of cross-sections of control and HIMW treatment specimens, taken at 80 kV. (**a**) Cell wall of control. (**b,c**) Cell wall of HIMW treatment (100 L). (**d**) Bordered pit (aspirated pit) of control. (**e,f**) Bordered pit of HIMW treatment (100 L). CCML: cell corner middle lamella between adjoining tracheids; CML: compound middle lamella (P and ML); S1: outer secondary wall; S2: middle secondary wall; S3: inner secondary wall; PT: pit torus; PM: pit margo; PB: pit border; PA: pit aperture. The arrows in the figures indicate the location of the damage.

3.3. FTIR Spectroscopy Analysis

Wood cell walls are primarily composed of cellulose, hemicellulose, and lignin, with the chemical component significantly influencing the properties of wood [29]. Degradation of wood material commences when its temperature exceeds 100 °C, a process that significantly affects the primary chemical composition of the cell wall [23]. HIMW treatment employs high-temperature steam heating of the wood, which can induce alterations in the chemical structure of the cell wall. Furthermore, the non-thermal effects of microwaves also affect the structure of the treated sample [30,31].

The chemical functional groups of wood and their corresponding FTIR peaks are listed in Table 3. The absorption peaks at 1605 cm^{-1}, 1510 cm^{-1}, and 1268 cm^{-1} are associated with characteristic stretching or bending vibrations in different lignin groups, and the absorption peaks at 1425 cm^{-1}, 1371 cm^{-1}, 1317 cm^{-1}, and 896 cm^{-1} correspond to cellulose groups. Additionally, the absorption peaks at 1736 cm^{-1} and 810 cm^{-1} are assigned to hemicellulose groups [32–34].

Table 3. Chemical functional group and corresponding FTIR peaks.

Assignment	FTIR Peak (cm^{-1})	Functional Group
Hemicellulose	810	Glucomannan in softwood
Cellulose	896	C−H bending and asymmetric out-of-plane ring stretching
Lignin	1268	C−O stretching vibrations of the methyl and phenyl propane units
Cellulose	1317	CH$_2$ wagging
Cellulose	1371	C−H bending vibrations in polysaccharides
Cellulose	1425	C=C aromatic skeleton vibration and CH$_2$ bending vibration
Lignin	1510	C=C stretching of the aromatic ring
Lignin	1605	C=O stretching vibration and C=C aromatic ring vibration
Hemicellulose and xylan	1736	C=O stretching of acetyl or carboxylic acid

Figure 7 displays the average FTIR absorbance spectra of HIMW treatment and control samples in the 1800−800 cm^{-1} fingerprint region. All spectra were normalized at 1025 cm^{-1}. Compared to the control samples, the chemical structure of the HIMW treatment samples exhibited slight changes in the wood cell wall. By analyzing the trends of characteristic peaks of cellulose, hemicellulose, and lignin, it was observed that different chemical components displayed different responses to HIMW treatment.

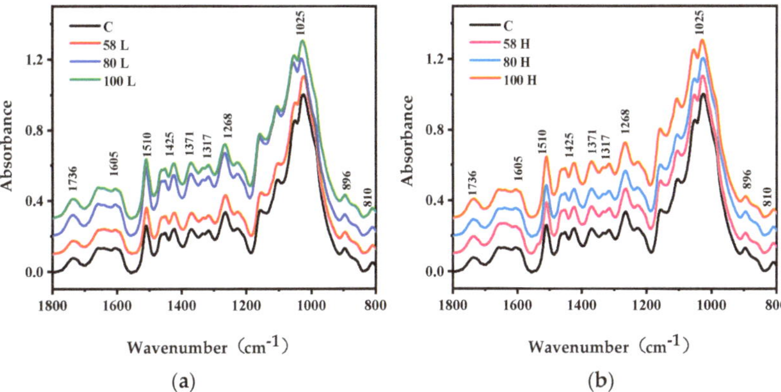

Figure 7. FTIR spectra of the wood cell wall of control and HIMW treatment specimens. (**a**) FTIR spectra of low MC and control specimens. (**b**) FTIR spectra of high MC and control specimens.

At 1736 cm^{-1} and 810 cm^{-1}, the absorbance intensity of the HIMW treatment samples was higher than that of the control samples, indicating the destruction of carbonyl and acetyl groups in xylan and glucomannan. These absorption peaks correspond to the stretching vibrations of the carbonyl group (C=O) in the xylan branch chain and glucomannan in softwood [30] (Table 3). When the MWED was 60 kWh/m^3 for both low and high MC samples, no notable changes in absorbance intensity were observed between the HIMW treatment and control samples. However, at MWEDs of 80 kWh/m^3 and 100 kWh/m^3, the absorbance intensity of the low MC hemicellulose peak showed an overall increase (Figure 7a), while the changing trend in the high MC sample was not evident (Figure 7b). This could be attributed to the negative impact of a large amount of free water on temperature rise.

Regarding cellulose, the absorbance peaks at 1317 cm^{-1}, 1371 cm^{-1}, 1425 cm^{-1}, and 896 cm^{-1} are associated with oscillating vibrations of CH$_2$ in crystalline cellulose, bending vibrations of C−H in polysaccharides, and plane bending vibrations of O−H in amorphous cellulose, respectively. These peaks are used to evaluate the structural changes in cellulose [35,36]. Due to its high thermal stability and relatively compact structure, cellulose experienced minimal damage, increasing cellulose adsorption slightly at 1317 cm^{-1} and 1425 cm^{-1}. However, with the increase in MWED, especially at 80 kWh/m^3 and 100 kWh/m^3, cellulose underwent a change under high-temperature conditions. This may be caused by the non-thermal effects of microwaves [30].

The absorbance peaks at 1268 cm^{-1}, 1605 cm^{-1}, and 1510 cm^{-1} in lignin exhibit elevated levels across most spectra. These changes in peak intensity are attributed to alterations in the relative content of lignin. The absorption peak at 1510 cm^{-1} shows a significant increase in HIMW treatment wood samples compared to the control, particularly at low MC, MWED is 80 kWh/m^3 and 100 kWh/m^3. This increase is due to the degradation of other components and the rise in the relative content of the aromatic skeleton, which is caused by a complex combination of microwave and hygrothermal effects [37]. On the other hand, the characteristic absorption peaks at 1268 cm^{-1} and 1605 cm^{-1} only show slight increases after HIMW treatment, indicating complex structural changes within guaiacyl lignin and side chains [38]. The change in the chemical composition of cellulose, hemicellulose, and lignin are caused by the HIMW treatment. These may affect the physical and mechanical properties of wood and the crosslinking of impregnation modifier. However, further research is needed to understand the specific structural components of cell wall chemistry affected by HIMW treatment.

3.4. Cellulose Crystalline Structure Analysis

The hemicellulose and lignin present in wood are characterized by their amorphous nature, whereas cellulose consists of both amorphous and crystalline regions. The crystalline structures of cellulose are relatively small in size. The thermal stability, mechanical strength, hardness, density, and hygroscopicity of wood are significantly impacted by the structure of crystalline and the amorphous regions within cellulose [39].

Crystallinity is a crucial parameter that measures the proportion of crystalline regions compared to amorphous content in a material [40]. In this study, XRD was employed to investigate changes in the crystallinity of wood, as well as the crystal width of cellulose crystallites [41]. Variations in the intensity of the diffraction bands indicated alterations relative to crystallinity. The peak near $2\theta = 22°$, corresponding to the crystallographic plane with a Miller index of (200), was commonly used to assess the crystallinity of the samples [35,42]. A moderate peak was observed at around 34.5°, which belonged to the (040) crystallographic plane, which was not the dominant contributor [43].

The HIMW treatment did not induce any changes in the crystal structure of Chinese fir cellulose, as evidenced by similar XRD patterns observed in both treated and control samples (Figure 8a). The HIMW treatment samples did not exhibit noticeable alterations in the number or position of reflection peaks compared to the control sample, indicating the absence of crystal-type transformations during the HIMW treatment. However, under different conditions, the intensity and width of the (200) reflection peaks in the treated samples changed, indicating some crystallinity modifications. Furthermore, a decrease in the diffraction intensity of amorphous regions near 2θ of 18.6° was observed in the samples after HIMW treatment, which is attributed to amorphous scattering in cell walls. Interestingly, the HIMW treatment significantly enhanced the diffraction of the amorphous regions of cellulose, resulting in a higher degree of crystallinity in the treated samples.

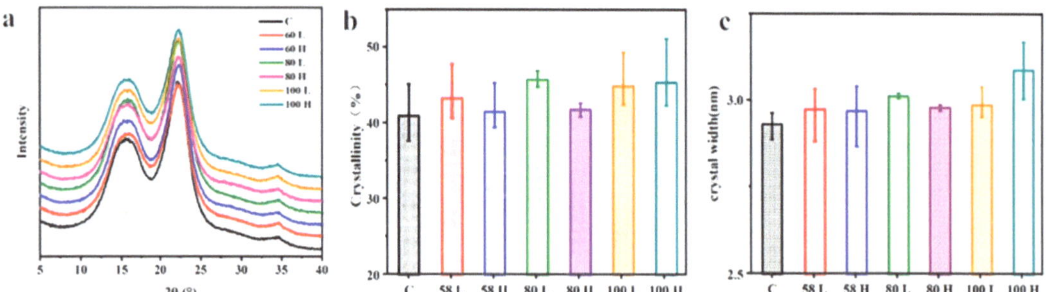

Figure 8. The cellulose crystalline structures of the control and HIMW treatment specimens. (**a**) X-ray diffractograms. (**b**) The relative crystallinities. (**c**) Crystal width.

Figure 8b illustrates that the control sample has a crystallinity index (*CI*) value of 41.03%. As the MWED increased, the *CI* value of the treated samples increased compared to the control samples. When the MWED is increased from 58 kWh/m^3 to 100 kWh/m^3, the mean *CI* of the low MC sample and that of the sample with the high MC sample exhibit an increase of 3.83% and 9.29%, respectively. The reduction in the amorphous cellulose region was the main reason for the increase in *CI* of HIMW treatment samples. At the same MWED, except for the group of 100 kWh/m^3, the crystallinity of the low MC sample is higher than that of the high MC sample because the sample with low MC obtains a higher temperature in a short time during HIMW treatment. This is a contribution to the thermal effect of microwaves. This increase in crystallinity can be attributed to the decomposition of amorphous regions in cellulose and hemicellulose, as well as the crystallization of amorphous cellulose resulting from thermal degradation [44,45]. Additionally, the non-thermal effect of dielectric relaxation induced by microwaves leads to a change in the initial distance between hydroxyl groups due to variations in hydrogen bonding degrees of freedom. This, in turn, promotes the recrystallization of cellulose to some extent [46].

Figure 8c depicts the changes in the crystal width of cellulose in the HIMW treatment and control samples. The average D_{200} value of the control sample is 2.93 nm and is lower than that of the HIMW treatment samples. When the MWED is 100 kWh/m^3, the average D_{200} values of the low and high MC samples increased to 2.99 nm and 3.09 nm, resulting in a 2.05% and 5.46% increase, respectively. This phenomenon can be attributed to hydrothermal processes that rapidly weaken the cellulose molecular chains and improve mobility [46]. The exposure of cellulose to HIMW facilitates the reorganization and recrystallization processes, enhancing the volume and degree of microfibril aggregation, ultimately increasing cellulose's crystallinity and crystal width [47,48].

4. Conclusions

This study focused on the crack morphology, damage position, and chemical composition changes in the cell wall of Chinese fir via HIMW treatment. The results showed the MC of the wood and the MWED played significant roles in influencing the crack morphology. When the MWED is 58 kWh/m^3 with low MC and high MC and 80 kWh/m^3 with low MC, the proportion of cracks with a length below 40 mm and width below 0.2 mm is more than 70% and 80%, respectively. Morphological observations revealed that as the MWED increased, so did the maximum crack's width and length. Additionally, the low MC in the wood caused more cracks, and the higher MC in the wood resulted in more uniform cracks. HIMW treatment induced various types of cracks in wood, including cracks between wood ray parenchyma and longitudinal tracheids and between longitudinal tracheids. In addition, tearing of the cell wall along the microfiber angle and rupturing of the pit torus and pit margo on the pit membrane were observed in the wood using the HIMW treatment. The primary mode of cell wall destruction involved tearing from the junction of the S1/S2 interface and destruction in the S2 layer. Furthermore, HIMW treatment led to a slight change in the cell wall composition, namely cellulose, hemicellulose, and lignin. With the increase in MWED, hemicellulose degradation occurs first. The recrystallization of cellulose and the lignin content increases because the aromatic C=O groups change. The non-thermal effects of microwaves may affect the relative cellulose content of HIMW treatment samples, even though cellulose has greater thermal stability. These changes in chemical composition and structure consequently affected the crystallinity and crystal size of the cell wall. In particular, higher MWED correlated with increased relative crystallinity and crystal width. Cracks in wood and damage to cell walls can improve the drying rate and drying quality of wood. Moreover, in the functional modification impregnation of Chinese fir wood, suitable cracks can provide favorable channels for the penetration of impregnant chemicals so that the wood modifier can penetrate deeper and be distributed uniformly, thus improving the high efficiency and high value-added utilization of Chinese fir wood.

Author Contributions: Conceptualization, X.L. and X.G.; methodology, X.L.; software, X.F.; validation, X.G.; formal analysis, X.L.; data curation, X.L. and X.F.; writing—original draft preparation, X.L.; writing—review and editing, X.L. and Y.Z.; supervision, Y.Z. and S.Y.; funding acquisition, Y.Z. All authors have read and agreed to the published version of the manuscript.

Funding: This research was funded by the National Key R&D Program of China (2023YFD2201402), the National Project to Promote S&T Achievements in Forestry and Grassland (2020133146) and the National Natural Science Foundation of China (32101461). The APC was funded by the National Key R&D Program of China.

Data Availability Statement: The raw data supporting the conclusions of this article will be made available by the authors on request.

Conflicts of Interest: The authors declare no conflicts of interest.

References

1. Ganguly, S.; Hom, S.K.; Tripathi, S.; Ghosh, S.; Kanyal, R.; Samani, A. Quantitative evaluation of microwave irradiation on short-rotation plantation wood species. *Maderas-Cienc. Tecnol.* **2021**, *23*, 1–14. [CrossRef]
2. Yu, X.Y.; Wei, N.F.; Liu, Q.S.; Wu, Z.Y.; Fan, M.Z.; Zhao, W.G.; Zeng, Q.Z. Study on microwave pretreatment technology to improve theeffect of shellac impregnation of fast-growing Chinese Fir. *J. Renew. Mater.* **2022**, *8*, 2041–2053. [CrossRef]
3. Okon, K.E.; Lin, F.C.; Lin, X.; Chen, C.X.; Chen, Y.D.; Huang, B. Modification of Chinese fir (*Cunninghamia lanceolata* L.) wood by silicone oil heat treatment with micro-wave pretreatment. *Eur. J. Wood Wood Prod.* **2017**, *76*, 221–228. [CrossRef]
4. Petty, J.A. Permeability and structure of the wood of Sitka spruce. *Proc. R. Lond.* **1970**, *175*, 149–166.
5. Hansmann, C.; Gindl, W.; Wimmer, R.; Teischinger, A. Permeability of wood—A Review. *Wood Res.* **2002**, *47*, 1–16.
6. Wardrop, A.B.; Davies, G.W. Morphological factors relating to the penetration of liquids into wood. *Holzforschung* **1961**, *15*, 129–141. [CrossRef]
7. Bailey, P.J.; Preston, R.D. Some aspects of softwood permeability I. Structural studies with Douglas fir sapwood and heartwood. *Holzforschung* **1969**, *23*, 113–120. [CrossRef]
8. Palin, M.A.; Petty, J.A. Permeability to water of the cell wall material of spruce heartwood. *Wood Sci. Technol.* **1981**, *15*, 161–166. [CrossRef]
9. Torgovnikov, G.; Vinden, P. High-intensity microwave wood modification for increasing permeability. *For. Prod. J.* **2009**, *59*, 84–92.
10. Torgovnikov, G.; Vinden, P. Microwave method for increasing the permeability of wood and its applications. *For. Prod. J.* **2010**, *60*, 303–311.
11. Fan, Z.Q.; Peng, L.M.; Liu, M.H.; Feng, Y.; He, J.R.; Wu, S.Q. Analysis of influencing factors on sound absorption capacity in microwave-treated *Pinus radiata* wood. *Eur. J. Wood Wood Prod.* **2022**, *80*, 985–995. [CrossRef]
12. Hermoso, E.; Vega, A. Effect of microwave treatment on the impregnability and mechanical properties of *Eucalyptus globulus* wood. *Maderas-Cienc. Tecnol.* **2016**, *18*, 55–64. [CrossRef]
13. Zhang, J.X.; Luo, Y.F.; Liao, C.R.; Xiong, F.; Li, X.; Sun, L.L.; Li, X.J. Theoretical investigation of temperature distribution uniformity in wood during microwave drying in three-port feeding circular resonant cavity. *Dry. Technol.* **2017**, *35*, 409–416. [CrossRef]
14. Weng, X. Study on the Influence Mechanism of Microwave Treatment on Drying Characteristics of Chinese Fir Wood. Ph.D. Thesis, Research Institute of Wood Industry, Chinese Academy of Forestry, Beijing, China, 2020.
15. Mascarenhas, F.J.R.; Dias, A.M.P.G.; Christoforo, A.L. State of the art of microwave treatment of wood: Literature review. *Forest* **2021**, *12*, 745. [CrossRef]
16. He, S.; Yu, H.; Wu, Z.X.; Chen, Y.H.; Fu, F. Effect of microwave treatment on liquid impregnate property of *Pinus Sylvestris*, L. var. Lumber. *J. Microw.* **2016**, *32*, 90–96.
17. Zhang, Y.L.; Jia, K.; Cai, L.P.; Shi, S.Q. Acceleration of moisture migration in Larch wood through microwave pre-treatments. *Dry. Technol.* **2013**, *31*, 666–671. [CrossRef]
18. Xing, X.F.; Li, S.M.; Jin, J.W.; Lin, L.Y.; Zhou, Y.D.; Peng, L.M.; Fu, F. Improving gas permeability and characterizing the multi-scale pore size distribution of radiata pine (*Pinus radiata* D. Don) treated via high-intensity microwave. *Wood Sci. Technol.* **2023**, *57*, 1345–1367. [CrossRef]
19. Terashima, N.; Fukushima, K.; He, L.F.; Takabe, K. Comprehensive model of the lignified plant cell wall. In *Forage Cell Wall Structure and Digestibility*; American Society of Agronomy, Inc.: Madison, WI, USA, 1993; pp. 247–270.
20. Salmén, L.; Burgret, I. Cell wall features with regard to mechanical performance. A review COST Action E35 2004-2008: Wood machining-micromechanics and fracture. *Holzforschung* **2009**, *63*, 121–129. [CrossRef]
21. Segal, L.; Creely, J.J.; Martin, A.E.; Conrad, C.M. An empirical method for estimating the degree of crystallinity of native cellulose using the X-ray diffractometer. *Text. Res. J.* **1962**, *29*, 786–794. [CrossRef]
22. Nishiyama, Y.; Kuga, S.; Okano, T. Mechanism of mercerization revealed by X-ray diffraction. *J. Wood Sci.* **2000**, *46*, 452–457. [CrossRef]
23. Saporiti, J. Effect of microwave treatment on oak compression strength. *Silva Lusit.* **2006**, *14*, 51–58.
24. Kol, H.S.; Çayır, B. The effects of increasing preservative uptake by microwave pre-treatment on the microstructure and mechanical properties of Oriental spruce. *Wood Mater. Sci. Eng.* **2023**, *18*, 732–738. [CrossRef]
25. Li, Z.; Zhan, T.; Eder, M.; Jiang, J.; Lyu, J.; Cao, J. Comparative studies on wood structure and microtensile properties between compression and opposite wood fibers of Chinese fir plantation. *J. Wood Sci.* **2021**, *67*, 12–18. [CrossRef]
26. Muzamal, M. Structural Modifications in Spruce Wood during Steam Explosion Pretreatment. Ph.D. Thesis, Chalmers University of Technology, Gothenburg, Sweden, 2016.
27. Tong, Y.P.; Zhao, G.J. Structure of bordered pit membrane of *Cunninghamia lanceolata* tracheid. *Sci. Silvae Sin.* **2007**, *43*, 151–153.
28. Bao, F.C.; Lyu, J.X.; Zhao, Y.K. Effect of different positions of bordered pit torus in Yezo Spruce on its permeability. *Acta Bot. Sin.* **2001**, *43*, 119–126.
29. Booker, R.E.; Sell, J. The nanostructure of the cell wall of softwoods and its functions in living tree. *Holz Roh Werkst.* **1998**, *56*, 1–8. [CrossRef]

0. Zhai, C.K.; Teng, N.; Pan, B.H.; Chen, J.; Liu, F.; Zhu, J.; Na, H.N. Revealing the importance of non-thermal effect to strengthen hydrolysis of cellulose by synchronous cooling assisted microwave driving. *Carbohydr. Polym.* **2018**, *197*, 414–421. [CrossRef] [PubMed]
1. Bichot, A.; Lerosty, M.; Radoiu, M.; Méchin, V.; Bernet, N.; Delgenès, J.-P.; García-Bernet, D. Decoupling thermal and non-thermal effects of the microwaves for lignocellulosic biomass pretreatment. *Energy Convers. Manag.* **2020**, *203*, 112220. [CrossRef]
2. Fackler, K.; Stevanic, J.S.; Ters, T.; Hinterstoisser, B.; Schwanninger, M.; Salmén, L. FTIR imaging microscopy to localise and characterise simultaneous and selective white-rot decay within spruce wood cells. *Holzforschung* **2011**, *65*, 411–420. [CrossRef]
3. Temiz, A.; Terziev, N.; Jacobsen, B.; Eikenes, M. Weathering, water absorption and durability of silicon, acetylated and heat-treated wood. *J. Appl. Polym. Sci.* **2006**, *102*, 4506–4513. [CrossRef]
4. Popescu, C.M.; Singurel, G.; Popescu, M.C.; Vasile, C.; Argyropoulos, D.S.; Willfor, S. Vibrational spectroscopy and X-ray diffraction methods to establish the differences between hardwood and softwood. *Carbohydr. Polym.* **2009**, *77*, 851–857. [CrossRef]
5. Lionetto, F.; Sole, R.D.; Cannoletta, D.; Vasapollo, G.; Maffezzoli, A. Monitoring wood degradation during weathering by cellulose crystallinity. *Materials* **2012**, *5*, 1910–1922. [CrossRef]
6. Huang, X.N.; Kocaefe, D.; Kocaefe, Y.; Boluk, Y.; Krause, C. Structural analysis of heat-treated birch (*Betule papyrifera*) surface during artificial weathering. *Appl. Surf. Sci.* **2013**, *264*, 117–127. [CrossRef]
7. Carrillo, F.; Colom, X.; Sunol, J.; Saurina, J. Structural FTIR analysis and thermal characterisation of lyocell and viscose-type fibres. *Eur. Polym. J.* **2004**, *40*, 2229–2234. [CrossRef]
8. Li, M.Y.; Cheng, S.C.; Li, D.; Wang, S.N.; Huang, A.M.; Sun, S.Q. Structural characterization of steam-heat treated *Tectona grandis* wood analyzed by FT-IR and 2D-IR correlation spectroscopy. *Chin. Chem. Lett.* **2015**, *26*, 221–225. [CrossRef]
9. Guo, J.; Song, K.L.; Salmén, L.; Yin, Y.F. Changes of wood cell walls in response to hygro-mechanical steam treatment. *Carbohydr. Polym.* **2015**, *115*, 207–214. [CrossRef]
0. Tarmian, A.; Mastouri, A. Changes in moisture exclusion efficiency and crystallinity of thermally modified wood with aging. *iFor.-Biogeosci. For.* **2019**, *12*, 92–97. [CrossRef]
1. Dwianto, W.; Tanaka, F.; Inoue, M.; Norimoto, M. Crystallinity changes of wood by heat or steam treatment. *Wood Res.* **1996**, *83*, 47–49.
2. Tribulová, T.; Kačík, F.; Evtuguin, D.V.; Čabalová, I.; Ďurkovič, J. The effects of transition metal sulfates on cellulose crystallinity during accelerated ageing of silver fir wood. *Cellulose* **2019**, *26*, 2625–2638. [CrossRef]
3. French, A.D. Idealized powder diffraction patterns for cellulose polymorphs. *Cellulose* **2014**, *21*, 885–896. [CrossRef]
4. Inagaki, T.; Siesler, H.W.; Mitsui, K.; Tsuchikawa, S. Difference of the crystal structure of cellulose in wood after hydrothermal and aging degradation: A NIR spectroscopy and XRD study. *Biomacromolecules* **2010**, *11*, 2300–2305. [CrossRef] [PubMed]
5. Rayirath, P.; Avramidis, S.; Mansfield, S.D. The effect of wood drying on crystallinity and microfibril angle in black spruce (*Picea mariana*). *J. Wood Chem. Technol.* **2008**, *28*, 167–179. [CrossRef]
6. Bhuiyan, M.T.R.; Hirai, N.; Sobue, N. Changes of crystallinity in wood cellulose by heat treatment under dried and moist conditions. *J. Wood Sci.* **2000**, *46*, 431–436. [CrossRef]
7. Guo, J.; Rennhofer, H.; Yin, Y.F.; Lichtenegger, H.C. The influence of thermo-hygro-mechanical treatment on the micro- and nanoscale architecture of wood cell walls using small- and wide-angle X-ray scattering. *Cellulose* **2016**, *23*, 2325–2340. [CrossRef]
8. Xing, D.; Li, J.; Wang, X.Z.; Wang, S.Q. In situ measurement of heat-treated wood cell wall at elevated temperature by nanoindentation. *Ind. Crops Prod.* **2016**, *87*, 142–149. [CrossRef]

Article

A Novel Bamboo–Wood Composite Utilizing High-Utilization, Easy-to-Manufacture Bamboo Units: Optimization of Mechanical Properties and Bonding Performance

Yifan Ma [1,2], Yu Luan [1,2], Lin Chen [1,2], Bin Huang [1,2], Xun Luo [1,2], Hu Miao [1,2,*] and Changhua Fang [1,2,*]

1   International Centre for Bamboo and Rattan (ICBR), Beijing 100102, China
2   Key Laboratory of National Forestry and Grassland Administration/Beijing for Bamboo & Rattan Science and Technology, Beijing 100102, China
*   Correspondence: miaohu@icbr.ac.cn (H.M.); cfang@icbr.ac.cn (C.F.)

Abstract: Bamboo–wood composites have found extensive applications in the container flooring, furniture, and construction industries. However, commonly utilized bamboo units such as four side-planed rectangular bamboo strips and bamboo scrimber suffer from either low utilization rates or high adhesive content. The recently developed bamboo-flattening technology, which employs softening methods with saturated high-pressure steam, may improve the utilization rate and reduce the adhesive content, but its complex processes and high cost restrict its widespread application. This study introduces a novel bamboo–wood composite utilizing high-utilization, easy-to-manufacture bamboo units processed through a straightforward flattening-and-grooving method. However, the stress concentration introduced by the grooving treatment may affect the mechanical properties and stability of the bamboo–wood composites. In order to optimize the mechanical properties and bonding performance, response surface methodology based on a central composite rotatable design was used to map the effects of hot-pressing parameters (time, temperature, and pressure) on the mechanical properties. The bamboo-woodbamboo–wood composites prepared with optimized conditions of 1.18 min/mm pressing time, 1.47 MPa pressure, and a 150 °C temperature had a 121.51 MPa modulus of rupture and an 11.85 GPa modulus of elasticity, which exhibited an error of only ~5% between the experimental and model predictions. Finite element analysis revealed that, in comparison to homogeneous flat bamboo composites, grooved bamboo composites exhibited distinct tensile ductility and toughness due to discontinuous stress fields and alternating rigid–soft layers, which alter the stress transmission and energy dissipation mechanisms. Additionally, grooving treatment not only effectively improved the surface wettability of the bamboo plants, thus enhancing the permeability of the adhesive, but also facilitated adhesive penetration into parenchymal cells and fibers. This led to the formation of a more robust glue–nail structure and chemical bonding.

Keywords: bamboo; bamboo–wood composite; bamboo-flattening and -grooving unit; response surface methodology; finite element analysis

Citation: Ma, Y.; Luan, Y.; Chen, L.; Huang, B.; Luo, X.; Miao, H.; Fang, C. A Novel Bamboo–Wood Composite Utilizing High-Utilization, Easy-to-Manufacture Bamboo Units: Optimization of Mechanical Properties and Bonding Performance. *Forests* **2024**, *15*, 716. https://doi.org/10.3390/f15040716

Academic Editor: Bruno Esteves

Received: 13 March 2024
Revised: 10 April 2024
Accepted: 12 April 2024
Published: 18 April 2024

1. Introduction

Bamboo's rapid growth, excellent mechanical properties, and abundant availability position it as an excellent alternative material for wood, which is in short supply [1,2]. Bamboo–wood composites have found extensive applications in container flooring, furniture, and construction [3]. Due to the hollow cylindrical structure of bamboo culms, currently, the most commonly used basic units of bamboo engineering materials are bamboo strips and bamboo scrimber [4]. Bamboo strips are produced by planing arc-shaped bamboo splits on all four sides, resulting in a low material utilization rate (approximately 30%) [5,6]. Bamboo scrimber is manufactured by crushing bamboo splits into bundles followed by adhesive soaking, drying, and pressing, leading to a higher adhesive content (approximately 15%–30%) [7,8]. These processing steps are numerous and contribute to high processing costs. Despite recent advancements in bamboo-flattening technology, which have increased the utilization rate and reduced adhesive content to some extent, the method's reliance on high-pressure saturated steam softening leads to complex processes and high costs, thereby limiting its widespread application.

To address the aforementioned challenges, this study introduces a novel bamboo–wood composite with high utilization and simple processing. The process involved the direct and rapid unfolding of arc-shaped bamboo strips, followed immediately by grooving to produce grooved–flattened bamboo (GFB). Both steps were seamlessly executed on the same equipment, ensuring efficiency. Grooving treatment not only enlarges the bonding area but also facilitates adhesive penetration. Subsequent to flattening and grooving, the bamboo fibers and parenchymal cells are fully exposed, leading to enhanced bonding capability. Since it is challenging to apply adhesive directly onto the flattened and grooved bamboo [9], this study employed adhesive application on wooden veneers, significantly enhancing adhesive efficiency. Poplar wood veneers were used in this study for bamboo–wood composite preparation. Due to their low density, they effectively fill the grooves on the bamboo units during hot pressing, enhancing the composite's structural integrity.

However, the stress concentration introduced by the grooving treatment significantly affects the mechanical properties and stability of bamboo–wood composites [10]. There-fore, this study proposes combining the respective advantages of wood and bamboo, utilizing the softer characteristics of wood to solve the stress concentration problem of the grooving process. The fabrication methods of bamboo–wood composites based on the grooved–flattened bamboo (GFB) unit were systematically examined. The material properties were evaluated by assessing the tensile strength and bonding ability of the bamboo-based composites. Initially, mathematical models were developed using response surface methodology (RSM) and central composite design (CCD) to predict overall material performance. A comprehensive investigation was conducted to understand how various processing parameters influence the mechanical and adhesive performance of the com-posite materials. Optimized process parameters were proposed to ensure the composite material achieved optimal performance. Additionally, mechanical verification was carried out for the optimized parameters. Scanning electron microscopy (SEM), confocal laser scanning microscopy (CLSM), and Fourier transform infrared spectroscopy (FTIR) were utilized to investigate the adhesive mechanism of the scored bamboo laminations.

2. Materials and Methods

2.1. Materials

Moso bamboo (*Phyllostachys edulis*) (4 years old) was obtained from Jinzhai County, Anhui Province, China. Poplar veneers with a nominal thickness of 3 mm were sourced from the Poplar Board Production Base in Heze City, Shandong Province. The average moisture content of the veneers was around 12%, and the average air-dry density was 0.37 g/cm^3. Phenolic resin (PF) was obtained from Jinzhai CIMC New Materials Technology Development Co., Ltd., Lu'an, China, with a viscosity of 2500 mPas, solid content exceeding 47%, and a pH value of 11.

2.2. Preparation of Bamboo–Wood Composites

2.2.1. Preparation of Grooved-Flattened Bamboo (GFB) Units

Figure 1a illustrates the manufacturing process of the GFB units. The fresh bamboo culms were cut into 3–4 strips with a thickness of 8 mm and then fed into a specific device for the integrated preparation of GFB units. In this equipment, bamboo units underwent three treatments, namely, flattening, removing the bamboo culm outer and inner layers, and grooving. Firstly, pressure rollers flattened the arc-shaped bamboo strips, and carving knives on these rollers were responsible for creating slots on the bamboo surface. This method was direct and rapid, allowing the flattening of bamboo strips with a large width (up to approximately 10 cm). Additionally, the internal stress of bamboo generated during the flattening process was released through slotting, which contributed to improved dimensional stability. Secondly, the inner layer and outer layer of the flattened bamboo strips were removed, exposing the porous structure, and thereby guaranteeing the material's bonding stability. Thirdly, the bamboo strips underwent grooving treatment to achieve surfaces with a consistent "V" configuration, enhancing the robustness of the gluing area. Finally, GFB units with a final thickness of 5.5–7.5 mm were obtained and then conditioned to achieve a moisture content of 8%–10%.

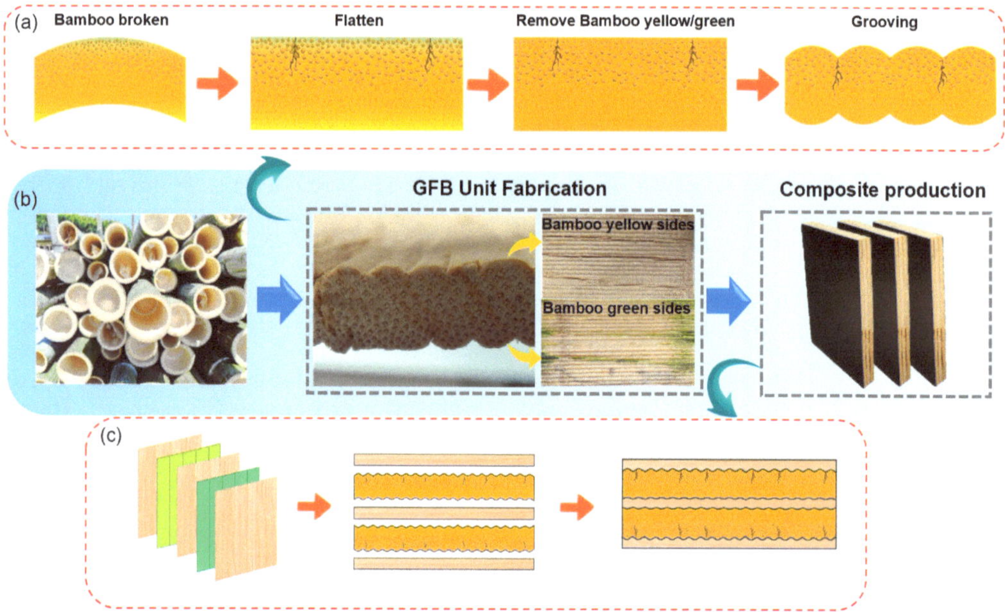

Figure 1. Preparation of bamboo–wood composite laminated boards: (**a**) GFB fabrication, (**b**) preparation of bamboo–wood composite, and (**c**) bamboo–wood composite structure.

2.2.2. Structural Design of Bamboo–Wood Composites

The prepared GFB and wood veneer were sawn into boards measuring 750 mm (length) × 34 mm (width) × 6.8–7 mm (thickness) and 750 mm (length) × 34 mm (width) × 1.5 mm (thickness), respectively. The layup structure of the bamboo–wood composite is shown in Figure 1c. To mitigate the stress concentration defects introduced by grooving in GFB, this study designed a five-layer composite structure in which three wood layers were alternated with two GFB layers, and their grain directions were perpendicular. The grooves on the GFB surface increased the contact area, and the softness of the wood layers facilitated a good fit with the wavy adhesive interface. Finally, PF-impregnated film papers covered the surfaces of the composites.

It is noteworthy that to facilitate a performance comparison, the as-received bamboo culms were divided into three sections. After undergoing high-temperature steam softening at 180 °C for 70 min, the bamboo slices were subsequently treated to remove the green and yellow layers. Then, these sections were flattened through rollers, resulting in homogeneous flat bamboo units. Finally, flat bamboo–wood composites were crafted, adhering to a consistent five-layer structural design.

2.3. Optimization of Process Parameters for Bamboo–Wood Composite Production

In this study, we employed the central composite rotatable design (CCRD), a widely recognized approach in response surface methodology (RSM), to optimize the hot-pressing parameters for bamboo–wood composite production. The CCRD methodology, acknowledged for its effectiveness in process modeling, analysis, and optimization, facilitates the efficient exploration of parameter effects while accommodating both linear and quadratic effects, along with interactions, within a minimal number of experiments [11,12]. To prepare bamboo–wood composites with a stable structure and controllable properties, we conducted a systematic study on three key hot-pressing parameters. The three parameters under scrutiny encompass the hot-pressing time, pressure, and hot-pressing temperature. To explore the relationship between these parameters and material performance more meticulously and accurately, we set three different levels for each parameter. Drawing from a wealth of experimental findings and literature references [13], Table 1 encapsulates the process parameter values and their corresponding tiers as addressed in this research.

Table 1. Experimental design scheme of the response surface.

Level	A (Temperature, °C)	B (Hot-Pressing Time, min/mm)	C (Pressure, MPa)
−1	110	1	1
0	130	1.2	1.5
1	150	1.4	2

The modulus of rupture (MOR) and modulus of elasticity (MOE) of the bamboo–wood composites, prepared under different process parameter conditions, were measured by a universal mechanical testing machine (5528, Instron Corporation, Norwood, MA, USA) according to GB/T17657-2013 [14]. They were taken as the measured response values. The relationship between the influencing factors and response values was established using Design-Expert software (v8.0.6, Stat-Ease, Inc., Minneapolis, MN, USA). During the model-building process, random experiments were conducted to avoid the introduction of systematic errors into the system. With the assistance of the Design-Expert software and operating at a 95% confidence interval, we deduced coefficients pertinent to the second-order polynomial regression model. Furthermore, to ensure the accuracy of the developed model, an analysis of variance (ANOVA) was performed, including significance tests for the regression model and its coefficients.

2.4. The Tensile Shear Properties of Bamboo–Wood Composites

The tensile shear test, employed to ascertain the adhesive strength of materials, adheres to the national standard GB/T17657-2013 [14]. Samples were meticulously crafted from the board, each measuring 100 mm in length and 25 mm in width. For each board, three specimens were prepared to evaluate the dry-bonding strength of the material. Utilizing a 100 kN sensor, an optimal load speed was chosen to ensure the specimens reached failure within a precise window of 60 ± 30 s. Upon the specimens' failure, a detailed examination of the fracture surfaces was conducted to gather insights into the bonding efficacy.

To compare the mechanical response between the grooved bamboo composite and homogenous bamboo composite, finite element analysis was employed. The tension specimens were modeled using Autodesk CAD v2023 software (Autodesk, San Francisco, CA, USA), and the strain distribution in the material during tension was studied using ANSYS v12.1 software (ANSYS Inc., Canonsburg, PA, USA). The MOR and MOE of the GFB were 45.96 MPa and 1557.95 MPa, respectively [10]. The average bending modulus of the wood veneer was 10,615 MPa, the tensile modulus was 11,339 MPa, and the maximum tensile stress was 28.42 MPa [15]. Cohesive elements were incorporated at the interface between the wood and bamboo layers to analyze the effectiveness of the adhesive bonding. The transverse shear behavior of the cohesive units was quantified by the changes in the position of the upper and lower surfaces perpendicular to the direction of thickness. Notably, the cohesive units employed a bilinear constitutive model. Thus, the damage process at the adhesive interface mainly encompassed two stages: initiation of damage and propagation. The damage propagation followed the Bishop–Kuwabara Failure Criterion [16]. The tensile process of the material was achieved by fixing the left end and applying tensile displacement at the right end.

2.5. Bonding Interface and Mechanism

The physical bonding mechanisms between bamboo and wood in composite materials were investigated using a scanning electron microscope (SEM) (GeminiSEM360, Zeiss, Oberkochen, Germany) and confocal laser scanning microscopy (CLSM) (LSM510Meta, Zeiss, Oberkochen, Germany). SEM was used to observe the microstructural characterization of the GFB unit and the bonding surface of the bamboo–wood composite. The acceleration voltage was 5 kV. CLSM was used to observe the glue distribution at the adhesive interface. The bamboo–wood composites were cut into 20 μm thick cross-sections using a sliding microtome, stained for 1 h with 0.5% toluidine blue, washed with deionized water, and sealed with a mixture of glycerol and water at a 1:1 ratio. The prepared slides were observed using CLSM, with the laser wavelength set at 488 nm and the laser current set at 6.5 A.

The chemical bonding mechanisms between the resin and bamboo unit were studied using Fourier transform infrared (FTIR) spectroscopy (Thermo Fisher, Perkin Elmer, Inc., Shelton, CT, USA). The FTIR spectra samples were obtained from 20 μm thick cross-sectional slices prepared by slicing. Specifically, samples were carefully selected from targeted areas within the bamboo–wood composites, such as the bamboo unit areas and areas close to the adhesive layers, to ensure the representativeness and relevance of the results. The FTIR spectra were recorded with a spectral resolution of 4 cm^{-1} in the range of 4000 cm^{-1} to 400 cm^{-1}.

2.6. Penetration of Water

Contact angles were used to assess the surface hydrophilicity of the pith ring side of natural bamboo, the inner side of flat bamboo without the pith ring, and the inner side of GFB. The dynamic changes in the contact angles of the water droplets on the bamboo's green side were measured with contact angle apparatus (OCA20, Dataphysics Ltd., Filderstadt, Germany) utilizing an automatic circle-fitting approach, and the droplet volume was set at 3 μL. For consistency of the experimental outcomes, the experiment was conducted three times. Once the water droplet formed and left the syringe needle, the image was captured for 20 s and recorded by a CCD camera.

3. Results

3.1. Numerical Modeling and Effectiveness Assessment

The optimal response of an empirical model was established through response surface methodology, as shown in Table 2. Employing the Design-Expert statistical software, the F test was used to ascertain the significance of all the coefficients at a 95% confidence interval. The mathematical relationship between each factor and developmental response was explained using the 2FI model [17]. After determining the significant coefficients, a mathematical model to estimate the MOR (1) and MOE (2) of the composite material based on coded factors can be obtained, as shown below:

$$MOR = 112.54 + 9.49A - 0.3994B - 0.2489C + 0.0475AB - 0.0525AC$$
$$+0.165BC - 0.4299A^2 - 1.97B^2 - 1.49C^2 \tag{1}$$

$$MOE\ (Gpa) = 10.89 + 0.6627A - 0.1269B - 0.0707C + 0.085AB + 0.0875AC$$
$$+0.0775BC - 0.2003A^2 - 0.1243B^2 - 0.3064C^2 \tag{2}$$

The appropriateness of the developed model was then evaluated using ANOVA. The statistical results of the ANOVA are presented in Table 3. At the 95% confidence level, the calculated F-ratio is greater than the tabulated value, which proves that the model is adequate. The coefficient of determination (R^2) is also widely used to assess the reliability of fitted regression models. A higher R^2 value (with a maximum of 1) and a lower standard error (SE) indicate that the regression model is very reliable and can be used to predict the response. In the context of these crafted models, both the computed and adjusted R^2 values surpass 90% and 80%, respectively, indicating the high reliability of the regression model

Additionally, the coefficient of variation (CV) indicates the degree of deviation of the unit output from the mean. The CV values in Table 3 are low, indicating the precision and reliability of the conducted experiments. A value of "Prob > F" less than 0.0500 indicates that the model also possesses statistical significance.

Table 2. Experimental design matrix and corresponding responses.

Std Order	Run Order	Point (a)	Input Parameters			Output Parameters	
			A	B	C	MOR, MPa	MOE, GPa
1	7	F1	110	1	1	100.12	10.16
2	8	F1	150	1	1	121.54	11.12
3	2	F1	110	1.4	1	99.89	9.55
4	12	F1	150	1.4	1	120.46	11.1
5	1	F1	110	1	2	98.72	9.43
6	16	F1	150	1	2	118.89	10.99
7	13	F1	110	1.4	2	98.11	9.38
8	14	F1	150	1.4	2	119.51	11.03
9	9	A1	96.3641	1.2	1.5	96.21	9.22
10	11	A1	163.636	1.2	1.5	123.56	11.2
11	4	A1	130	0.863641	1.5	106.75	10.75
12	6	A1	130	1.53636	1.5	104.28	10.1
13	3	A1	130	1.2	0.659104	105.88	9.87
14	17	A1	130	1.2	2.3409	107.89	9.95
15	10	C1	130	1.2	1.5	113.1	10.89
16	15	C1	130	1.2	1.5	112.44	10.91
17	5	C1	130	1.2	1.5	112.56	10.92

(a) F1 factorial point, A1 axial point, C1 center point.

Table 3. ANOVA table for the response surface model.

Terms	Responses	
	MOR, MPa	MOE, Gpa
Sum of squares		
Regression	1288.17	7.66
Residual	47.2	0.3503
Lack of fit	46.95	0.3498
Pure error	0.2472	0.0005
Mean squares		
Regression	143.13	0.8511
Residual	6.74	0.05
Lack of fit	9.39	0.07
Pure error	0.1236	0.0002
Degrees of freedom		
Regression	9	9
Residual	7	7
Lack of fit	5	5
Pure error	2	2
Std. Dev	2.6	0.2237
Mean	109.41	10.39
C.V.%	2.37	2.15
PRESS	357.17	2.7
R^2	0.9647	0.9563
Adj R^2	0.9192	0.9
Adeq Precision	16.0218	12.9913
F-ratio (calculated)	21.23	17.01
Prob F	0.0003	0.0006
Remark	Significant	Significant

3.2. Effects of Process Parameters on Responses

The duration of hot pressing, applied pressure, and hot-pressing temperature are key process parameters in the production of bamboo–wood composites [18]. They directly affec the effectiveness of bonding curing and thereby the properties of the composite materia. In general, these three processing parameters interact to collectively define the overal properties of bamboo–wood composite materials. To obtain the best performance, these parameters need to be meticulously fine-tuned and optimized during experiments. In the following sections, the other parameters are thought to be at their center levels wheneve: an interaction effect or a comparison between any two input parameters is being discussed

3.2.1. The Response of MOR

Figure 2 illustrates the 2D contour interactions among any two treatment parameters affecting the MOR. A notable positive correlation was observed between the MOR and the temperature during hot pressing (Figure 2a,b). It can be inferred that higher hot-pressing temperatures enhance the flowability and penetrability of the adhesive and effectively im prove the physical bonding and chemical bonding ability between PFs and bamboo–wood composites. Conversely, the effects of hot-pressing time and pressure on the MOR are less significant. Intriguingly, as the hot-pressing time and pressure increase, the MOR first increases and then tends to decrease. This is presumed to be due to the inadequate dispersion of the adhesive under shorter durations and lower pressures of hot pressing resulting in weaker bonding. On the other hand, grooving treatment leaves numerou stress concentration areas on the bamboo surface, and higher pressure and prolonged ho pressing may reduce the stability of GFB, leading to localized crack initiation. Therefore, to achieve optimal bending strength, it is necessary to consider the appropriate pressure and hot-pressing time to ensure the uniform distribution of the adhesive and the densificatio of the material.

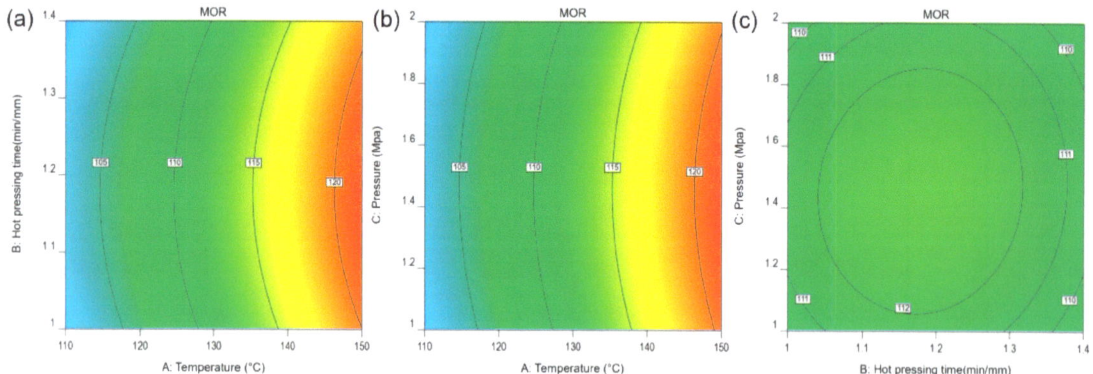

Figure 2. Response of the MOR to the process parameters: (**a**) hot-pressing temperature and time (**b**) hot-pressing temperature and pressure; (**c**) hot-pressing time and pressure.

3.2.2. The Response of the MOE

Figure 3 clearly shows the effects of the different processing parameter combinations on the MOE of the bamboo–wood composite materials. As observed, the range of MOE variation is primarily between 9 and 11 GPa. The influence mechanism of the processing parameters on the MOE is similar to that on the MOR. Specifically, higher temperatures during hot pressing aid in enhancing the MOE of bamboo–wood composites (Figure 3a,b An increase in hot-pressing time and pressure similarly promoted an initial increase fol lowed by a decrease in the MOE. Distinctly, within the experimental parameter range o this study, the effect of pressure on the MOE is more pronounced than that of hot-pressing time, as depicted in Figure 3c. This greater variability, introduced by pressure, is due to the instability of the GFB caused by the scoring treatment. Hence, to optimize the MOE of bamboo–wood composite materials, careful selection of the appropriate pressure is necessary to ensure consistency during the hot-pressing process.

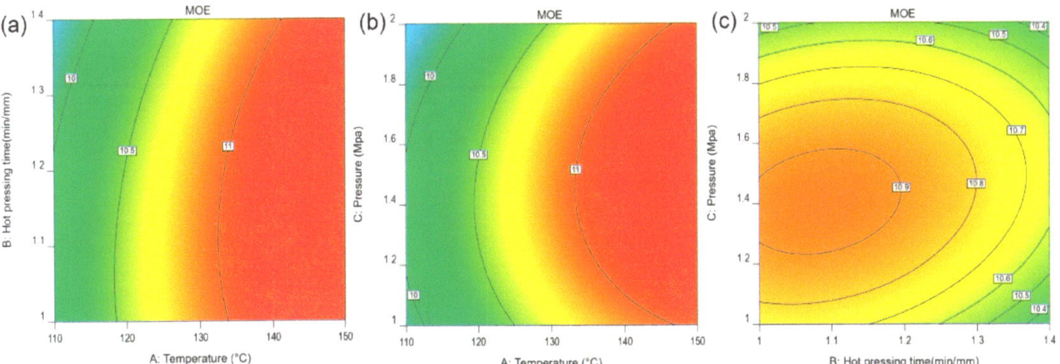

Figure 3. Response of the MOE to the process parameters: (**a**) hot-pressing temperature and time; (**b**) hot-pressing temperature and pressure; (**c**) hot-pressing time and pressure.

3.2.3. Optimization and Verification of the Effects of the Process Parameters on the Responses

The findings from the previous section reveal that the process parameters exhibit a more complex response to the material's mechanical properties. Therefore, it is imperative to select the appropriate process parameters to endow bamboo–wood composites with the desired performance combination. As mentioned above, the primary advantage of the RSM is optimizing the response by manipulating the independent variables [13]. Optimization aims to determine the best mechanical performance within the experimental realm of processing parameters [19]. For the sake of optimization, numerical and graphical methods were employed by selecting the favored goals for the parameters and responses, as demonstrated in Table 4.

Table 4. Goals and limitations considered in the optimization process.

Name	Goal	Lower Limit	Upper Limit
A: Temperature	In range	110	150
B: Hot-pressing time	In range	1	1.4
C: Pressure	In range	1	2
MOR	Maximize	96.21	123.56
MOE	Maximize	9.22	11.2

A desirability-based approach was employed to evaluate the optimization results. The desirability of each factor, which ideally ought to closely approach 1, contributes to comprehensive optimization [20]. The computed findings revealed an optimized comprehensive desirability of 0.962 in this research, signifying congruence between the optimized input and target output values (Figure 4a). It has been established in earlier discussions that the temperature during hot pressing is positively correlated with the overall mechanical performance of the material. Consequently, Figure 4b,c depicts the response surface and corresponding contour map of the interaction effects between hot-pressing time and pressure on the response surface at 150 °C, which illustrates a graphical search for potential optimal cooperation. The higher composite desirability further confirms the consistency between the mechanical properties and bonding strength. Specifically, the quantified results also suggest that the optimal lamination conditions for bamboo–wood composites are a hot-pressing duration of 1.18 min/mm, 1.47 MPa of pressure, and a hot-pressing temperature of 150 °C. Under these parameters, the optimal values for the MOE and MOE of the bamboo–wood composite material are 121.625 MPa and 11.354 GPa, respectively.

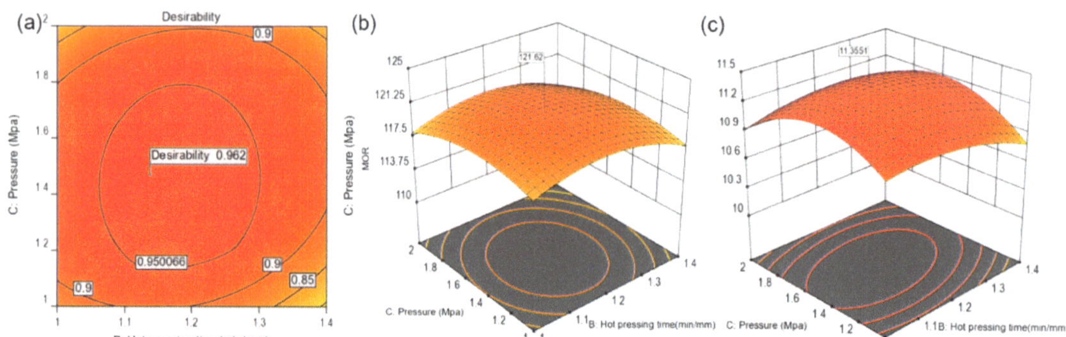

Figure 4. Response surface and corresponding contour plots of 2D optimization diagrams determined by desirability (**a**) and the interaction between hot-pressing time and pressure at 150 °C (**b,c**).

Subsequent verification experiments were further conducted, as shown in Figure 5a. The mechanical properties of the grooved bamboo composites prepared under the optimized conditions were tested three times, and exhibited an error of only ~5% for the mechanical properties predicted by RSM, with an average MOR and MOE of 121.51 MPa and 11.85 GPa, respectively. To further verify the adhesive capability of the material, we compared the three-point bending and mechanical responses of the grooved bamboo and homogeneous flat bamboo composite materials. The grooved bamboo composite material displayed a mechanical response similar to that of the flat bamboo composite. If the highest point of the stress–strain curve in Figure 5a is regarded as the start of crack initiation, the curve can be divided into two stages: the elastic–plastic stage (the E&P stage) and the crack propagation stage (the C-stage). During the elastic–plastic stage, both the grooved and flat bamboo composite materials exhibited almost identical performances, indicating that the grooving treatment had little effect on the MOE of the composite material. This might be because the elastic deformation of the side-bonded bamboo integrated material is related only to the performance of the unit, and the bonding performance between the units has little impact on it. However, the grooving treatment had some effect on the MOR of the material. The MOR of the grooved bamboo composite material was slightly less than that of the flat bamboo composite. This decrease could be due to the preset cracks causing the initiation of cracks. For the C-stage, the crack propagation stage exhibits noticeable fluctuations. This is primarily because, unlike single materials, the structural characteristic of laminated composite materials determine their ability to suppress the generation of local stress during the plastic deformation process and delay the phenomenon of strain localization. Interestingly, the C-stage of the grooved bamboo laminate is significantly longer, which is related to the unique milling structure of the device, which will be further discussed later in the text. Furthermore, we calculated the areas enclosed by the curves of the different stages and assessed the energy absorption capacity of the material in each stage based on the formula [21]

$$W = \int_0^d F dx$$

where F is the applied force and d is the total displacement at different stages. The specific calculation results can be seen in Figure 5b. Through this quantitative evaluation method, we found that homogeneous flat bamboo composites demonstrated superior performance during the elastic–plastic deformation stage. However, during the crack propagation stage, the energy absorption capacity of the grooved bamboo composites reached twice that of the composite structure without grooving.

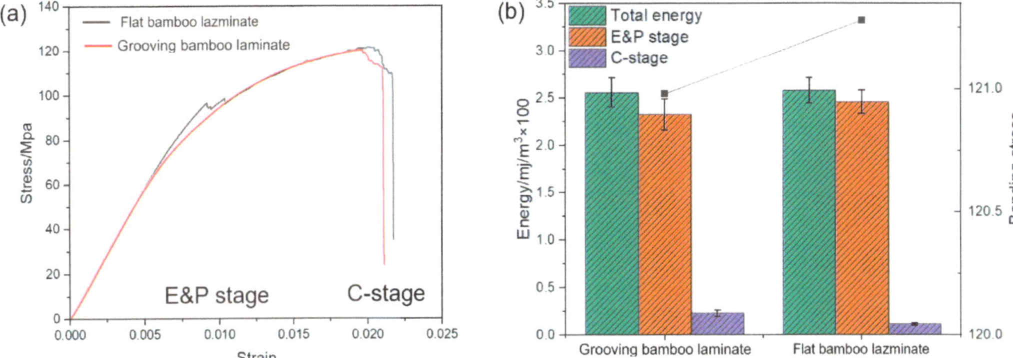

Figure 5. Three-point bending mechanical responses of bamboo–wood composites based on grooved bamboo and flat bamboo: (**a**) stress—strain curve; (**b**) energy absorption capacity at different stages of the tensile process.

3.3. The Tensile Shear Properties of Bamboo–Wood Composites

Enhancing our understanding of the ductile behavior of cracks in grooved bamboo composite materials necessitated conducting experiments on samples with pre-introduced grooves with a 3 mm width, subjected to double-ended stretching. The structures of these experiments are depicted in Figure 6. As observed, cracks predominantly originate within the bamboo–wood adhesive layer. Observations indicate that the propagation of cracks is significantly influenced by the softer layer's structure, attributable to the lower strength of the wood side. This influence steers the crack paths, causing deviations that culminate in the emergence of three distinct fracture morphologies: interlayer, fracture, and shear cracks.

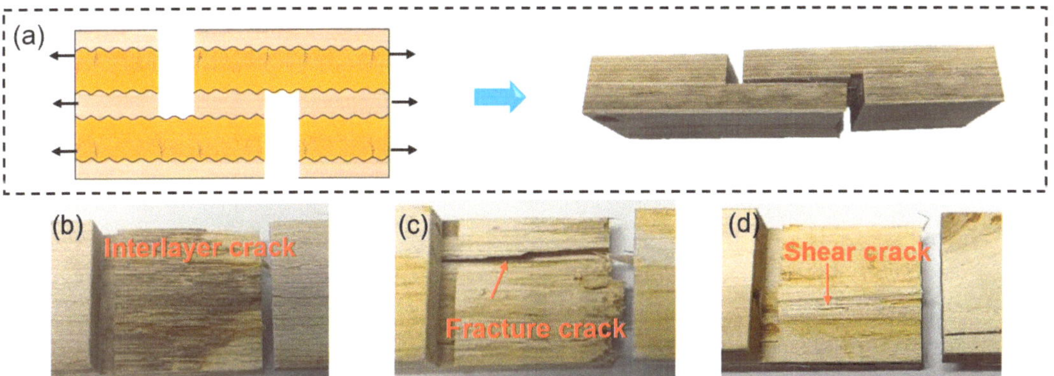

Figure 6. Tensile shear responses of bamboo–wood composites: (**a**) tensile diagram and fracture morphology; (**b**–**d**) typical fracture mechanism.

We further conducted tensile simulation tests on bamboo–wood composite materials using ANSYS to further explore the mechanical response of the material. As depicted in Figure 7, the evolutions of the stress distribution during the tensile process in both homogeneous flat bamboo and grooved bamboo composites can be observed. As observed, different structures under tensile action exhibited similar macroscopic stress distributions, which essentially corresponded to the maximum shear force direction at 45°. However, the locations of stress concentration between the two are noticeably different. For the flat bamboo composite material, the stress is more uniformly distributed around the notches, while for the grooved bamboo composite material, the stress is more concentrated at

the wavy milling grooves in the notches, where there is a smaller cross-sectional area.
This phenomenon reveals that preset grooves would introduce a degree of stress concentration. Interestingly, the grooved bamboo composite material could withstand larger
displacements. At a displacement of 0.14 mm, the adhesive junction of the flat bamboo
composite exhibited significant stress concentration and crack propagation. In contrast,
for the grooved bamboo composite material, delamination was observed only when the
displacement reached 0.19 mm, where the stress peaked at 95.96 MPa. It can be inferred that
grooved bamboo, to a certain extent, possesses greater elasticity and load-bearing capacity,
enabling it to remain intact under greater displacements without breaking easily. These
characteristics are directly related to the smaller bearing area and preset engraving cracks
in the grooved bamboo. To substantiate this hypothesis, as shown in Figure 7d, we further
studied the stress evolution during the delamination process. The stress transmission in
the flat bamboo structure is more continuous. The homogeneous structural features of
the flat bamboo suggest that the crack propagation encounters less resistance. However,
the stress transmission in the grooved structure clearly exhibits discontinuity. During
the crack propagation process, the stress transmission undergoes additional orientation
rotations. On the other hand, the irregular wavy features in the grooved structure introduce
an alternating change between the soft and hard layers, leading to increased resistance
during crack propagation. Therefore, it can be inferred that the wavy bamboo morphology
formed by milling increases the bonding area of bamboo, which can effectively increase the
energy absorption and comprehensive mechanical properties at the fracture stage.

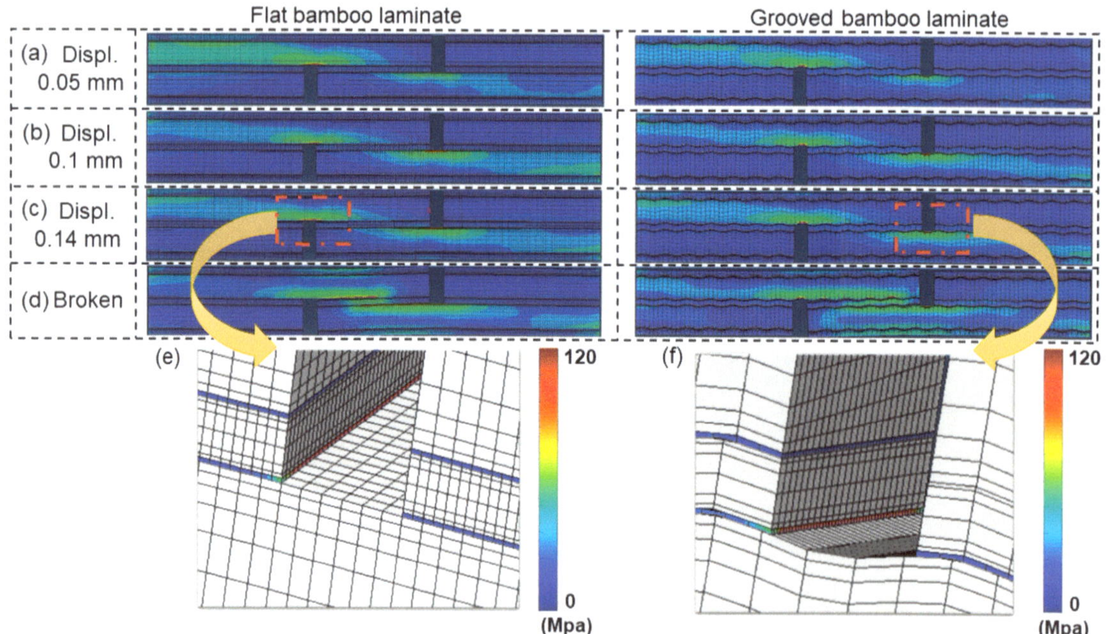

Figure 7. Strain distribution of grooved bamboo laminates and flat bamboo laminates in tension using
FEM: (**a–d**) the stress distribution at different displacements until failure; (**e,f**) the stress distribution
in the adhesive layer at a displacement of 0.14.

3.4. Bonding Mechanism

The most crucial aspect of bamboo–wood composite material performance lies in the
bonding capacity between its various layers. Figure 8a–c shows the SEM images of the GFB
surface, where Figure 8a shows the inner part of bamboo and Figure 8c shows the outer
part of bamboo. The grooving treatment led to the disappearance of the cortex and pith

ring and the exposure of fine pores. Larger grooving depths essentially expose a greater proportion of the parenchymal cells of bamboo, while also reducing the anisotropy between the outer and inner sides of the bamboo. This process plays a positive role in enhancing the stability of bamboo–wood composite materials. To determine the impact of grooving treatment on the material permeability, we further compared the wettability of the pith ring side of natural bamboo (NB), the inner side of flat bamboo without the pith ring (FB), and the inner side of GFB (Figure 8f). The initial contact angle of the bamboo pith ring after sanding decreased from 95° to 73°. However, compared to that on the flat bamboo which had the pith ring sanded, the change on the GFB surface was not significant and only decreased by 2°, indicating that both processes effectively disrupted the smooth surface of the bamboo. Nonetheless, there was a marked difference in the final wettability of the three materials, with contact angles of 79.87°, 65.13°, and 27.66°. The liquid on the surface of the GFB almost entirely permeated into the bamboo, demonstrating that grooving could improve the permeability of the bamboo more effectively.

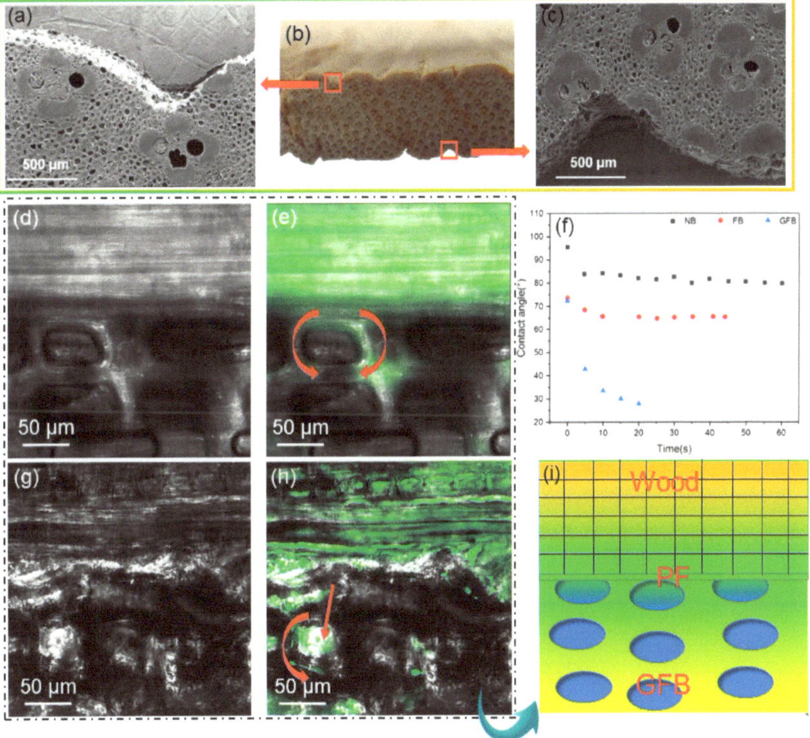

Figure 8. Mechanism of physical bonding. (**a**–**c**) SEM image of the grooved bamboo's inner side and the bamboo's outer side. (**d**,**e**) CLSM images of the composite interface of non-grooved bamboo before and after gluing. (**f**) The contact angles of the NB, FB, and GFB. (**g**,**h**) CLSM images of the composite interface of grooved bamboo before and after gluing. (**i**) The physical bonding mechanism of gluing.

Figure 8d,g displays the composite interface of the grooved bamboo and non-grooved bamboo. The parenchymal cells in the non-grooved bamboo still maintained their complete cell wall structure during the hot-pressing process. After bonding, CLSM revealed that the substance penetrating the cavities and cell walls of parenchymal cells was PF (Figure 8e). Compared to those of fiber cells, the cell walls of parenchymal cells are thinner, the cell cavities are larger, and the cell walls have dense pits, which are more conducive to the penetration of glue [22]. Furthermore, the damaged parenchymal cells of grooved bamboo at the bonding interface exhibited significant fracturing due to hot pressing. The cavities of cells at the adhesive interface were filled with a substance identified as PF based on CLSM (Figure 8h). It can be inferred that PF can penetrate cell walls via the pits present on the cells. Overall, PF could penetrate parenchymal and fiber cells more effectively after grooving, forming a more stable glue–nail network structure (Figure 8i).

Chemical bonding is an essential reason for the formation of gluing forces [23]. Figure illustrates the infrared spectral differences between grooved bamboo composites and homogeneous flat bamboo composites, revealing the chemical reaction mechanism of PF with the cells of bamboo. In the respective infrared spectra, the peak at 1738 cm^{-1} is attributed to the stretching vibration of the C=O ester carbonyl (hemicellulose) in bamboo. The peak at 1601 cm^{-1} is likely the result of the stretching vibration of the aromatic ring in PF and the bending vibration of C-H. Moreover, the chemical reaction between PF and bamboo results in a more complex infrared spectral response, which significantly fluctuates in the 1300–1500 cm^{-1} intensity range, where the peak at 1516 cm^{-1} is likely due to benzene ring stretching. The peak at 1460 cm^{-1} is attributed to the asymmetric stretching vibration of C-H/symmetric deformation vibration of CH. The peak at 1426 cm^{-1} is attributed to the scissoring vibration of CH2. The peak at 1368 cm^{-1} is attributed to the bending of aliphatic CH in cellulose and hemicellulose. The most prominent peaks at 1242 cm^{-1}, 1159 cm^{-1}, and 1032 cm^{-1} are associated with the stretching vibrations of C-O in cellulose and hemicellulose. Compared with the flat bamboo laminates, grooved bamboo laminates show a similar intensity of the above peaks. It can be indicated that the grooving treatment will not cause deterioration of the chemical reaction at the interface.

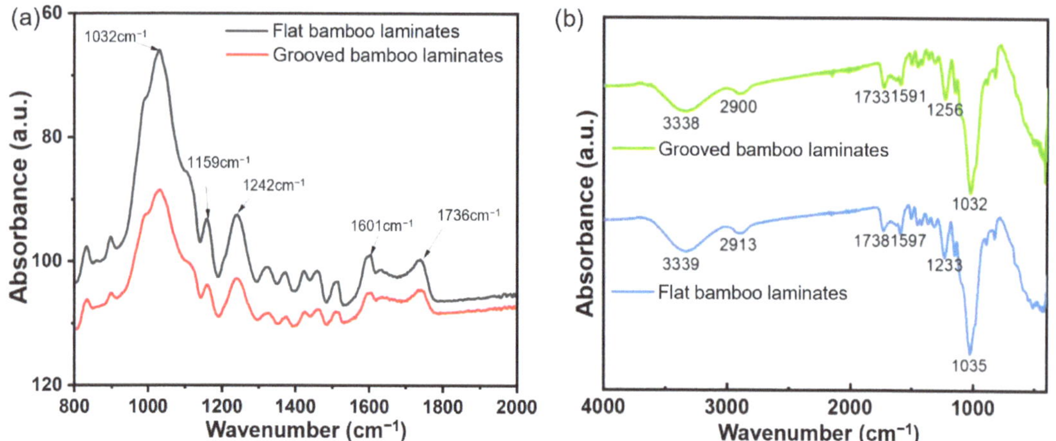

Figure 9. (**a**) Infrared spectra and (**b**) infrared difference spectrum of grooved bamboo laminates and flat bamboo laminates near the adhesive layer.

4. Application

The bonding area is an important factor affecting adhesion. Research shows that increasing the bonding area can effectively improve bonding stability [17,24]. In the present study, we also found that milling can effectively increase the energy absorption and mechanical properties of bamboo at the fracture stage by increasing the gluing area of bamboo. Hence, by manipulating the cutting blades and modifying the number and shape of the arc-shaped protrusions, we were able to adjust the ratio of the bonding area, thus improving the designability of the system. As shown in Figure 10a–c, the wavy surface milling pattern resembles a cycloid. Therefore, by controlling the blades of the milling cutter, the shape and number of arc-shaped protrusions are altered, thereby increasing the proportion of the gluing area. In this study, to simplify the calculations, an approximate circular method is employed to analyze the area changes introduced by scoring. The increased gluing area ratio due to the surface milling pattern can be estimated using the following formula:

$$P = \frac{L_2}{L_1} = \frac{\arccos(1 - \Delta)}{\sqrt{1 - \left(1 - \Delta\right)^2}}$$

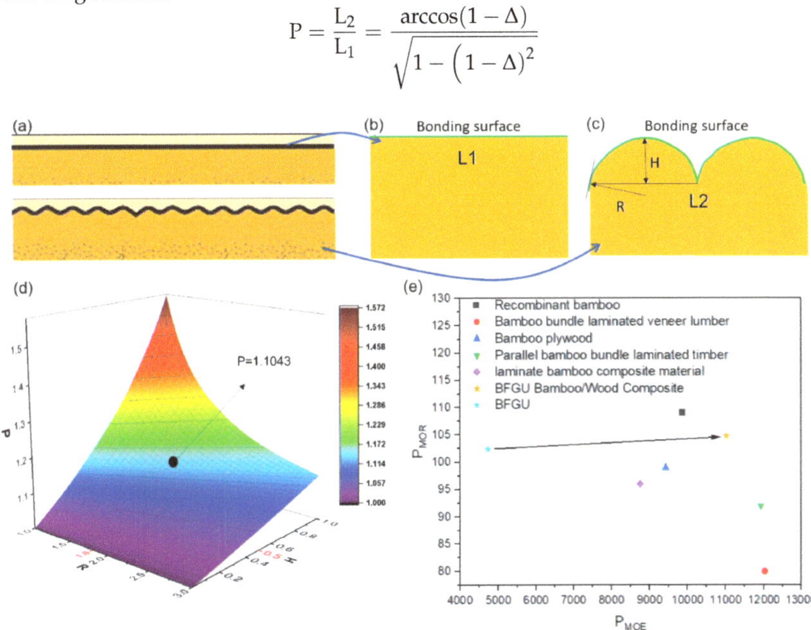

Figure 10. (**a**–**c**) Increased gluing area ratio due to surface milling pattern. (**d**) Impact of arc-shaped milling pattern on gluing area ratio. (**e**) Comparison of comprehensive properties of different bamboo materials.

Here, Δ represents the ratio of the height H of the arc-shaped milling pattern to the radius R of the circular arc-shaped milling pattern. Figure 10d shows the impact of the height H and the radius R of the arc-shaped milling pattern on the gluing area ratio. As observed, the milling height H is negatively correlated with the gluing area ratio, while the other factors are the opposite. Considering the decrease in bamboo volume due to scoring, a moderate score was chosen for this study, with R being 1.8 mm and H 0.5 mm. The corresponding gluing area ratio is 1.1043, indicating that the area of the wavy milling pattern is ~10% greater than that of ordinary flattened bamboo. To integrate the mechanical properties and adhesive stability and to provide a comprehensive mechanical assessment of the bamboo–wood composite materials, an area-weighted performance index was introduced:

$$P_{MOE} = MOE \times P$$

$$P_{MOR} = MOR \times P$$

Based on the above equation, we compared the modification properties of several different bamboos [7,25–30], as shown in Figure 10e. It is evident that bamboo–wood composites utilizing GFB units demonstrate excellent comprehensive performance. Specifically, compared to that of GFB units, the composite performance of bamboo–wood materials was significantly improved. After the flattened bamboo sheets and wood veneers were combined, the surface defects were filled, and due to the increased bonding area, the mechanical properties of the final laminated material were guaranteed. Additionally, although the mechanical performance did not significantly improve compared to that of regular flattened bamboo sheets, this difference could be attributed to the elimination of certain mechanically significant tissues (such as fiber cells) during the etching process, as well as the reintroduction of cracks in the cross-section. However, a higher bonding area contributes to greater material stability.

Bio-based composites based on wood and bamboo are widely utilized in building, furniture, transportation, etc. In terms of processing, the innovative grooving and flattening unit eliminates the need for softening in traditional bamboo flattening, integrating standardization and flattening and thereby providing greater economic value. The average production of 1 cubic meter of board saved 65 kg of PF resin adhesive relative to that of bamboo scrimber. In comparison to bamboo composites, this new type of bamboo–wood laminated material has more than twice the efficiency of usage and similar adhesive requirements and needs only half as much bamboo material to produce an average of 1 m of board [31]. Overall, the grooving structure does not have a significant impact on the mechanical properties of bamboo laminate materials. The bamboo square composite material can reach the level of conventional bamboo square composites and can be partially replaced. Furthermore, compared to the bamboo composites made by the planing unit, the bamboo composites prepared by the grooving unit may achieve a more dimensional design and are suited to more usage situations with larger size specifications. As a result, the bamboo/wood composite based on the GFB unit in this study may augment and replace existing indoor bamboo–wood composite materials and has a wide range of applications.

5. Conclusions

This work successfully developed a novel bamboo–wood composite material using flattened and grooved bamboo units prepared by an efficient integrated machining process. Response surface methodology was applied to model the effects of hot-pressing parameters on the mechanical and bonding performance. Optimization of the parameters produced composites with an excellent MOR and MOE. Furthermore, the physical and chemical bonding mechanisms of the optimized samples were also studied. The following conclusions can be drawn:

Based on the RSM, the optimal hot-pressing process parameters were determined to be 1.18 min/mm hot-pressing time, 1.47 MPa pressure, and a 150 °C temperature. Under these conditions, the bamboo–wood composite exhibited a 121.51 MPa modulus of rupture and an 11.85 GPa modulus of elasticity, which represented only a ~5% error between the experimental and model predictions.

Finite element analysis revealed that, in comparison to homogeneous flat bamboo composites, grooved bamboo composites exhibit distinct tensile ductility and toughness.

Grooving treatment not only effectively improved the wettability of the bamboo surface and increased the permeability of the adhesive on the bamboo surface, but also augmented the bonding area.

Author Contributions: Writing—review and editing, Y.M., Y.L., L.C. and C.F.; performing the experiments, Y.M., X.L., L.C., B.H. and H.M.; project administration, C.F. All authors have read and agreed to the published version of the manuscript.

Funding: This work was funded by the "14th Five-year" National Key R&D Program of China (2022YFD2200902) and the Foundation of the International Centre for Bamboo and Rattan (No 1632022021).

Data Availability Statement: The data presented in this study are available on request from the corresponding author. The data are not publicly available due to privacy/ethical restrictions.

Acknowledgments: The authors wish to thank Jinzhai CIMC New Materials Technology Development Co., Ltd., Lu'an, China for for providing materials used for experiments. We also thank Renzhong Tao for his technical support.

Conflicts of Interest: The authors declare no conflicts of interest.

References

1. Hong, C.; Li, H.; Xiong, Z.; Lorenzo, R.; Corbi, I.; Corbi, O.; Wei, D.; Yuan, C.; Yang, D.; Zhang, H. Review of connections for engineered bamboo structures. *J. Build. Eng.* **2020**, *30*, 101324. [CrossRef]
2. Hailemariam, E.K.; Hailemariam, L.M.; Amede, E.A.; Nuramo, D.A. Identification of barriers, benefits and opportunities of using bamboo materials for structural purposes. *Eng. Constr. Archit. Manag.* **2023**, *30*, 2716–2738. [CrossRef]
3. Chen, J.; Guagliano, M.; Shi, M.; Jiang, X.; Zhou, H. A comprehensive overview of bamboo scrimber and its new development in China. *Eur. J. Wood Wood Prod.* **2021**, *79*, 363–379. [CrossRef]
4. Bala, A.; Gupta, S. Engineered bamboo and bamboo-reinforced concrete elements as sustainable building materials: A review. *Constr. Build. Mater.* **2023**, *394*, 132116. [CrossRef]
5. Khajouei-Nezhad, M.; Semple, K.; Nasir, V.; Marggraf, G.; Hauptman, J.; Dai, C. Advances in engineered bamboo processing: Material conversion and structure. *Adv. Bamboo Sci.* **2023**, *5*, 100045. [CrossRef]
6. Wang, J.; Shi, D.; Zhou, C.; Zhang, Q.; Li, Z.; Marmo, F.; Demartino, C. An active-bending sheltered pathway based on bamboo strips for indoor temporary applications: Design and construction. *Eng. Struct.* **2024**, *307*, 117863. [CrossRef]
7. Huang, Y.; Qi, Y.; Zhang, Y.; Yu, W. Progress of bamboo recombination technology in China. *Adv. Polym. Technol.* **2019**, *2019*, 2723191. [CrossRef]
8. Huang, Y.; Ji, Y.; Yu, W. Development of bamboo scrimber: A literature review. *J. Wood Sci.* **2019**, *65*, 25. [CrossRef]
9. Lou, Z.; Li, Y.; Zhao, Y. Bamboo Flattening Technique. In *Bamboo and Sustainable Construction*; Springer: Berlin/Heidelberg, Germany, 2023; pp. 185–210.
10. Chen, L.; Luo, X.; Huang, B.; Ma, Y.; Fang, C.; Liu, H.; Fei, B. Properties and bonding interface characteristics of an innovative bamboo flattening and grooving unit (BFGU) for laminated bamboo lumber. *Colloids Surf. A Physicochem. Eng. Asp.* **2023**, *676*, 132185. [CrossRef]
11. Liu, Y.; Zhou, J.; Fu, W.; Zhang, B.; Chang, F.; Jiang, P. Study on the effect of cutting parameters on bamboo surface quality using response surface methodology. *Measurement* **2021**, *174*, 109002. [CrossRef]
12. Adamu, M.; Rahman, M.R.; Bakri, M.K.B.; Md Yusof, F.A.B.; Khan, A. Characterization and optimization of mechanical properties of bamboo/nanoclay/polyvinyl alcohol/styrene nanocomposites using response surface methodology. *J. Vinyl Addit. Technol.* **2021**, *27*, 147–160. [CrossRef]
13. Chen, L.; Yuan, J.; Wang, X.; Huang, B.; Ma, X.; Fang, C.; Zhang, X.; Sun, F.; Fei, B. Fine gluing of bamboo skin and bamboo pith ring based on sanding. *Ind. Crops Prod.* **2022**, *188*, 115555. [CrossRef]
14. *GB/T17657-2013*; Test Methods of Evaluating the Properties of Wood-Based Panels and Surface Decorated Wood-Based Panels. National Standards of the People's Republic of China: Beijing, China, 2013.
15. Uzcategui, M.G.; Franca, F.J.; Seale, R.D.; Senalik, C.A.; Ross, R.J. Flexural and tensile properties of 2 × 6 and 2 × 10 southern pine lumber: Southenr pine lumber properties. *Wood Fiber Sci.* **2022**, *54*, 257–269. [CrossRef]
16. Baltic, S.; Magnien, J.; Gänser, H.P.; Antretter, T.; Hammer, R. Coupled damage variable based on fracture locus: Prediction of ductile failure in a complex structure. *Int. J. Solids Struct.* **2020**, *207*, 132–144. [CrossRef]
17. Das, A.; Mohanty, K. Optimization of lignin extraction from bamboo by ultrasound-assisted organosolv pretreatment. *Bioresour. Technol.* **2023**, *376*, 128884. [CrossRef] [PubMed]
18. Nkeuwa, W.N.; Zhang, J.; Semple, K.E.; Chen, M.; Xia, Y.; Dai, C. Bamboo-based composites: A review on fundamentals and processes of bamboo bonding. *Compos. Part B Eng.* **2022**, *235*, 109776. [CrossRef]
19. Veza, I.; Spraggon, M.; Fattah, I.R.; Idris, M. Response surface methodology (RSM) for optimizing engine performance and emissions fueled with biofuel: Review of RSM for sustainability energy transition. *Results Eng.* **2023**, *18*, 101213. [CrossRef]
20. Garg, S.; Nayyar, A.; Buradi, A.; Shadangi, K.P.; Sharma, P.; Bora, B.J.; Jain, A.; Shah, M.A. A novel investigation using thermal modeling and optimization of waste pyrolysis reactor using finite element analysis and response surface methodology. *Sci. Rep.* **2023**, *13*, 10931. [CrossRef] [PubMed]
21. Mo, T.; Chen, Z.; Zhou, D.; Lu, G.; Huang, Y.; Liu, Q. Effect of lamellar structural parameters on the bending fracture behavior of AA1100/AA7075 laminated metal composites. *J. Mater. Sci. Technol.* **2022**, *99*, 28–38. [CrossRef]
22. Shi, J.; Liang, Y.; Yu, H.; Ban, Z.; Zhang, Y.; Yang, W.; Yu, W. A new strategy for bamboo ultra-long radial slice preparation and novel composite fabrication. *Ind. Crops Prod.* **2023**, *203*, 117232. [CrossRef]
23. Wang, X.; Yao, Y.; Xie, X.; Yuan, Z.; Li, W.; Yuan, T.; Huang, Y.; Li, Y. Investigation of the microstructure, chemical structure, and bonding interfacial properties of thermal-treated bamboo. *Int. J. Adhes. Adhes.* **2023**, *125*, 103400. [CrossRef]
24. Su, H.; Du, G.; Ren, X.; Liu, C.; Wu, Y.; Zhang, H.; Ni, K.; Yin, C.; Yang, H.; Ran, X. High-performance bamboo composites based on the chemical bonding of active bamboo interface and chitosan. *Int. J. Biol. Macromol.* **2023**, *244*, 125345. [CrossRef] [PubMed]
25. Zhou, H.; Wei, X.; Smith, L.M.; Wang, G.; Chen, F. Evaluation of uniformity of bamboo bundle veneer and bamboo bundle laminated veneer lumber (BLVL). *Forests* **2019**, *10*, 921. [CrossRef]
26. Zhang, J.; Ren, H.; Zhong, Y.; Zhao, R. Analysis of compressive and tensile mechanical properties of recombinant bamboo. *J. Nanjing For. Univ.* **2012**, *36*, 107–111.
27. Han, J.; Gao, X.T. A study on tensile mechanical properties of bamboo plywood. *Adv. Mater. Res.* **2011**, *291*, 1009–1014. [CrossRef]
28. Qi, J.Q.; Xie, J.L.; Huang, X.Y.; Yu, W.J.; Chen, S.M. Influence of characteristic inhomogeneity of bamboo culm on mechanical properties of bamboo plywood: Effect of culm height. *J. Wood Sci.* **2014**, *60*, 396–402. [CrossRef]

29. Chen, G.; Yu, Y.; Li, X.; He, B. Mechanical behavior of laminated bamboo lumber for structural application: An experimental investigation. *Eur. J. Wood Wood Prod.* **2020**, *78*, 53–63. [CrossRef]
30. Jain, S.; Kumar, R.; Jindal, U. Mechanical behaviour of bamboo and bamboo composite. *J. Mater. Sci.* **1992**, *27*, 4598–4604.
31. Lou, Z.; Zheng, Z.; Yan, N.; Jiang, X.; Zhang, X.; Chen, S.; Xu, R.; Liu, C.; Xu, L. Modification and Application of Bamboo-Based Materials: A Review—Part II: Application of Bamboo-Based Materials. *Forests* **2023**, *14*, 2266. [CrossRef]

Article

The Effect of Wet and Dry Cycles on the Strength and the Surface Characteristics of Coromandel Lacquer Coatings

Wenjia Liu [1,2], Ling Zhu [1,2], Anca Maria Varodi [3], Xinyou Liu [1,2,3,]* and Jiufang Lv [1,2,]*

1   College of Furnishing and Industrial Design, Nanjing Forestry University, Nanjing 210037, China; liuwenjiajia@hotmail.com (W.L.); zlwaxinge@njfu.edu.cn (L.Z.)
2   Co-Innovation Center of Efficient Processing and Utilization of Forest Resources, Nanjing Forestry University, Nanjing 210037, China
3   Faculty of Furniture Design and Wood Engineering, Transilvania University of Brasov, 500036 Brasov, Romania; anca.varodi@unitbv.ro
*   Correspondence: liu.xinyou@njfu.edu.cn (X.L.); lvjiufang8189@njfu.edu.cn (J.L.); Tel.: +86-25-8542-7408 (X.L.)

Abstract: Research on the degradation mechanism of coating materials is crucial for the preservation of cultural heritage. The purpose of this study was to evaluate the protective effect of Coromandel coatings on wooden substrates by analyzing their dimensions, weight, adhesion strength, hydrophobicity, and glossiness. The results indicate that after five cycles, the radial moisture expansion rate of the wood specimen is 0.332%, while that of the lacquer specimen is 0.079%, representing 23.8% of the radial moisture expansion rate of untreated wood specimens. This performance is superior to that of the ash and pigment specimens. Across different experimental conditions, the change in the mass of the Coromandel specimens aligns with the trend in their dimensional changes, indicating that moisture absorption and desorption are the primary reasons for dimensional changes. The influence of temperature on mass and dimensional stability is significant only in terms of dry shrinkage rate. After wet and dry cycles at 40 °C, the adhesion strength of the Coromandel specimens decreases the most, with the ash specimens decreasing by 7.2%, the lacquer specimens by 3.2%, and the pigment specimens by 4.5%. Following wet and dry cycles at three different temperatures, the contact angle of the lacquer layers changes by less than 5%, with their contact angle values exceeding 120°. These data indicate that among the Coromandel coatings, the lacquer layer provides the best protection for the wooden substrate, while the ash coating is the most fragile. The degradation rate of the Coromandel specimens increases with rising temperatures. These findings emphasize the critical roles of humidity and temperature in protecting wooden coatings and aim to provide theoretical insights and practical significance for the preservation of wooden artifacts and the assessment of coating performance.

Keywords: wet and dry cycles; lacquer coating; dimensional stability; adhesion

Citation: Liu, W.; Zhu, L.; Varodi, A.M.; Liu, X.; Lv, J. The Effect of Wet and Dry Cycles on the Strength and the Surface Characteristics of Coromandel Lacquer Coatings. *Forests* **2024**, *15*, 770. https://doi.org/10.3390/f15050770

Academic Editor: Anuj Kumar

Received: 18 March 2024
Revised: 23 April 2024
Accepted: 25 April 2024
Published: 27 April 2024

1. Introduction

The history of Chinese lacquer painting in China spans over 8000 years [1,2], serving as a profound symbol of Chinese civilization. As documented in the "Record of Painting and Decorating" and interpreted by relevant experts [3–5], Coromandel lacquer involves a meticulous process. This includes applying an ash base onto wooden surfaces, coating it with black lacquer, intricately carving and selectively removing lacquer layers, and filling them with vibrant pigments to create colorful and intricate patterns. This technique flourished during the Ming and Qing dynasties, with its works commonly found on screens and other artifacts. The term "Coromandel" derives from the export of such artistry along the southeastern coast of India [6–8].

Based on the documented evidence and examination of heritage artifacts, as illustrated in Figure 1, the structure of Coromandel lacquer can be summarized as comprising wood layers, ground layers (ash layers), lacquer layers, and pigment layers. Compared to other lacquer layer structures, the composition of Coromandel lacquer layers is more diverse, with the ground layer being relatively soft, easy to carve and decorate, and cost effective. However, precisely because of this, the adhesion of Coromandel lacquer layers is lower. In heritage artifacts, issues such as cracking and peeling of Coromandel lacquer coatings often occur, closely linked to the environmental conditions [9]. Specifically, when rainy days are followed by sunny weather, as the humidity decreases, the moisture within the coatings permeates and evaporates, causing the coating material to contract [10,11]. This affects the cohesion between the wood, ash layers, lacquer, and pigment coatings, resulting in cracking or peeling. This phenomenon is attributed to the hygroscopic nature of wood's chemical components and the plasticity of the capillary system, leading to deformations, cracks, and a significant decrease in various strength indicators of the wood [12–14]. Repeated cycles of wetting and drying further exacerbate these issues, causing a greater decrease in wood's strength indicators and affecting its fundamental properties and functionality [15,16]. The phenomenon of hygroscopic aging explains how repeated wet and dry cycles reduce wood's hygroscopicity, reflected in dimensional changes, indicating the loss of a hygroscopic reaction after the weakening of the wood's cell wall polarity [17]. Even repair materials like glutinous rice pulp lime mortar undergo slight irreversible dimensional changes under wet and dry cycling [18].

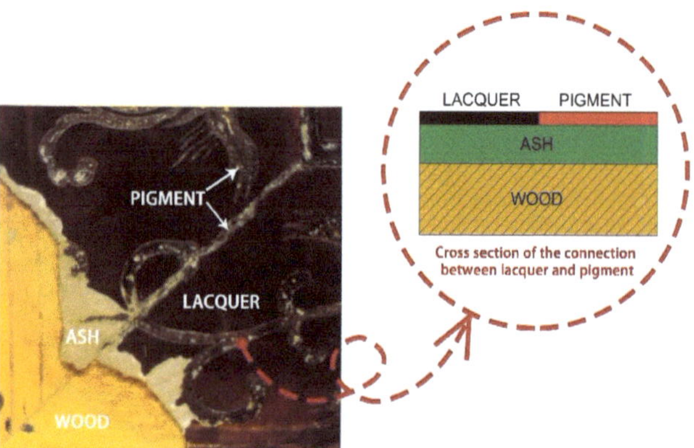

Figure 1. Structure diagram of damaged Coromandel screen [19].

The method of using wet and dry cycles and temperature to accelerate aging can reflect the influence of moisture on wood. Research indicates that during wet and dry cycles, wood undergoes minimal chemical changes [20]. However, repeated wet and dry cycles can cause aging effects on wood [21,22]. The commonly used wood wet and dry cycling aging conditions refer to standards such as ASTM G154 and ASTM D1037 [23,24], involving methods like immersion in water, drying, or wet and dry cycling at room temperature [25,26]. Studies have employed five cycles of alternating aging using a temperature of 50 °C and a relative humidity of 90%, establishing a relationship between moisture absorption and the aging dimensions. Additionally, research has explored the correlation between the radial and tangential shrinkage rates of wood and its thermal conductivities. Furthermore, the use of wet and dry cycles involving immersion in water and subsequent drying has been employed to verify changes in the dimensional stability of thermally modified wood. These studies collectively demonstrate that wet and dry cycles are the primary factors causing changes in the dimensional stability and moisture absorption of wood.

Surface coatings on wood exert a restraining effect on its swelling and shrinking due to moisture. This occurs because these coatings encapsulate the surface of the wood fibers, minimizing their direct exposure to moisture-laden air [10,27,28]. However, research on the changes in coating properties using wet and dry cycling methods is limited. This limit tation arises from the complexity of factors influencing surface coatings, which extend beyond humidity and drying conditions to include factors such as light exposure and chemical erosion. In contrast to the development of new materials, research on material degradation in cultural heritage aims to identify weaknesses in the traditional materials and the environmental factors causing material deterioration. This research aims to effectively reduce the damage caused by environmental factors to cultural heritage and to undertake targeted repair and preservation efforts.

In this study, wooden specimens, ash specimens, lacquer specimens, and pigment specimens were prepared according to the traditional processes. Artificial wet and dry cycles were induced using saturated solutions to control humidity and heating in an oven. After five wet and dry cycles, changes in the longitudinal and radial dry shrinkage rates, weight change rates, glossiness, contact angle, adhesion, and other values of the four specimens were measured systematically. The influence of the wet and dry cycle conditions on the wood substrate coatings was comprehensively investigated to provide data for studying the aging of Coromandel lacquer coating and similar coatings, offering valuable insights for the protection and restoration of relevant artifacts.

2. Materials and Methods

2.1. Materials

The raw lacquer was derived from the sap of a Chinese lacquer tree (*Toxicodendron vernicifluum* (Stokes) F. A. Barkley) located in Maoba, Hubei Province, China, and was procured from the Yangzhou Lacquerware Factory. To shorten the curing time of the raw lacquer, a refinement process was executed, following the guidelines established in the existing literature [29]. Ferrous hydroxide was prepared within the paint laboratory at Nanjing Forestry University. The natural mineral pigment ochre (Fe_2O_3 in chemical composition) was sourced from Lin Shan Tang, Suzhou. Tung oil was obtained from a retailer in Nanjing. Furthermore, pig's blood was acquired from the Yangzhou Lacquerware Factory. This material was concocted by blending fresh pig's blood with a small quantity of lime and water, resulting in a gelatinous pig blood substance. This substance was subsequently utilized as a binding agent for the ground layer [30].

The Chinese fir wood (*Cunninghamia lanceolata* (Lamb.) Hook.) was purchased from Zhejiang Jeson Wood products Co., Ltd. (Huzhou, China) and possessed a moisture content of $10 \pm 2\%$. The wood was then cut into 96 pieces of diameter-cut boards measuring 100 (longitudinal) $\times$ 100 (radial) $\times$ 10 (tangential) mm each. These boards underwent sanding using 400# sandpaper and were conditioned within an artificial climate chamber at $(20 \pm 2)\,°C$ and $(55 \pm 5)\,\%$ relative humidity until reaching a consistent weight, a process that spanned 14 days, before being readied for specimen preparation.

2.2. Material Preparation for the Black Lacquer and Oil Pigment

Drawing insights from both Xiu Shi Lu and pertinent research, the customary additive utilized in the formulation of black lacquer is generally ferrous hydroxide [4,31]. However, due to its susceptibility to oxidation in the presence of air, the preservation of ferrous hydroxide poses a challenge [32]. To ensure its efficacy, ferrous hydroxide is derived from the reaction between ferrous sulfate and sodium bicarbonate, as depicted using the chemical equation:

$$FeSO_4 + 2NaHCO_3 = Fe(OH)_2\downarrow + Na_2SO_4 + 2CO_2\uparrow$$

The amalgamation of 1 mol of $FeSO_4$ solution with 2 mol of $NaHCO_3$ solution results in the formation of a precipitate, $Fe(OH)_2$. This substance takes the form of a finely textured paste, obtainable using a filtration process. Gradually, the $Fe(OH)_2$ paste is incorporated into the refined raw lacquer at a mass ratio of 20%, with thorough agitation yielding the desired black lacquer.

Furthermore, ochre powder and Tung oil are meticulously blended at a 1:1 mass ratio to create Tung oil pigment [33], ensuring an even distribution through thorough stirring.

2.3. Preparation of the Coromandel Specimens

Within the framework of traditional Chinese finishing techniques, the incorporation of a ground base layer between the lacquer and the wood is a customary practice aimed at enhancing the lacquer's environmental stability [19,34]. Three distinct pig blood putties were concocted by blending gel-like pig blood with three varieties of tile ash, classified by their coarseness (80-mesh for coarse ash, 100-mesh for medium ash, and 120-mesh for fine ash [35]). This blending was achieved at a 1:1 ratio. Subsequently, the fir specimens were consecutively coated with the coarse putty, followed by the medium putty, and eventually the fine putty. Each layer of putty underwent a drying period of 7 days. The cumulative thickness of these three putty layers was approximately 1 mm. The base putty layer served as a transitional structure that bridges the gap between the substrate and the painted layer. This layer mends any gaps in the substrate, safeguards it, and prevents paint moisture from permeating into the wood, thereby averting wood deformation [19]. Concurrently, it ensures a seamless surface, priming it for the final finish. Eventually, the wood specimen underwent sanding with 400# sandpaper. A total of 72 pig's blood–ash specimens were meticulously prepared.

For the specimens coated with the ground base layer of pig's blood and ash, a layer of black lacquer was uniformly applied to their surface. These painted specimens were then placed within a climatic chamber set at 15–20 °C and 70–85% relative humidity for over 21 days to facilitate drying. The outcome was 24 black lacquer specimens. As for the oil pigment application, it was evenly administered onto the surface of the specimen coated with the pig blood ground layer. Subsequently, these specimens were placed in a climate chamber maintained at 15–20 °C and 60–65% RH, allowing them to dry for a span surpassing two weeks. This process yielded a set of 24 oil pigment specimens.

2.4. Wet and Dry Cycles

The 96 specimens were categorized based on their distinct specimen structures and experimental parameters, detailed in Table 1. Out of these, each set of 24 specimens featured 6 as the control group, while the remaining 18 underwent the wet and dry cycle experiment. These 18 specimens were further divided into three subgroups, each placed within an oven at temperatures of 40 °C, 30 °C, and 20 °C, while maintaining a temperature fluctuation of ±2 °C. Within each temperature subgroup, there were 6 wood specimens, 6 ash specimens, 6 black lacquer specimens, and 6 oil pigment specimens. To replicate the desired humidity conditions, a desiccator was employed in conjunction with a saturated magnesium chloride hexahydrate solution for the dry cycle, maintaining humidity levels between 30 and 33%. Similarly, the wet cycle utilized a saturated potassium chloride solution, maintaining humidity levels within the range of 80–85%. The experimental procedure consisted of initially subjecting the specimens to wet cycle conditions for a duration of 7 days, followed by a transition to dry cycle conditions for another 7 days. This complete sequence constitutes a singular wet and dry cycle, and this entire process was iterated for a total of 5 cycles. A visual representation of the flow of the wet and dry cycle experiments is presented in Figure 2.

Table 1. Design of various specimen codes, additives, and experimental conditions.

Code	Specimen Name	Specimen Structure	Temperature	Number of Specimens
W40	Wood specimen	wood	40 °C	6
W30	Wood specimen	wood	30 °C	6
W20	Wood specimen	wood	20 °C	6
W + A40	Ash specimen	wood + ash	40 °C	6
W + A30	Ash specimen	wood + ash	30 °C	6
W + A20	Ash specimen	wood + ash	20 °C	6

Table 1. *Cont.*

Code	Specimen Name	Specimen Structure	Temperature	Number of Specimens
W + A + L40	Lacquer specimen	wood + ash + lacquer	40 °C	6
W + A + L30	Lacquer specimen	wood + ash + lacquer	30 °C	6
W + A + L20	Lacquer specimen	wood + ash + lacquer	20 °C	6
W + A + P40	Pigment specimen	wood + ash + pigment	40 °C	6
W + A + P30	Pigment specimen	wood + ash + pigment	30 °C	6
W + A + P20	Pigment specimen	wood + ash + pigment	20 °C	6

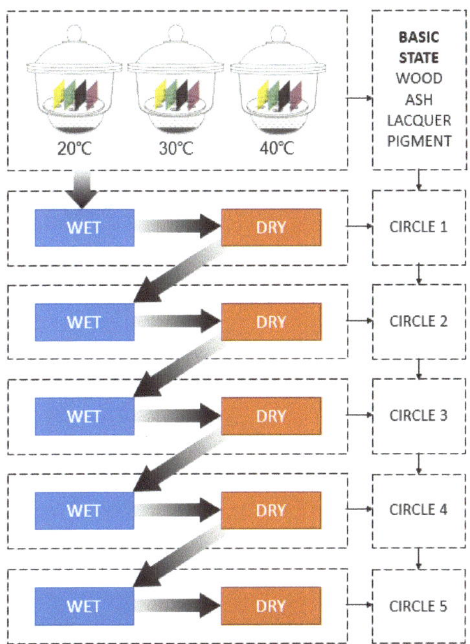

Figure 2. Flow chart of wet and dry cycle experiment.

2.5. Physical Tests

The tangible outcome of moisture absorption and release in Coromandel coatings is the alteration of the specimen's mass and dimensions. The practice of measuring these attributes before and during distinct cycles effectively captures the influence of wet and dry cycle conditions on a specimen's dimensions and their temporal trends. To this end, reference was made to the GB/T 1927.4-2021 standards titled "Test Method for Physical and Mechanical Properties of Small Clear Wood Specimens, Part 8: Determination of Wet Swelling" and "Test Method for Physical and Mechanical Properties of Small Clear Wood Specimens, Part 6: Determination of Dry Shrinkage" [36]. For measurement purposes, vernier calipers were utilized to assess the specimens' dimensions, while electronic balance scales were employed to gauge the specimens' weights.

The equations for determining the rates of longitudinal and radial wet swelling as well as dry shrinkage are as follows:

$$\alpha_l = \frac{L_{li} - L_{lo}}{L_{l0}} \times 100\% \tag{1}$$

$$\alpha_r = \frac{L_{ri} - L_{ro}}{L_{r0}} \times 100\% \tag{2}$$

In these formulas, α_l and α_r represent the wet expansion and dry shrinkage of the specimen, respectively, in both the longitudinal and radial directions, expressed as percentages. L_{li}, L_{ri} denote the longitudinal and radial dimensions of the specimen after a single wet cycle or dry cycle, measured in millimeters (mm), while L_{lo} and L_{ro} indicate the initial longitudinal and radial dimensions of the specimen, also measured in millimeters (mm).

When both the α_l and α_r values are positive, the specimen undergoes wet expansion. Conversely, when both the α_l and α_r values are negative, the specimen experiences dry shrinkage.

Upon the specimens achieving equilibrium moisture content, the initial weight of each specimen is measured. Throughout the progression of the wet and dry cycles, the weight change rate of each specimen during one complete cycle is documented and expressed using the subsequent equation.

The equation for calculating the weight change rate [37,38] is as follows:

$$W = \frac{m - m_0}{m_0} \times 100\% \tag{3}$$

where W represents the rate of weight change, encompassing both the rate of weight decrease from high humidity to low humidity and the rate of weight increase from low humidity to high humidity, expressed as a percentage. m_0 denotes the initial weight of the specimen, measured in grams (g). m signifies the weight of the specimen maintained under specific humidity conditions for a duration of 7 days, also measured in grams (g).

When the calculated value of W is positive, the specimen undergoes wet swelling. Conversely, if the calculated value of W is negative, the specimen undergoes dry shrinkage.

Subsequent to the completion of all the wet and dry cycles, the adhesion of each coating was evaluated using a BEVS adhesion meter equipped with a 20 mm diameter test column. This evaluation was conducted in accordance with the national standard "Color and Varnish Pull-off Method Adhesion Test" (GB/T 5210-2006) [39]. The aim was to identify the weakest interface within the Coromandel coating.

In accordance with the standards ASTM D523-2014 "Standard Test Method for Mirror Gloss" [40] and GBT4893.6-2013 "Physical and Chemical Properties Test of Furniture Surface Paint Film, Part 6 Gloss Determination Method" [41], a three-angle gloss meter (Guangdong San En Shi Intelligent Technology Co., Ltd., HG268, Guangdong, China) was utilized to measure the gloss of the specimens throughout the wet and dry cycles. For these measurements, the 60° angle condition was chosen. Prior to the commencement of the aging experiment and at each 7-day interval during the aging process, the glossiness data were recorded, with the same point consistently measured on the same specimen. The light loss rate of the fixed points on each test specimen was computed, and the average value yielded the light loss rate of the respective test specimen.

The formula for calculating the light loss rate is as follows:

$$H = \frac{H_0 - H_i}{H_0} \times 100\% \tag{4}$$

where H_0 denotes the initial gloss, expressed as a percentage. H_i signifies the gloss after aging, expressed as a percentage.

A DSA100S droplet shape analyzer from KRUSS in Germany was employed to assess the water contact angle of the surfaces coated with Coromandel. For the testing process, the test specimen (measuring 100×100 mm) was subdivided into smaller 10 mm wide segments. Each coating underwent parallel testing at five distinct locations, utilizing a droplet volume of 2 µL. The initial water contact angle was recorded at each of these locations.

The core objective of this study was to illuminate the stability and aging resistance attributes of the four coatings. This was accomplished by examining the performance changes exhibited by the four Coromandel specimens across varying cycle durations and temperatures.

2.6. Statistical Analysis

The statistical analysis was carried out using SPSS software (IBM Corp., IBM SPSS Statistics for Windows, v. 25, Armonk, NY, USA). To assess the significant effects among different specimens and temperatures during the wet and dry cycling process, an analysis of variance (ANOVA) was executed at a significance level of 0.05. This analysis encompassed the rate of wet swelling, dry shrinkage, and weight change in the radial direction of the Coromandel specimens. Prior to conducting the one-way ANOVA tests, two essential prerequisites were addressed. The homogeneity of variance was evaluated using the Levene test, while the normal distribution of data was verified using the Shapiro–Wilk test. Both these tests were essential to establish the suitability of the data for the ANOVA. Subsequently, the mean values were distinguished using Fisher's protected minimum significant difference (LSD) test at a significance level of $\alpha = 0.05$. This statistical methodology enabled the differentiation of meaningful variations among the data sets.

3. Results

3.1. Dimensional Stability

The moisture-related swelling and shrinking rates of wood are important parameters describing its dimensional changes. For Coromandel specimens using Chinese fir wood as the base material, the main source of dimensional variation lies in the fir wood itself, while the ash layer, lacquer layer, and pigment layer of the Coromandel affect it. In order to better study the dimensional stability of the Coromandel layers, the experiment first verified the longitudinal and radial dimension change rates due to the swelling and shrinking of the fir wood specimens under different temperature conditions during wet and dry cycling, as shown in Figure 3. The longitudinal wet swelling and shrinkage rates of the fir wood specimens are lower than the radial wet swelling and shrinkage rates [40,41], and the radial wet swelling and shrinkage rates are approximately ten times those of the longitudinal rates, consistent with the research findings in literature [26,42,43]. The underlying cause is attributed to the structural and fiber orientation disparities within the wood. Longitudinally oriented fibers exhibit greater stability due to their robust interconnections, resulting in minimal longitudinal shrinkage or swelling. Conversely, fibers oriented in the radial direction are comparatively weaker, featuring more fragile interconnections. Consequently, they are more susceptible to moisture-induced changes, resulting in pronounced shrinkage or swelling in the radial direction.

Throughout the aging process, the Coromandel specimens underwent five cycles of wet and dry conditions. Notably, the rate of wet and dry shrinkage exhibited a gradual reduction over time. This phenomenon can be attributed to the "moisture-absorption fatigue synthesis mechanism" intrinsic to wood. Hygroscopic aging refers to the partial loss of wood's hygroscopic response, occurring as a consequence of a process wherein the cell wall polarity diminishes [44]. This mechanism entails a partial saturation of the cell wall polar groups, rendering them less responsive to water vapor fixation. This is reflected in the decrease in the swelling and shrinking rates observed in the experimental data, which conforms to this pattern under different temperature conditions.

The maximum longitudinal and radial wet swelling and dry shrinkage rates of the Coromandel specimens under wet and dry cycling conditions are presented in Table 2. At room temperature (20 °C), the radial wet swelling rate of 0.565% significantly exceeded the longitudinal wet swelling rate of 0.061% during the wet cycle. Similarly, the radial dry shrinkage rate of −0.292% was notably higher than the longitudinal dry shrinkage rate of −0.033% observed during the dry cycle. This discrepancy arises from the mechanism of water absorption in the longitudinal direction of wood [12,45]. In this scenario, water primarily traverses continuous pathways composed of tubular cells, driven by potent capillary forces. This facilitates rapid water transport within the wood. In contrast, when water is conveyed transversely across the wood, its movement is constrained by the necessity to navigate through grain pores, slowing down the transport process.

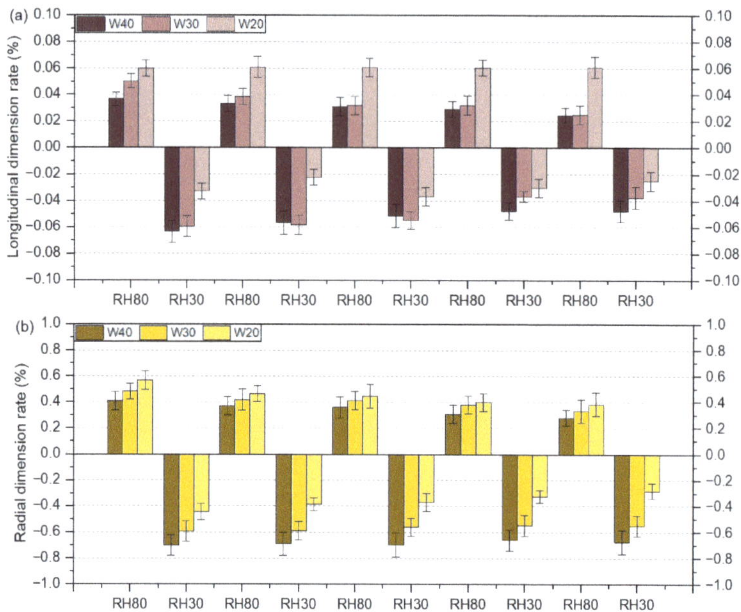

Figure 3. Longitudinal and radial dimension rates of the wood specimen under wet and dry cycling conditions. (**a**) Longitudinal dimension rate of the wood specimen; (**b**) radial dimension rate of the wood specimen.

Table 2. Maximum values of longitudinal and radial direction wet expansion and dry shrinkage of the Coromandel specimens under wet and dry cycling conditions.

Code		Wood 20 °C (%)	Ash 20 °C (%)	Lacquer 20 °C (%)	Pigment 20 °C (%)
Wet circle	α_l	0.061 ± 0.008	0.058 ± 0.012	0.049 ± 0.008	0.048 ± 0.011
	α_r	0.565 ± 0.071	0.633 ± 0.102	0.272 ± 0.059	0.489 ± 0.064
Dry circle	α_l	-0.037 ± 0.007	-0.051 ± 0.013	-0.050 ± 0.006	-0.046 ± 0.007
	α_r	-0.445 ± 0.062	-0.208 ± 0.057	-0.425 ± 0.063	-0.184 ± 0.032

The ash, lacquer, and pigment specimens likewise exhibited greater radial wet and dry shrinkage rates compared to their longitudinal counterparts. Pig's blood–ash, black lacquer, and oil pigment finishes collectively influence the wet swelling and dry shrinkage behavior of the underlying fir wood base material. During the wet cycle, both the black lacquer and the oil pigment significantly curbed the wood's wet swelling. Conversely, in the dry cycle, the pig's blood–ash and oil pigment exerted significant inhibitory effect on the wood's dry shrinkage. Given that radial wet swelling and dry shrinkage are the pivotal factors contributing to the dimensional deformation of Coromandel furniture, a comprehensive analysis of the radial dimensional behavior is presented below.

The radial dimension change rates of the Coromandel specimens under wet and dry cycling conditions are shown in Table 3. In the wet cycle, the moisture absorption rate of the ash specimens is slightly higher than that of the wood specimens, while the moisture absorption rate of the pigment specimens is slightly lower than that of the wood specimens and the moisture absorption rate of the lacquer specimens is significantly lower than that of the wood specimens. This is because in the three-layer structure of Coromandel specimens, the lacquer effectively prevents moisture from penetrating through the surface layer and the ash layer into the wood layer [46]. After five cycles, the radial moisture expansion rates for the wood specimens, ash specimens, lacquer specimens, and pigment specimens are 0.332%, 0.427%, 0.079%, and 0.368%, respectively. The lacquer specimen notably outperforms the ash and pigment specimens, representing only 23.8% of the radial moisture expansion rate of the untreated wood specimens. The moisture expansion data on the Coromandel coatings directly reflect the dimensional changes caused by moisture absorption under natural conditions, which are particularly important. In the dry cycle, the drying rate of the wood specimens and the lacquer specimens is higher than that of the ash specimens and the pigment specimens. This is because the hygroscopic nature of the ash and pigment specimens affects the drying properties of the base wood. When the type of specimen is used as a factor, all the p-values are less than 0.05, indicating that the type of specimen has a significant effect on Coromandel specimens [47].

Table 3. Radial dimension change rate of the Coromandel specimens under wet and dry cycling conditions.

Code	Cycle1		Cycle2		Cycle3		Cycle4		Cycle5	
	Wet	Dry	Wet	Dry	Wet	Dry	Wet	Dry	Wet	Dry
Specimen type										
W	0.484 [a]	−0.580 [b]	0.417 [a]	−0.556 [b]	0.406 [a]	−0.542 [b]	0.361 [a]	−0.512 [b]	0.332 [a]	−0.503 [b]
W + A	0.507 [a]	−0.268 [a]	0.488 [a]	−0.258 [a]	0.470 [a]	−0.260 [a]	0.435 [a]	−0.248 [a]	0.427 [a]	−0.240 [a]
W + A + L	0.164 [b]	−0.516 [b]	0.120 [b]	−0.506 [b]	0.099 [b]	−0.499 [b]	0.088 [b]	−0.475 [b]	0.079 [b]	−0.460 [b]
W + A + P	0.409 [a]	−0.350 [a]	0.393 [a]	−0.322 [a]	0.383 [a]	−0.306 [a]	0.379 [a]	−0.274 [a]	0.368 [a]	−0.240 [a]
Temperature										
40 °C	0.348 [a]	−0.549 [b]	0.326 [a]	−0.535 [c]	0.313 [a]	−0.526 [b]	0.282 [a]	−0.495 [b]	0.268 [a]	−0.472 [a]
30 °C	0.375 [a]	−0.421 [a]	0.356 [a]	−0.406 [b]	0.354 [a]	−0.399 [a]	0.343 [a]	−0.383 [ab]	0.324 [a]	−0.375 [ab]
20 °C	0.450 [a]	−0.315 [a]	0.381 [a]	−0.291 [a]	0.353 [a]	−0.281 [a]	0.322 [a]	−0.253 [a]	0.313 [a]	−0.237 [a]
p Values										
Specimen type	<0.001	<0.001	<0.001	<0.001	<0.001	<0.001	<0.001	<0.001	<0.001	0.001
Temperature	0.267	<0.001	0.653	<0.001	0.783	<0.001	0.572	0.001	0.454	0.003
Specimen type × Temperature	0.435	0.698	0.505	0.434	0.722	0.679	0.496	0.741	0.279	0.672

Note: Mean values of H followed by the same small superscript letters ([a–c]) within a group are not significantly different based on Fisher's protected LSD test at the 0.05 significance level. The p-value indicates the significance of the influencing factor. A smaller p-value indicates stronger significance. Generally, when the p-value is less than 0.05, the influencing factor is considered significant, and when the p-value is greater than 0.05, it is considered insignificant.

The moisture expansion and contraction rates of the Coromandel specimens are also influenced by different temperatures under wet and dry cycle conditions. The moisture expansion rates of the Coromandel specimens in the radial direction at different temperatures are shown in Figure 4. Under wet cycle conditions, at 40 °C, the radial moisture expansion rate of the four types of Coromandel specimens is lower than at 30 °C, which, in turn, is lower than at 20 °C. Under dry cycle conditions, at 40 °C, the radial drying shrinkage rate of the four types of Coromandel specimens is greater than at 30 °C, which, in turn, is greater than at 20 °C. Temperature promotes the drying process of the Coromandel specimens, leading to an increase in the radial drying shrinkage rate of the Coromandel specimens. Some scholars have also conducted drying under high-temperature and low-humidity conditions [48], indicating that this condition has the highest aging efficiency. Meanwhile, low temperatures are more conducive to the moisture absorption of the Coromandel specimens. It can be inferred that under conditions of a low temperature and high humidity or a high temperature and low humidity, the dimensional variation in the Coromandel coatings is maximal, and their dimensional stability is poorest. Through significance analysis, it can be observed that during the dry cycle process, the p-value is less than 0.05, indicating significance; however, during the wet cycle, it is not significant.

In terms of the significance analysis of the wet and dry cycles, the type of specimen wielded a significant impact on the wet swelling and dry shrinkage rates, while temperature significantly affected dry shrinkage. However, the interaction between the Coromandel specimen type and temperature was not significant. This suggests that specimen type and temperature independently influence the wet swelling and dry shrinkage rates. Consequently, when assessing and regulating these rates, considering specimen type and temperature as distinct factors, rather than focusing on their interaction, is advisable.

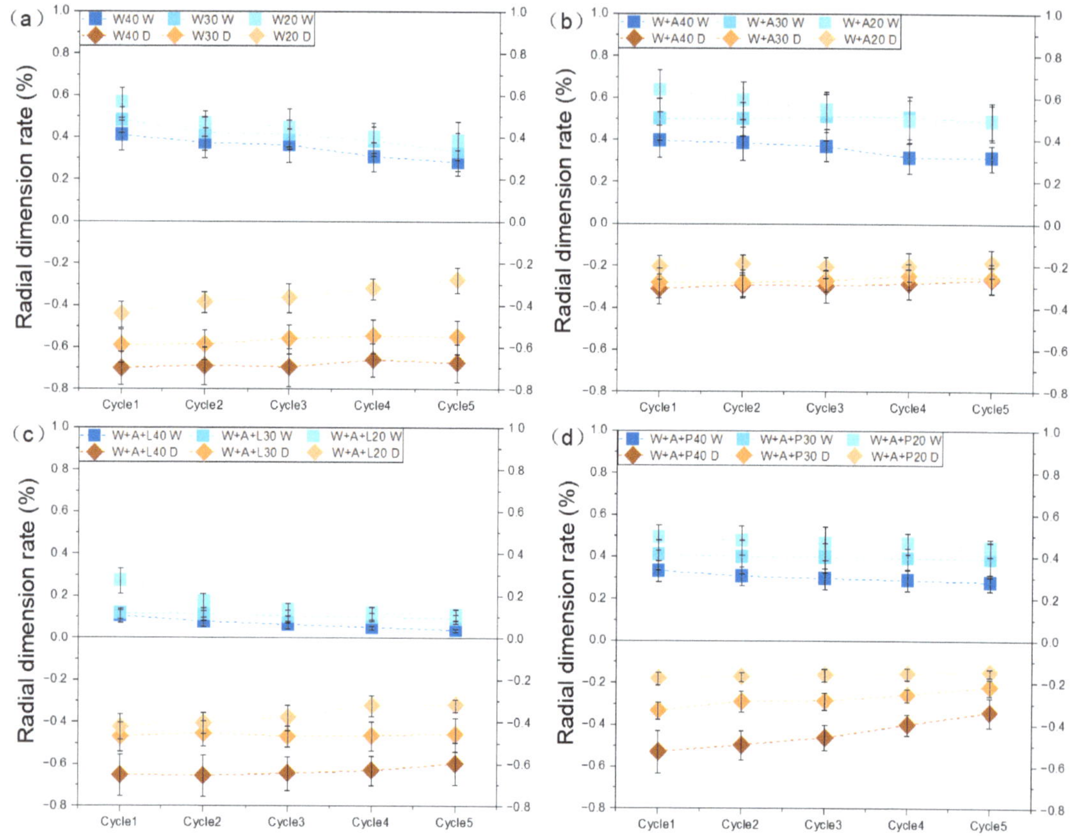

Figure 4. Radial dimension change rate of the Coromandel specimens at different temperature (**a**) Wood specimen; (**b**) ash specimen; (**c**) lacquer specimen; (**d**) pigment specimen.

3.2. Weight Change

The weight change rate of the Coromandel specimens under the wet and dry cycles is shown in Table 4 and Figure 5. During wet and dry cycles, changes in specimen weight are mainly due to changes in the moisture absorption and desorption of the specimens, reflecting primarily changes in the moisture absorption and desorption of the base material fir wood, under different temperatures and wet and dry cycle conditions. Simultaneously, they reflect the effects of the ash layer, the black lacquer layer, and the oil pigment layer on the moisture absorption and desorption of the base fir wood. Under wet and cycle conditions, the quality of the Coromandel coatings fluctuates with humidity. The influence of oil pigment and ash on the moisture absorption rate of fir wood is relatively weak, but the protective effect of the black lacquer is evident. This is because the dense black lacquer blocks the exchange of moisture between the ash and the surface of the wood and the external humidity environment [49]. Under dry cycle conditions, the weight change in the wood specimens and the lacquer specimens is greater than that of the ash specimens and the pigment specimens. Regarding the influence of temperature on the Coromandel specimens, the weight change rate of the Coromandel coatings is highest at 40 °C and lowest at 20 °C. The type of specimen has a significant effect on weight change, but temperature and the interaction between specimen type and temperature are not significant. These patterns are consistent with the dimensional stability of the Coromandel specimens, indicating that moisture absorption and desorption are the primary factors influencing dimensional changes.

Table 4. Weight change rate of the Coromandel specimens under wet and dry cycling conditions.

Code	Cycle1		Cycle2		Cycle3		Cycle4		Cycle5	
	Wet	Dry	Wet	Dry	Wet	Dry	Wet	Dry	Wet	Dry
Specimen type										
W	2.513 [a]	−2.697 [bc]	2.421 [a]	−2.677 [b]	2.362 [a]	−2.692 [b]	2.218 [a]	−2.594 [b]	1.978 [a]	−2.623 [c]
W + A	2.262 [ab]	−1.843 [ab]	2.187 [a]	−1.817 [a]	2.188 [a]	−1.789 [a]	2.163 [a]	−1.671 [a]	2.085 [a]	−1.767 [ab]
W + A + L	1.383 [b]	−2.746 [c]	1.191 [b]	−2.721 [b]	1.020 [b]	−2.674 [b]	0.891 [b]	−2.578 [b]	0.778 [b]	−2.570 [bc]
W + A + P	2.526 [a]	−1.559 [a]	2.490 [a]	−1.480 [a]	2.389 [a]	−1.482 [a]	2.254 [a]	−1.417 [a]	2.314 [a]	−1.414 [a]
Temperature										
40 °C	1.634 [a]	−3.256 [c]	1.614 [b]	−3.224 [c]	1.586 [a]	−3.261 [c]	1.508 [a]	−3.160 [b]	1.342 [a]	−3.213 [c]
30 °C	2.212 [ab]	−2.401 [b]	2.143 [ab]	−2.320 [b]	2.040 [ab]	−2.266 [b]	1.972 [a]	−2.105 [b]	1.863 [ab]	−2.132 [b]
20 °C	2.667 [a]	−0.977 [a]	2.460 [a]	−0.978 [a]	2.344 [a]	−0.951 [a]	2.166 [a]	−0.931 [a]	2.161 [a]	−0.936 [a]
_p_ Values										
Specimen type	0.009	0.002	0.002	<0.001	<0.001	0.001	0.001	<0.001	<0.001	0.001
Temperature	0.008	<0.001	0.019	<0.001	0.014	<0.001	0.088	<0.001	0.004	<0.001
Specimen type × Temperature	0.977	0.987	0.976	0.947	0.998	0.979	0.994	0.977	0.987	0.973

Note: Mean values of H followed by the same small superscript letters ([a–c]) within a group are not significantly different based on Fisher's protected LSD test at the 0.05 significance level. The *p*-value indicates the significance of the influencing factor. A smaller *p*-value indicates stronger significance. Generally, when the *p*-value is less than 0.05, the influencing factor is considered significant, and when the *p*-value is greater than 0.05, it is considered insignificant.

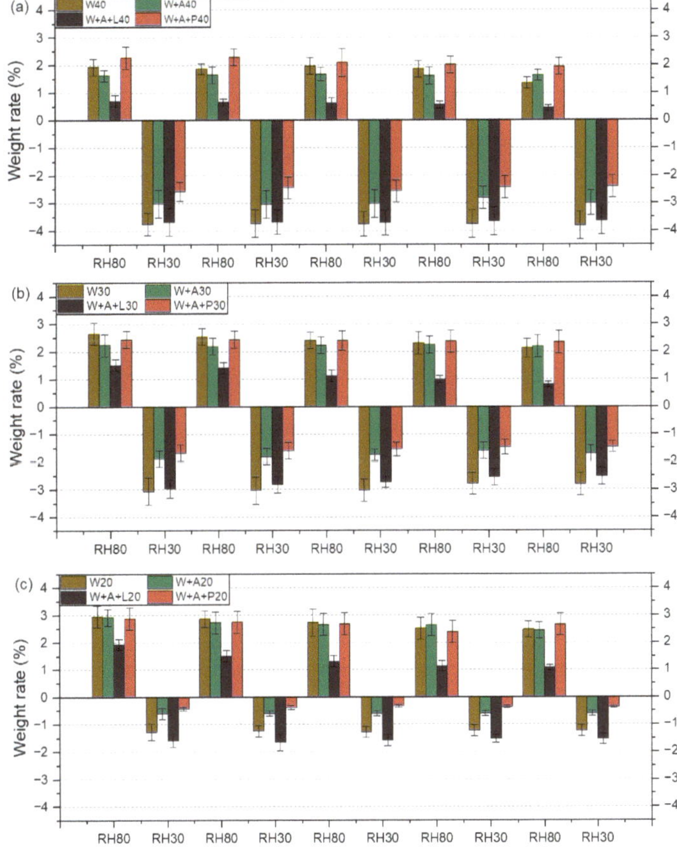

Figure 5. Weight change rate of the Coromandel specimens under wet and dry cycling conditions at different temperatures. (**a**) Weight change rate of the Coromandel specimens at 40 °C; (**b**) weight change rate of the Coromandel specimens at 30 °C; (**c**) weight change rate of the Coromandel specimens at 20 °C.

3.3. Adhesion

In the realm of organic coatings, the adhesion to the substrate stands as a pivotal characteristic [50]. In general, increased humidity leads to loss of the coating's material strength, as well as reduced adhesion between the wood and the coating [51]. To delve deeper into the impact of wet and dry cycles on Coromandel coatings, an evaluation of the adhesion alterations of the Coromandel specimens prior to and following the wet and dry cycles was conducted, as illustrated in Figure 6. Notably, the ash, lacquer, and pigment specimens each exhibited a marginal reduction in their adhesion after undergoing the wet and dry cycles. It is worth highlighting that the adhesion of the lacquer and pigment specimens slightly surpassed that of the ash specimen. The greatest decrease was observed at 40 °C, with the ash specimens decreasing by 7.2%, the lacquer specimens by 3.2%, and the pigment specimens by 4.5%.

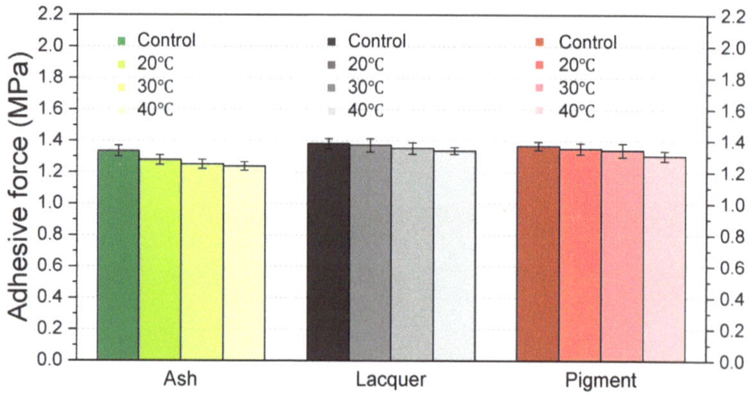

Figure 6. Changes in adhesion of Coromandel specimens after wet and dry cycles.

When subjected to tension, there was a fracture between the ash layer and the wood as shown in Figure 7. Additionally, it can be observed from the figures that a small amount of the ash layer still remained on the wood substrate after the adhesion test, indicating that the mechanical strength of the ash layer itself is not high. The significant variations in the radial moisture expansion and contraction rates and the weight change rates of the ash specimens in the dimensional stability experiments also confirm this result. This suggests that the ash layer is a weak interface in Coromandel coatings, and the interface between the ash layer and the wood layer is the weakest layer.

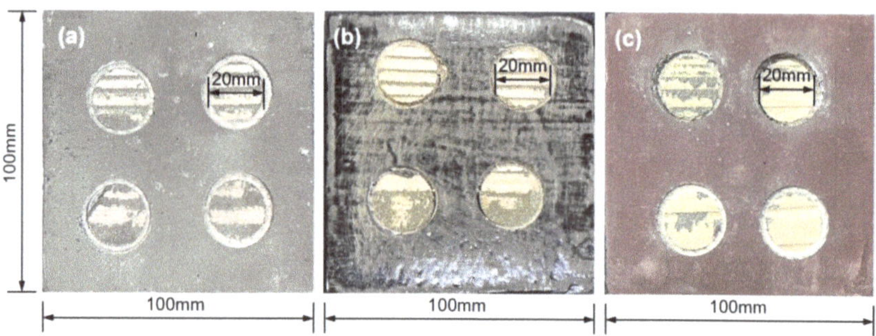

Figure 7. Part of Coromandel specimens after the adhesion test. (**a**) Ash specimen; (**b**) lacquer specimen; (**c**) pigment specimen.

The fracture of the lacquer and pigment specimens still occurs between the ash layer and the wood substrate. This is because the adhesion between the ash layer and the wood substrate is still weaker than the adhesion between the ash layer and the lacquer or pigment layers. However, the adhesion of the lacquer and pigment specimens is higher than that of the ash specimen. This is because during the wet and dry cyclic aging process, the encapsulation of the lacquer and pigment layers partially prevents moisture from penetrating, delaying the degradation of the ash layer and wood substrate [52,53], thereby improving the adhesion to a certain extent.

3.4. Contact Angle

Figure 8 depicts the contact angle values of the Coromandel specimens subjected to wet and dry cycling, at the initial stage, after cycling at 20 °C, 30 °C, and 40 °C. Aging has a decreasing effect on the contact angle of the coating for all the specimens [54]. This is because the humidity variation during the cycles causes the expansion and contraction of the specimens, resulting in microscale damage on the surface, leading to an increase in the roughness of the Coromandel coatings. Liquids tend to spread more easily on rough surfaces, causing an increase in wetting and a decrease in hydrophobicity, resulting in decreased contact angle values. As the temperature during the wet and dry cycles increases, the decrease in the contact angle values becomes more pronounced. This is because temperature accelerates the aging of the specimens, further reducing their hydrophobicity.

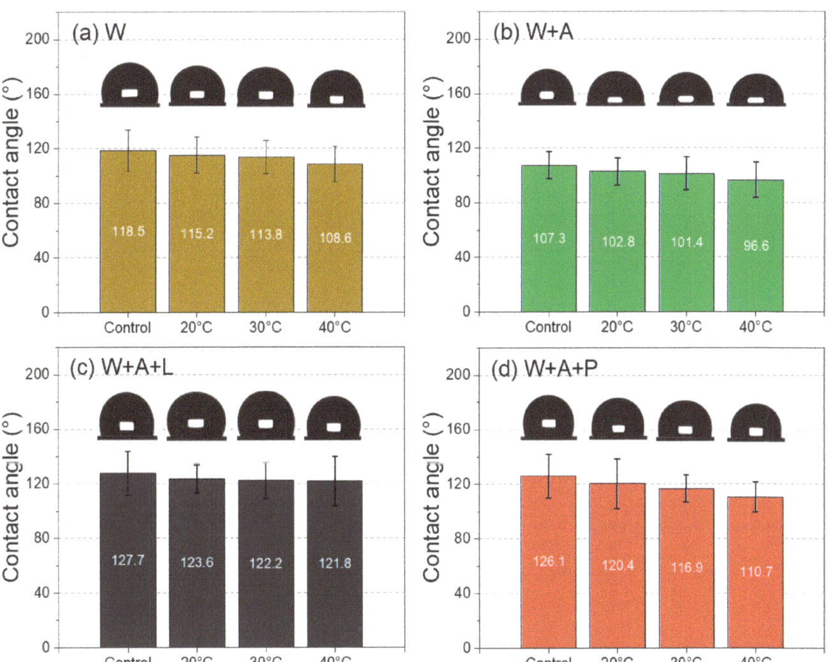

Figure 8. Contact angle of Coromandel specimens under wet and dry cycling conditions. (**a**) Wood specimen; (**b**) ash specimen; (**c**) lacquer specimen; (**d**) pigment specimen.

Using the contact angle values of the fir wood substrate before it underwent wet and dry cycling as a benchmark, it was observed that both the lacquer and pigment specimens exhibited higher contact angles compared to the wood specimens, while the ash specimens displayed lower contact angles. This reflects the protective effect of the lacquer and pigment coatings on the wood. After undergoing the wet and dry cycles, the contact angle values of all four types of specimens decreased. At 20 °C, the decrease rates of the wood layer, ash layer, lacquer layer, and pigment layer are 2.8%, 4.2%, 3.2%, and 4.5%, respectively; at 30 °C, the decrease rates are 4.0%, 5.5%, 4.3%, and 7.3%, respectively; at 40 °C, the decrease rates are 8.4%, 10.0%, 4.6%, and 12.2%, respectively. It can be observed that the decrease rates of the ash and pigment layers are higher than those of the wood and lacquer layers.

The change in the contact angle of the lacquer layer is less than 5% under all three temperature conditions, and the contact angle values are greater than 120°. This is because the main component of lacquer, phenolic compounds, contains a benzene ring structure which provides strong π-π stacking interactions between molecules [55], forming a closely packed molecular structure on the lacquer film surface, making the surface smooth and the contact angle value high. In the oil pigment layer, the fatty acids and oleic acids of the binding agent Tung oil undergo oxidation reactions during the drying process, forming a dense cross-linked network structure [56]. The surface of the coating has no obvious pores or irregularities, making the surface of the coating smoother, resulting in a contact angle value of 126.1. Under high-temperature conditions, the unsaturated fatty acids in Tung oil may undergo oxidation due to oxygen, and volatile organic solvents may be released [57], accelerating the deterioration of the pigment coating surface, resulting in a decrease in the contact angle value of the pigment specimen to 110.7° at 40 °C. These data emphasize the protective effect of lacquer and pigment coatings on wood substrates, while the ash coating as a transitional structure does not provide protection.

3.5. Glossiness of the Black Lacquer and Oil Pigment Coatings

Glossiness is an important indicator for evaluating surface physical property changes in color coatings [58]. The gloss loss of the lacquer specimens under wet and dry cycling conditions is shown in Figure 9. With an increase in the number of cycles, the glossiness damage to the lacquer specimens gradually intensifies. This is due to the microstructural changes on the surface of the lacquer film caused by the wet and dry cycles [59], resulting in an increase in its surface roughness, leading to a decrease in its glossiness. After five wet and dry cycles under 40 °C conditions, the gloss loss rate of the black lacquer specimen reached up to 6.83%, indicating that the temperature accelerated the aging effect of moisture on the lacquer film. The gloss loss rate is directly proportional to the temperature, with the gloss loss rate at 40 °C significantly higher than those at 30 °C and 20 °C. This may be because the high temperature promotes oxidation reactions on the surface of the black lacquer [60], leading to the formation of oxides and consequently causing a decrease in glossiness.

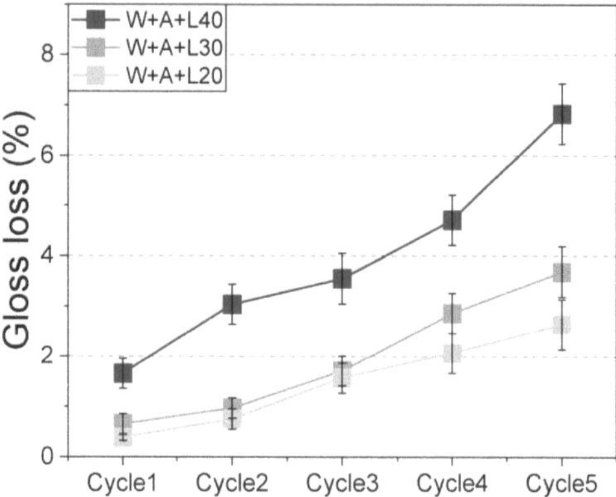

Figure 9. Gloss loss of black lacquer specimen under wet and dry cycling conditions.

As the experimental conditions were designed to simulate temperature and humidity variations in natural environments with a moderate acceleration intensity, and considering that the glossiness of the pigment specimens was inherently low, significant differences in the glossiness values after the aging of the pigment specimens were not observed.

4. Conclusions

In conclusion, using alternating wet and dry cycle experiments on four types of Coromandel coatings, namely no coating, a pig blood–ash layer, a black lacquer layer, and an oil pigment layer, the protective effects of these coatings on the wood substrate were studied. The research findings are as follows:

(1) Under wet and dry cycle conditions, all four types of Coromandel specimens experienced moisture expansion and contraction. The results indicate that under conditions of 40 °C and 30% RH, the dry shrinkage rate of the Coromandel specimens was maximal, while under conditions of 20 °C and 80% RH, the wet expansion rate was maximal. The ash, lacquer, and pigment layers all provided a certain degree of protection to the wood substrate. From the radial moisture expansion rate, it could be seen that after five wet cycles, the radial moisture expansion rate of the lacquer specimens was 0.079%, which was 23.8% of the radial moisture expansion rate of the wood specimens, indicating the best protective effect.

(2) Specimen type significantly influences the moisture swelling and dry shrinkage rates, while temperature significantly affects the dry shrinkage rates. Across different experimental conditions, changes in the mass of the Coromandel specimens align with their dimensional changes, indicating that moisture absorption and desorption are the primary reasons for dimensional changes.

(3) With an increase in the number of wet and dry cycles, the adhesion of the Coromandel specimens showed a slight decrease. The ash layer was identified as a weak interface, representing the weakest interface layer between the ash layer and the wood layer. After wet and dry cycles at 40 °C, the adhesion strength of the Coromandel specimens decreased the most, with the ash specimens decreasing by 7.2%, the lacquer specimens by 3.2%, and the pigment specimens by 4.5%.

(4) After wet and dry cycles, the surface glossiness and hydrophobicity of the Coromandel coatings decreased to a certain extent with an increasing number of cycles, further exacerbated by induced temperature. Following wet and dry cycles at three different temperatures, the contact angle of the lacquer layers changed by less than 5%, with their contact angle values exceeding 120°, while the loss of glossiness reaches up to 6.83%.

In summary, this study provides a comprehensive understanding of the complex interactions between Coromandel coatings, wood substrates, and environmental conditions. These findings emphasize the importance of temperature and humidity in the stability and degradation of Coromandel coatings. Furthermore, future research could delve deeper into the durability and protective effects of different types of Coromandel coatings, as well as conducting a more thorough analysis of the mechanisms according to which various environmental factors affect Coromandel lacquer artifacts.

Author Contributions: Conceptualization, W.L. and X.L.; software, W.L.; investigation, A.M.V.; data curation, L.Z.; writing—original draft preparation, W.L.; writing—review and editing, X.L.; project administration, J.L.; funding acquisition, J.L. All authors have read and agreed to the published version of the manuscript.

Funding: Funding was received from the Ministry of Education's Humanities and Social Sciences Research Fund project (21YJC760017), China.

Data Availability Statement: Data is contained within the article.

Acknowledgments: The authors would like to thank the College of Furnishings and Industrial Design, Nanjing Forestry University, for facilitating the experimental conditions.

Conflicts of Interest: The authors declare no conflicts of interest.

References

1. Wu, M.; Zhang, B.; Jiang, L.; Wu, J.; Sun, G. Natural lacquer was used as a coating and an adhesive 8000 years ago, by early humans at Kuahuqiao, determined by ELISA. *J. Archaeol. Sci.* **2018**, *100*, 80–87. [CrossRef]
2. Zhai, K.; Sun, G.; Zheng, Y.; Wu, M.; Zhang, B.; Zhu, L.; Hu, Q. The earliest lacquerwares of China were discovered at Jingtoushan site in the Yangtze River Delta. *Archaeometry* **2021**, *64*, 218–226. [CrossRef]
3. Yuming, S. *Explanatory Commentary on the 'Xiu Shi Lu' of Jianjia Tang*; Taiwan Commercial Press: Taipei, Taiwan, 1974.
4. Shixiang, W. *Explanatory Commentary on Xiu Shi Lu*; Sanlian Bookstore for Life, Reading, and New Knowledge: Beijing, China, 2013.
5. Bei, C. *Illustrated Explanation of Xiu Shi Lu*; Shandong Pictorial Publishing House: Jinan, China, 2021.
6. Wenfu, D. *A Study of Ancient Chinese Lacquered Furniture: Evidence from the 10th to 18th Centuries*; Cultural Relics Publishing House: Beijing, China, 2012.
7. Brugier, N. *Les laques de Coromandel*; La Bibliothèque des Arts: Lausanne, Switzerland, 2015.
8. Fei, N. Flocks of Birds Flying Westward: The Western Transmission of Chinese Kuancai Lacquer Screens—Starting from the Exhibition of Qing Kangxi's 'Bai Niao Chao Feng' Twelve-Panel Lacquer Screen with Black Lacquer and Polychrome at the Muenscher Lackkunst Museum, Germany. In Proceedings of the Guangzhou Cultural and Creative Industry Exposition; Guangzhou Cultural Museum: Guangzhou, China, 2017; pp. 112–117+194–228.
9. Zhu, Y.; He, J.; Zhang, L. Restoration of Coromandel Lacquer Screens in Jinhua Museum. In Proceedings of the ICOM-CC 20th Triennial Conference, Valencia, Spain, 18–22 September 2023.
10. Qu, S.; Ju, P.; Zuo, Y.; Zhao, X.; Tang, Y. The effect of various cyclic Wet-Dry exposure cycles on the Failure Process of Organic Coatings. *Int. J. Electrochem. Sci.* **2019**, *14*, 10754–10762. [CrossRef]
11. Wang, Q.; Feng, X.; Liu, X. Advances in historical wood consolidation and conservation materials. *Bioresources* **2023**, *1*, 6680–6723. [CrossRef]
12. Cheng, J. *Wood Science*; China Forestry Publishing House: Beijing, China, 1985.
13. Gereke, T.; Anheuser, K.; Lehmann, E.; Kranitz, K.; Niemz, P. Moisture behaviour of recent and natural aged wood. *Wood Re.* **2011**, *56*, 33–42.
14. Bai, R.; Wang, W.; Chen, M.; Wu, Y. Study of ternary deep eutectic solvents to enhance the bending properties of ash wood. *RSC Adv.* **2024**, *14*, 8090–8099. [CrossRef] [PubMed]
15. Liu, J.; Ji, Y.; Lu, J.; Li, Z. Mechanical Behavior of Treated Timber Boardwalk Decks under Cyclic Moisture Changes. *J. Korean Wood Sci. Technol.* **2022**, *50*, 68–80. [CrossRef]
16. Wang, J.; Cao, X.; Yang, Q.; You, H.; Qi, X. Experimental Study on Accelerating Aging of Larch Glued with Dry-Wet Cycling. *J. Hunan Univ.* **2022**, *50*, 156–164.
17. García Esteban, L.; Gril, J.; de Palacios de Palacios, P.; Guindeo Casasús, A. Reduction of wood hygroscopicity and associated dimensional response by repeated humidity cycles. *Ann. For. Sci.* **2005**, *62*, 275–284. [CrossRef]
18. Yang, R.; Li, K.; Wang, L.; Bornert, M.; Zhang, Z.; Hu, T. A micro-experimental insight into the mechanical behavior of sticky rice slurry-lime mortar subject to wetting-drying cycles. *J. Mater. Sci.* **2016**, *51*, 8422–8433. [CrossRef]
19. Kesel, W.D.; Dhondt, G. *Coromandel Lacquer Screens*; Art Media Resources Ltd.: Chicago, IL, USA, 2002.
20. Qing, Y.; Liu, M.; Wu, Y.; Jia, S.; Wang, S.; Li, X. Investigation on stability and moisture absorption of superhydrophobic wood under alternating humidity and temperature conditions. *Results Phys.* **2017**, *7*, 1705–1711. [CrossRef]
21. Cha, J.K. Effect of Cyclic Moisture Content Changes on Shrinkage and Thermal Conductivity in Domestic Quercus acutissim Carr. and Larix Kaempferi Carr. *J. Korean Wood Sci. Technol.* **2002**, *30*, 41–50.
22. Čermák, P.; Rautkari, L.; Horáček, P.; Saake, B.; Rademacher, P.; Sablík, P. Analysis of Dimensional Stability of Thermally Modified Wood Affected by Re-Wetting Cycles. *Bioresources* **2015**, *10*, 3242–3253. [CrossRef]
23. *ASTM G154*; Standard Practice for Operating Fluorescent Ultraviolet (UV) Lamp Apparatus for Exposure of Non-Metallic Materials. MICOM Inc.: Baie-d'Urfé, QC, Canada, 2023.
24. *ASTM D1037*; The ASTM International site is currently unavailable for scheduled maintenance. ASTM: West Conshohocken, PN, USA, 2020.
25. Akahoshi, H.; Obataya, E. Effects of wet–dry cycling on the mechanical properties of *Arundo donax* L. used for the vibrating reed in woodwind instruments. *Wood Sci. Technol.* **2015**, *49*, 1171–1183. [CrossRef]
26. Biblis, E.J.; Lee, W.C. *Effect of Repeated Humidity Cycling on Properties of Southern Yellow Pine Particleboard*; Auburn University: Auburn, AL, USA, 1975; pp. 1–12.
27. Yang, Z.; Zhang, M.; Liu, Y.; Lv, B.; Sun, X. Review of Literature on the Failure of Wood Coating. *Sci. Silvae Sin.* **2014**, *50*, 127–13.
28. Na, B.; Wang, H.; Ding, T.; Lu, X. Influence of Wet and Dry Cycle on Properties of Magnesia-Bonded Wood-Wool Panel. *Woo Res.* **2020**, *65*, 271–282. [CrossRef]
29. Haoliang, H. Methods of Refining Chinese Lacquer and Categories of Refined Chinese Lacquer. *Chin. Raw Lacq.* **2003**, *2*, 19–30. [CrossRef]
30. Miklin-Kniefacz, S.; Pitthard, V.; Parson, W.; Berger, C.; Stanek, S.; Griesser, M.; Kučková, Š.H. Searching for blood in Chinese lacquerware: Zhū xiě huī猪血灰. *Stud. Conserv.* **2016**, *61*, 45–51. [CrossRef] [PubMed]
31. Hwang, I.S.; Park, J.H.; Kim, S.-C. A Study on the Optical Characteristics According to the Lacquer Drying Conditions for the Conservation of Lacquerwares. *J. Korean Wood Sci. Technol.* **2018**, *46*, 610–621. [CrossRef]

2. Zheng, L.P.; Wang, L.Q.; Zhao, X.; Yang, J.L.; Zhang, M.X.; Wang, Y.F. Characterization of the materials and techniques of a birthday inscribed lacquer plaque of the qing dynasty. *Herit. Sci.* **2020**, *8*, 10. [CrossRef]

3. He, L. *Analysis and Identification of Ancient Chinese Polychromy*; Science Publishing House: Beijing, China, 2017.

4. Yang, L.; Zheng, J.T.; Huang, N. The Characteristics of Moisture and Shrinkage of *Eucalyptus urophylla* × E. Grandis Wood during Conventional Drying. *Materials* **2022**, *15*, 3386. [CrossRef]

5. Chang, B. *Collection of Chinese Traditional Crafts. Natural Lacquer Painting*; China Science and Technology Press: Beijing, China, 2018.

6. *3GB/T 1927.4-2021*; Test Method for Physical and Mechanical Properties of Flawless Small Specimens of Wood. State Administration for Market Regulation and the Standardization Administration of the People's Republic of China: Beijing, China, 2021.

7. Zhan, T.; Jiang, T.; Shi, T.; Peng, H.; Lyu, J. Heat Modification of Chinese Fir Wood Catalyzed by Fly Ash under Mild Temperature. *ACS Sustain. Chem. Eng.* **2023**, *11*, 14487–14496. [CrossRef]

8. Tao, M.; Liu, X.; Xu, W. Effect of the Vacuum Impregnation Process on Water Absorption and Nail-Holding Power of Silica Sol-Modified Chinese Fir. *Forests* **2024**, *15*, 270. [CrossRef]

9. *GB/T 5210-2006*; Color and Varnish Pull-off Method Adhesion Test. The General Administration of Quality Supervision, Inspection and Quarantine and the Standardization Administration of the People's Republic of China: Beijing, China, 2006.

10. *ASTM D523-2014*; Standard Test Method for Specular Gloss. ASTM International, W.C.: West Conshohocken, PA, USA, 2014.

11. *GB/T 4893.6-2013*; Physical and Chemical Properties Test of Furniture Surface Paint Film. Standardization Administration of the People's Republic of China: Beijing, China, 2013.

12. Liu, Y.; Zhao, G. *Wood Science*; China Forestry Press: Beijing, China, 2012.

13. Liu, H.; Ke, M.; Zhou, T.; Sun, X. Effect of Samples Length on the Characteristics of Moisture Transfer and Shrinkage of Eucalyptus urophylla Wood during Conventional Drying. *Forests* **2023**, *14*, 1218. [CrossRef]

14. Rowell, R.M.; Youngs, R.L. *Dimensional Stabilization of Wood in Use*; Forest Products Laboratory: Madison, WI, USA, 1981.

15. Inagaki, T.; Yonenobu, H.; Tsuchikawa, S. Near-infrared spectroscopic monitoring of the water adsorption/desorption process in modern and archaeological wood. *Appl. Spectrosc.* **2008**, *62*, 860–865. [CrossRef] [PubMed]

16. Obataya, E.; Ohno, Y.; Tomita, B. Changes in the vibrational properties of wood coated with urushi [Japanese] lacquer during moisture adsorption and desorption. *J. Jpn. Wood Res. Soc.* **2001**, *47*, 440–446.

17. Hastie, T.; Tibshirani, R.; Friedman, J.H. *The Elements of Statistical Learning: Data Mining, Inference, and Prediction*; Springer: Berlin/Heidelberg, Germany, 2009; Volume 2.

18. Lee, C.-J.; Lee, N.-H.; Park, M.-J.; Park, J.-S.; Eom, C.-D. Effect of Reserve Air-Drying of Korean Pine Heavy Timbers on High-temperature and Low-humidity Drying Characteristics. *J. Korean Wood Sci. Technol.* **2014**, *42*, 49–57. [CrossRef]

19. Thei, J. *Artificial Ageing of Japanese Lacquerware and Comparison of Conservation Treatments for Photodegraded Japanese Lacquer Surfaces*; Imperial College London: London, UK, 2012.

20. Montazeri, S.; Ranjbar, Z.; Rastegar, S.; Deflorian, F. A new approach to estimates the adhesion durability of an epoxy coating through wet and dry cycles using creep-recovery modeling. *Prog. Org. Coat.* **2021**, *159*, 106442. [CrossRef]

21. Volkmer, T.; Glaunsinger, M.; Mannes, D.; Niemz, P.J.B. Investigations on the influence of moisture on the adhesion strength of surface coating of wood. *Bauphysik* **2014**, *36*, 337–342. [CrossRef]

22. Zou, W.; Yeo, S.Y. Characterization of the Influence of the Particle Size of Pigments on Proteinaceous Binders Used in Ancient Wall Paintings Following Photo and Thermal Ageing. *Anal. Lett.* **2024**, *57*, 1078–1091. [CrossRef]

23. McSharry, C.; Faulkner, R.; Rivers, S.; Shaffer, M.S.P.; Welton, T. The chemistry of East Asian lacquer: A review of the scientific literature. *Stud. Conserv.* **2013**, *52*, 29–40. [CrossRef]

24. Gholamiyan, H.; Tarmian, A. Weathering Resistance of Poplar Wood Coated by Organosilane Water Soluble Nanomaterials. *Drv. Ind.* **2018**, *69*, 371–378. [CrossRef]

25. Zhang, F.L. *Chinese Lacquer Craft and Lacquerware Conservation*; Science Press: Beijing, China, 2010.

26. Na, W.; Ling, H. Analysis of Tung Oil Aging and Research on the Influence of Pigments on Its Aging. In Proceedings of the 9th Annual Academic Conference of the China Institute of Cultural Relics Preservation Technology, Chongqing, China, 22 November 2016; pp. 435–443.

27. dePolo, G. *Chemical and Mechanical Properties of Drying Oils During Polymerization*; Northwestern University: Evanston, IL, USA, 2023.

28. Peng, X.; Zhang, Z.; Wang, B.; Du, W. Research on the Aging Properties of Exterior Wood Coatings. In Proceedings of the 2012 International Conference on Biobase Material Science and Engineering, Changsha, China, 21–23 October 2012; pp. 121–126.

29. Shimadzu, Y.; Kawanobe, W. Deterioration Phenomena of Urushi Films:Combined Effects of Ultraviolet Radiation and Water. *Shikizai* **2003**, *76*, 385–390.

30. Liu, X. *Modelling the Mechanical Response of Japanese Lacquer (Urushi) to Varying Environmental Conditions*; Loughborough University: Loughborough, UK, 2012.

forests

Article

Highly Mechanical Strength, Flexible and Stretchable Wood-Based Elastomers without Chemical Cross-Linking

Yongyue Zhang [1], Jiayao Li [1], Yun Lu [2] and Jiangtao Shi [1,3,*]

[1] Department of Wood Science and Engineering, College of Materials Science and Engineering, Nanjing Forestry University, Nanjing 210037, China; 18195486182@163.com (Y.Z.); lijiayao@njfu.edu.cn (J.L.)
[2] Research Institute of Wood Industry, Chinese Academy of Forestry, Beijing 100091, China; y.lu@caf.ac.cn
[3] Co-Innovation Center of Efficient Processing and Utilization of Forest Resources, Nanjing Forestry University, Nanjing 210037, China
* Correspondence: shijt@njfu.edu.cn

Abstract: Wood exhibits a limited elastic deformation capacity under external forces due to its small range of elastic limit, which restricts its widespread use as an elastic material. This study presents the development of a stretchable wood-based elastomer (SWE) that is highly mechanical and flexible, achieved without the use of chemical cross-linking. Balsa wood was utilized as a raw material, which was chemically pretreated to remove the majority of the lignin and create a more abundant pore structure, while exposing the active hydroxyl groups on the cellulose surface. The polyvinyl alcohol (PVA) solution was impregnated into delignified wood, resulting in the formation of a cross-linked structure through multiple freeze–thaw cycles. After eight cycles, the tensile strength in the longitudinal direction reached up to 25.68 MPa with a strain of ~463%. This excellent mechanical strength is superior to that of most wood-based elastomers reported to date. The SWE can also perform complex deformations such as winding and knotting, and SWE soaked in salt solution exhibits excellent sensing characteristics and can be used to detect human finger bending. Stretchable wood-based elastomers with high mechanical strength and toughness have potential future applications in biomedicine, flexible electronics, and other fields.

Keywords: polymer cross-linking; anisotropic structure

Citation: Zhang, Y.; Li, J.; Lu, Y.; Shi, J. Highly Mechanical Strength, Flexible and Stretchable Wood-Based Elastomers without Chemical Cross-Linking. *Forests* **2024**, *15*, 836. https://doi.org/10.3390/f15050836

Academic Editor: Alain Cloutier

Received: 17 April 2024
Revised: 2 May 2024
Accepted: 8 May 2024
Published: 10 May 2024

1. Introduction

Natural wood, with its small strain limit in the elastic range (~10%), has long been used as an engineering material such as beams and posts in timber-framed buildings or as a rigid material such as furniture, flooring, doors, and windows [1,2]. However, as a biomaterial that has evolved over millions of years, wood has developed a structural basis for responding to environmental stimuli and is a "smart material" that responds and adapts to its environment [3–5]. Although it exhibits some elastic deformation when subjected to external forces, its elastic limit range is relatively small, which limits its use as an elastic material [6,7]. Research on the structural control of wood to prepare elastic materials is limited. If the elastic properties of wood can be accurately regulated, it can be used as a high-performance green material in various fields including machinery, automobiles, sports and leisure, aerospace, home packaging, and other applications [8–10].

The elasticity of wood is greatly limited by its structure and composition. Macroscopically, wood has a complex cellular structure arranged in different directions to form an interwoven network that is susceptible to damage from stress concentration during the tensile process; at the molecular level, the lignin content of wood is positively correlated with cell wall rigidity, and the thermoplastic behavior of lignin is not conducive to the elastic deformation of wood [11–13]. Consequently, one of the most effective methods for the preparation of elastic materials from wood is the removal of lignin.

Balsa wood, which is produced in tropical America, is one of the lightest woods in the world. It is characterized by low density, high porosity, and rapid growth. Its superior axial stiffness, strength, and energy absorption make it one of the most attractive core materials [14,15]. Due to its low lignin content and rich pore structure, balsa is often used in the construction of wood-based elastomers. For example, Gao et al. used balsa wood as a raw material, hollowed out the wood ray tissue by the chemical cropping strategy, and obtained a superhydrophobic wood-based elastomer by chemical heat treatment [16]. Wang et al. chemically delignified and silylated balsa wood to obtain wood-based elastomers with high elasticity [17]. Song et al. obtained highly elastic wood-based elastic sensors from balsa wood by delignification treatment and MXene modification [18]. However, the elastic deformation ability of elastic wood is limited because it is inflexible and cannot be stretched. This limits its application in fields such as biomedical and flexible electronics. Therefore, it is important to develop wood-based elastic materials with stretchable properties.

In elastomers, cross-linking and entanglement between long-chain molecules can confer excellent tensile properties to the material, which provides new ideas for the preparation of flexible wood with tensile properties [19–21]. In recent works, flexible hydrogel materials based on wood can be produced by incorporating a network of polymer molecules into the wood fiber backbone [22,23]. For example, Dong et al. constructed a wood-based composite gel sensor by introducing PAM into the delignified lignin skeleton [24]; Wang et al. constructed composite wood-based gels using gelatin [25]; and Shen et al. utilized ionic liquid impregnation to obtain wood/polyionic liquid (WA/PIL) hydrogels [26]. However, the elastic tensile properties of natural wood are affected by its anisotropic structure, which has been overlooked in this process, and the three-dimensional network structure formed by the polymer molecular chains and the fiber skeleton in the wood is too weak to withstand higher stresses and larger strains, so there is a need for the further development of tougher wood-based elastic materials.

Herein, we have prepared a high mechanical strength, high flexibility, and stretchable wood elastomer through delignification treatment and polymer impregnation crosslinking. The removal of lignin can effectively expose highly active hydroxyl groups on cellulose molecular chains and form rich pore structures. Physical freezing and thawing can facilitate the formation of a stable cross-linking structure between the polyvinyl alcohol solution and the lignocellulosic fiber skeleton [27,28]. To explore the potential applications of this material, we proceeded to combine SWE with salt solutions, thus obtaining wood-based anisotropic elastomers (E-SWE) with conductivity. The directional arrangement of channel structures within the lignocellulosic framework can be exploited for ion transport, thereby conferring an excellent sensing performance for E-SWE. Compared to traditional fossil-based elastomers (rubber, polyurethane, polydimethylsiloxane), SWE is expected to become a new generation of biomass elastomers due to its simple preparation process, excellent performance, and environmental friendliness based on natural wood.

2. Materials and Methods

2.1. Materials

Balsa wood (*Ochroma pyramidale*, from South America, density~0.1 gcm^3) was cut into slices along its growth direction with the dimensions of $60 \times 20 \times 2$ mm^3 (longitudinal $\times$ radial $\times$ tangential). Polyvinyl alcohol (PVA), sodium hydroxide (NaOH), sodium chlorite (NaClO$_2$), and glacial acetic acid were purchased from Nanjing Aladdin Biological Technology Co. Ltd. (Nanjing, China).

2.2. Delignification of the Balsa Wood

Balsa wood was immersed in a 6 wt% NaOH solution for 6 h at 90 °C in a water bath. The pretreated wood was washed 3~5 times with deionized water, and then the pretreated wood was delignified with 3 wt%NaClO$_2$ solution (pH~4.6 adjusted with acetic acid). The treatment time was about 6~8 h at 80 °C [29,30]. The obtained delignified wood samples were washed 3~5 times with deionized water and then freeze-dried.

2.3. Preparation of Super Stretched Wood Based Elastic Materials

The hydrogel solutions were first prepared as follows. Dissolve the PVA powder in deionized water at 95 °C and stir continuously with a magnetic stirrer to prepare PVA solutions with concentrations of 5% wt, 10% wt, and 15% wt. The PVA solution was poured into the delignified wood sample and vacuum impregnated at 200 Pa, with the vacuum being released at 30-min intervals to allow the hydrogel solution to fill the wood structure. This process was repeated three to four times. The impregnated samples were frozen in an ultra-low temperature refrigerator at a temperature of −60 °C for a period of one hour. Thereafter, the frozen samples were thawed naturally at room temperature for a period of two hours until they were completely thawed. The freeze–thaw process was repeated six to eight times in order to obtain stretchable wood-based elastomers (Figure 1). E-SWE with conductive properties was obtained by impregnating SWE into a salt (NaCl) solution for 24 h.

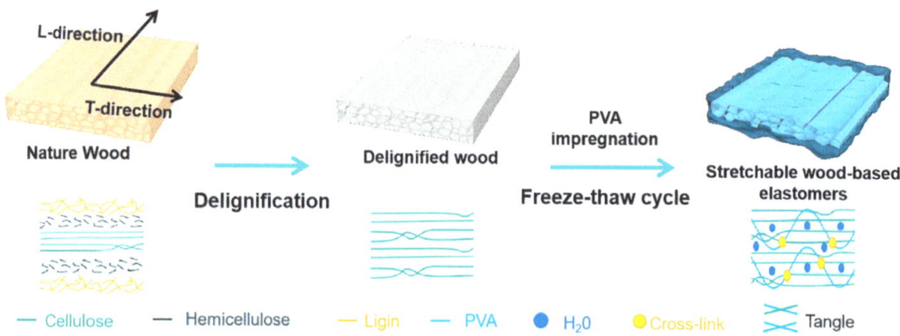

Figure 1. Construction method and mechanism of SWE.

2.4. Characterization

2.4.1. Morphology

Field emission scanning electron microscopy was used to observe the microscopic morphology of the cross section and chord section of natural wood and prepared SWE, respectively. The test voltage was 4 KV.

2.4.2. X-ray Diffraction Measurement

The cellulose crystal structure of the SWE was characterized using X-ray diffraction (Bruker D2 Phaser, Karlsruhe, Germany) with a scanning range of 5 ° to 90 ° (2θ); the speed was 2 s per degree. The crystallinity of the wood samples was calculated according to the Segal method. The crystallinity index (CrI) was determined according to Equation (1), where I_{200} is the maximum intensity of the 200 lattice diffraction around 22°, and I_{am} is the intensity of the amorphous diffraction at 18°.

$$CrI = \frac{I_{200} - I_{am}}{I_{200}} \times 100 \tag{1}$$

2.4.3. Small-Angle X-ray Scattering Measurement

Small-angle X-ray scattering (SAXS) measurements were performed at an X-ray wavelength (λ) of 0.154 nm, and the sample-to-detector distance was set to 1045 mm (Xeuss 2.0, Grenoble, France).

2.4.4. Fourier Transform Infrared (FTIR) Characterization

FTIR spectra were recorded in the range of 500−4000 cm^{-1} using a Nicolet iS10 FTIR spectrometer (Thermo Scientific Nicolet iS20, Waltham, MA, YSA).

2.4.5. Chemical Composition Analysis

Composition analysis of the wood samples was carried out according to the NREL (National Renewable Energy Laboratory, Golden, CO, USA) standard procedure.

2.4.6. Tensile Test

The tensile tests were performed on a universal material-testing machine (KXWW-01C, Chengde, China) utilizing a 100 N load cell at a crosshead speed of 5 mm/min. All specimens had a size of 30 mm × 5 mm with a thickness of around 1 mm.

2.4.7. Sensor Performance

An electrochemical workstation (CHI760E, Beijing Huake Putian Technology Co. Ltd., Beijing, China) was used to detect the resistance change of E-SWE. The AC voltage of the test resistor was 0.1 V, and the AC frequency was 1000 Hz. According to the resistance change of the hydrogel during the test, the resistance change rate trend of the entire process was calculated. $\Delta R/R_0 = (R - R_0)/R_0 \times 100\%$ (R_0 is the initial resistance during no-change operation during the test, R is the real-time resistance during the test).

The experimental data were analyzed using Excel (Microsoft Office Home and Student 2019) and Origin (OriginPro 2021).

3. Results and Discussion

3.1. The Physical Properties of SWE

The construction of the SWE was accomplished through a simple chemical pretreatment and physical freeze–thaw process. Various cells (wood fibers, rays, and conduits) constitute the anisotropic structure of natural wood. The fibers are oriented along the direction of growth for wood in the L-direction and perpendicular to the fiber direction in the T-direction. (Figure 1). In the cross-section shown in Figure 2a, honeycomb wood fibers were interlocked with each other, while the ray tissue was horizontally aligned along the radial direction [31]. In the delignification process, most of the lignin can be extracted in situ by pretreatment of the wood using NaOH solution, followed by the further removal of residual lignin using $NaClO_2$ solution (pH~4.6) to obtain white wood (Figure 2e) [32,33]. With the removal of lignin, the ray structure in natural wood was destroyed and replaced by a large cleavage structure (Figure 2f,g). Some of the cell wall structures in the cross-section were disrupted, resulting in a more porous and less dense channel structure. As a result of these treatments, the cellulose fibers exposed many highly reactive hydroxyl groups. However, the mechanical strength of the wood was weakened.

To reconstruct the loose and porous structure of delignified wood, filling the internal nanochannels with polymers is the preferred strategy. Therefore, biodegradable PVA (a commonly used polymer) was introduced into the system [34,35]. Polyvinyl alcohol, which contains abundant hydroxyl groups, can effectively reconstruct wood by forming hydrogen bonds with cellulose fibers. In our process, delignified wood was immersed in the PVA solution and vacuum impregnation was used to achieve complete penetration. The samples were frozen at $-20\ ^\circ$C for 6 h, followed by gelation. This process was repeated 6 to 8 times. During the freezing process, the ice crystals will phase-separate from PVA and keep squeezing the PVA molecular chains and cellulose molecular chains, which is similar to a continuous forging process. By continuously repeating this process, the pore structure of the wood in the chordal section is basically covered (Figure 2j), and the interwoven PVA network structure is formed within the wood fiber channels in the transverse section (Figure 2k–l) [27,28]. This indicates that PVA formed a stable cross-linking structure with the wood fiber backbone.

SAXS is a powerful technique for characterizing the cell wall structure without affecting the internal structure [36]. Figure 3 shows the SAXS 2D scattering images of natural wood and SWE similarly presenting an elliptical shape, showing the oriented arrangement of cellulose nanofibrils (CNFs). The similar CNF orientation indicates that the wood retained its anisotropic structure after crosslinking by PVA impregnation. We then probed the

nanostructural changes of SWE using SAXS. The 1D maps derived from the 2D SAXS mode
were fitted, as shown in Figure 3a,b. The interfiber correlation length (d) was calculated
from the peak centers of the Gaussian fits. In natural wood, the original fibers were densely
arranged in a hemicellulose–lignin matrix with a fiber spacing of ≈3.04 nm, which was
similar to the values reported in other studies (~3–4 nm) [37]. The SWE fibers obtained
after polymer impregnation and crosslinking had a spacing of ≈7.34 nm; the increase in
fiber spacing was conducive to the penetration of long-chain polymers into the fibers and
the formation of a denser crosslinking network structure with the fibers, which could more
easily withstand a greater amount of stress and strain.

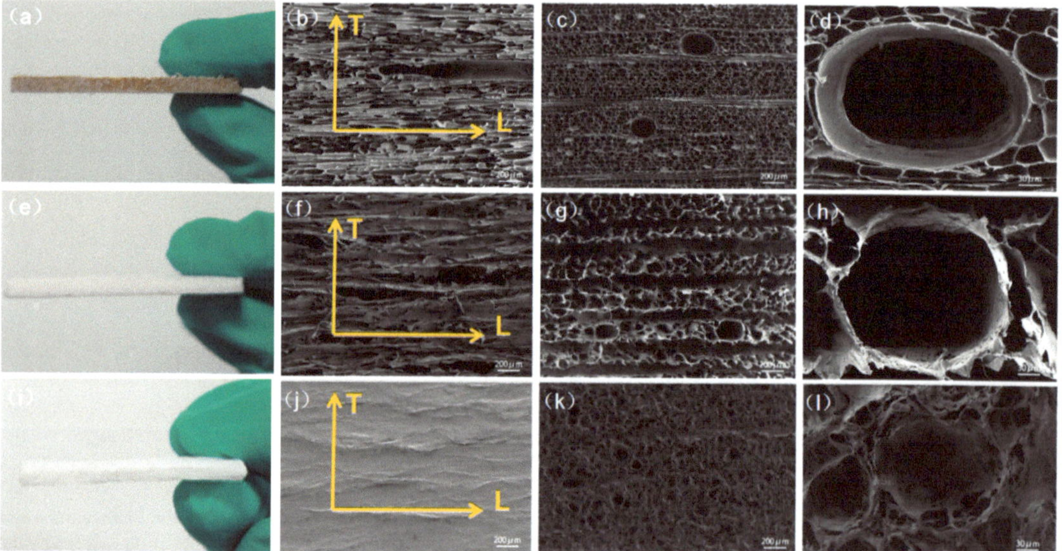

Figure 2. Macroscopic morphology and SEM images of natural wood, delignified wood, and SWE.
(**a**,**e**,**i**) Optical photos of natural wood, delignified wood, and SWE. (**b**,**f**,**j**) SEM image of natural wood,
delignified wood, and SWE tangential section; (**c**,**g**,**k**) SEM image of natural wood, delignified wood,
and SWE cross section; (**d**,**h**,**l**) SEM image of natural wood, delignified wood, and SWE cellular structure.

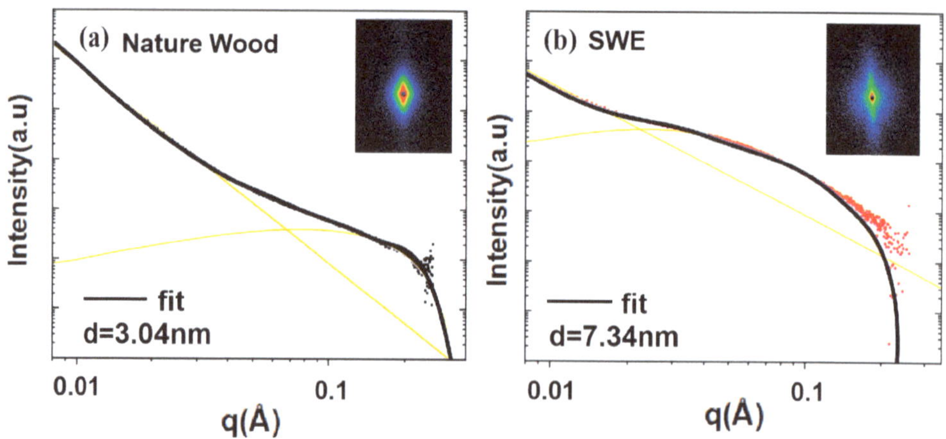

Figure 3. (**a**) 1D and 2D SAXS images of natural wood; (**b**) 1D and 2D SAXS images of SWE.

3.2. The Chemical Properties of SWE

The presence of rigid lignin in wood is an important cause of its elasticity, so the effective removal of lignin is important for the SWE to achieve high elasticity. We tested the contents of the three major elements in virgin and delignified wood (Table 1 and Figure 4). The natural wood contained 33.6% cellulose, 7.3% hemicellulose, and 20.6% lignin. After the delignification treatment, the lignin content decreased from 20.6% to 2.6%, the cellulose content increased from 33.6% to 72.6%, and the hemicellulose content remained almost unchanged, which indicates that this occurred after the NaOH pretreatment as well as $NaClO_2$ (pH ~4.6). This indicates that lignin was mostly removed and that cellulose was retained during the NaOH pretreatment and $NaClO_2$ (pH ~4.6) treatment. The mechanism of lignin removal from wood is that the $NaClO_2$ solution produces ClO_2 during heating, and the phenol radicals in the lignin molecules are oxidized by ClO_2 to form phenoxy radicals. The generated phenoxy radicals further react with ClO_2, and these two successive steps cause the lignin molecules to be converted into muconic acid, or the side chain is broken to form quinone [33].

Table 1. Contents of cellulose, lignin, and hemicellulose in natural wood and delignified wood.

Name	Cellulose (%)	Hemicellulose (%)	Lignin (%)
Nature wood	33.6 ± 0.8	7.3 ± 0.7	20.6 ± 1.7
Delignified wood	72.6 ± 1.2	9.0 ± 0.2	2.6 ± 0.3

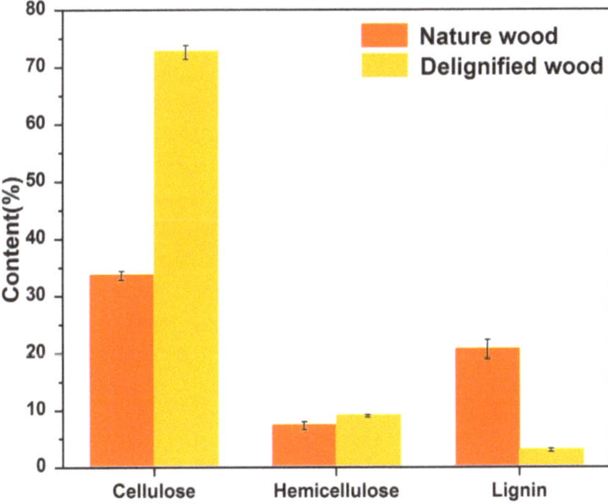

Figure 4. Contents of cellulose, lignin, and hemicellulose in natural wood and delignified wood.

Fourier transform infrared spectroscopy (FTIR) was utilized to further investigate the chemical structure changes of SWE (Figure 5). Compared with the natural wood, the delignified wood showed the absence of peaks of lignin (C-O stretching vibrational absorption peak) at 1232 cm^{-1}, and hemicellulose (non-conjugated C=O vibrational absorption peak) at 1734 cm^{-1} completely disappeared in the delignified wood, indicating that the lignin and hemicellulose in the wood had been completely removed. In the SWE, the absorption peak at 1417 cm^{-1} was attributed to the CH_2 bending vibration of the polyvinyl alcohol molecule, and the absorption peak at 1088 cm^{-1} was attributed to the C-O stretching vibrational mode of the polyvinyl alcohol molecule, which represents the formation of a crosslinked network of PVA molecules with cellulose molecule chains in the SWE.

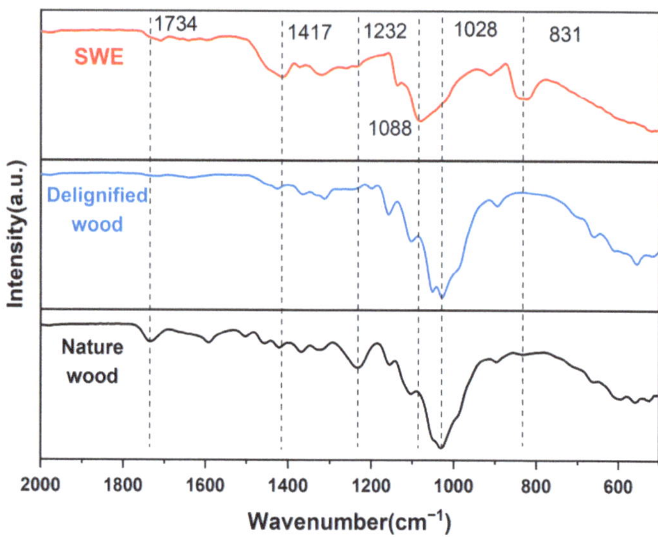

Figure 5. FTIR images of natural wood, delignified wood, and the SWE.

The XRD images indicate that the wood had a typical cellulose I structure, as evidenced by the three characteristic peaks at 34.67°, 21.98°, and 15.35° corresponding to the (040), (002), and (101) lattice planes of cellulose I, respectively (Figure 6). The peaks at the 002 lattice plane reflect the width of the crystalline zone, while the peaks at the 040 lattice plane reflect the length of the crystalline zone. The position of the characteristic peaks after delignification was basically the same as that of the original wood, but the crystallinity increased significantly. This is because amorphous hemicellulose and lignin are selectively removed during delignification, and cellulose is more likely to form a crystal structure through hydrogen bonding interactions. The XRD pattern of SWE retained the I-type structure of cellulose and showed the characteristic peaks of PVA, indicating that PVA molecules were well-filled into the pores of the wood.

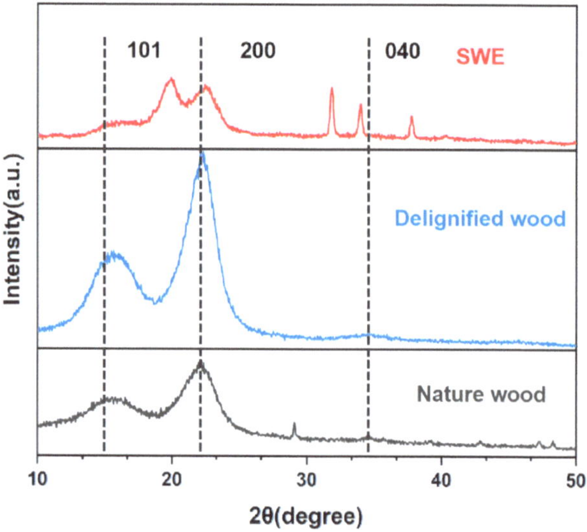

Figure 6. XRD images of natural wood, delignified wood, and the SWE.

3.3. Strengthening and Toughening of SWE

The SWE retained the natural anisotropic structure of wood and exhibited better mechanical properties than virgin wood. We investigated and compared the mechanical properties of the SWE in different directions. In the T direction, the tensile strength of virgin wood was about 2.3 MPa, the strain was about 10%, and the mechanical properties of the SWE were significantly reduced after delignification treatment (Figure 7a,b). The mechanical strength of natural wood in the L direction (~21.3 MPa) was significantly higher than that in the T direction, and the higher mechanical strength was attributed to the oriented arrangement of the cellulose molecular chains. The SWE obtained after reconstitution by PVA impregnation had excellent mechanical properties and the tensile strength increased with increasing PVA concentration. However, the strain of SWE will decrease at higher concentrations (15%PVA), mainly due to the high viscosity and poor fluidity of the high concentration PVA solution, which results in poor permeability in wood (Figure 7a,b). We further improved the mechanical properties of the SWE by using physical freezing and thawing. After eight freeze–thaw cycles, the mechanical properties in the T- and L-directions significantly improved. The mechanical strength in the T-direction increased to 5.63 MPa with a strain of 234% (Figure 7c). Similarly, the tensile strength in the L-direction increased to 25.68 MPa with a strain of 463% (Figure 7f). The freeze–thaw process is comparable to the physical forging of elastomers. During the freezing process, ice crystals repel water molecules from the solvent around them, while dispersing the polymer between them. The ice crystals melt when the temperature rises, releasing the polymers dispersed in them. Due to the interaction forces between the polymers, the polymer molecules are re-crosslinked to form a homogeneous gel structure. By repeating this process, it is possible to continuously squeeze the PVA molecules and cellulose nanofibers against each other to form an entangled network structure [38].

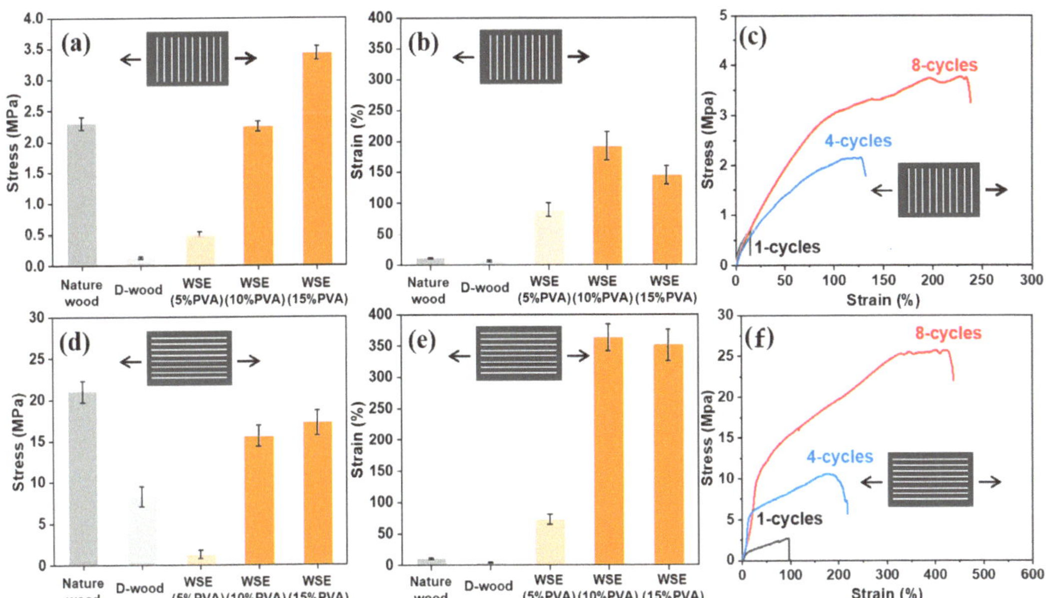

Figure 7. Comparison of the mechanical properties of natural wood, delignified wood, and the SWE (5% PVA, 10% PVA, 15% PVA). (**a**,**b**) Tensile strength and strain of three materials in the T-direction; (**c**) the effect of freeze–thaw cycles on the mechanical properties of the SWE T-direction; (**d**,**e**) the tensile strength and strain of three materials in the L-direction; (**f**) the effect of freeze–thaw cycles on the mechanical properties of the SWE L-direction.

We further used the microscopic in situ stretching technique to observe the micro morphological changes of delignified wood and SWE during stretching in the T-direction (Figure 8). The lignocellulosic fiber backbone was prone to fracture during external stretching (Figure 8a–c), which was related to the weak hydrogen bonding connections between fibers. The PVA in the SWE penetrated into the lignocellulosic fiber backbone and formed a crosslinked network structure, which would mean that it did not fracture during the stretching process (Figure 8e), and that the SWE would return to its original shape when the external force disappeared (Figure 8f).

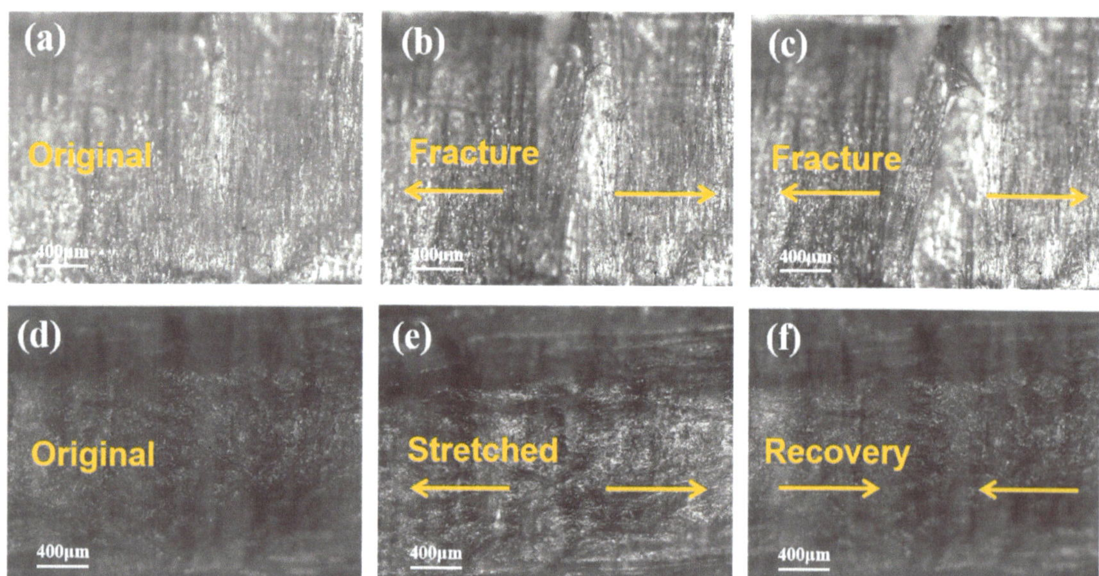

Figure 8. In situ tensile images of delignified light wood and the SWE. (**a–c**) Microscopic images of the tensile fracture of delignified wood; (**d–f**) microscopic images during the SWE stretching process.

3.4. The Flexible Deformation Capability of SWE

After the delignification and gelation processes, the cellulose fibers in the gel network were cross-linked and entangled with the PVA molecular chains, and the SWE showed a strong flexible deformation ability. Figure 9 shows the digital images of the SWE under different stresses. The basic structure of the SWE was not damaged when it was bent (Figure 9a), and we could also deform the SWE in complex ways such as winding and knotting (Figure 9b). When we stretched the SWE along the T-direction, the SWE showed excellent tensile ability (Figure 9c). Partially open microchannels created space in the SWE (Figure 9d), and the abundant pore channels allowed the SWE to bend and withstand severe deformations without rupturing in order to accommodate various deformations. This unique wrinkled cell wall structure gives the all-wood hydrogel excellent flexibility. The excellent mechanical properties of the SWE can mainly be attributed to the strong hydrogen bonding, physical entanglement, and van der Waals forces between the cellulose nanofibers and the PVA chains as well as the toughening effect of the cellulose nanofibers.

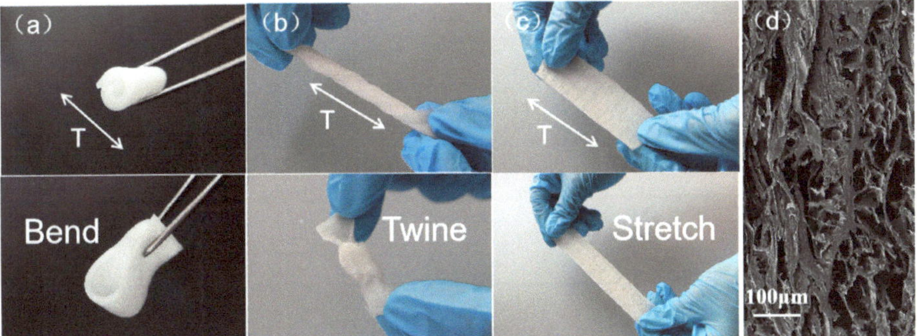

Figure 9. Macro image and micro morphology of the SWE undergoing flexible deformation. (**a**) Bending; (**b**) wrapping and knotting; (**c**) stretching; (**d**) SWE rich pore structure.

3.5. The Application in the Field of Flexible Electronics

In recent reports, cellulose nanofibers in wood are usually used as nanochannels for ion transport. To expand the application area of SWEs, we immersed it in a salt (NaCl) solution for approximately 24 h to allow for sufficient ion permeation. This resulted in the creation of an elastomeric material (E-SWE) with conductive properties. We then connected E-SWE to an electrochemical workstation and recorded the relative resistance change (R/R_0) through repeated stimulation. The well-arranged cellulose nanofibers contain nanochannels that function as fast channels for the ion transport sensor. As shown in Figure 10b, the fluctuation range of the current change value (($R - R_0$)/R_0) in the L-direction was smaller than that in the T-direction, indicating that the E-SWE has significant anisotropic electro sensitivity.

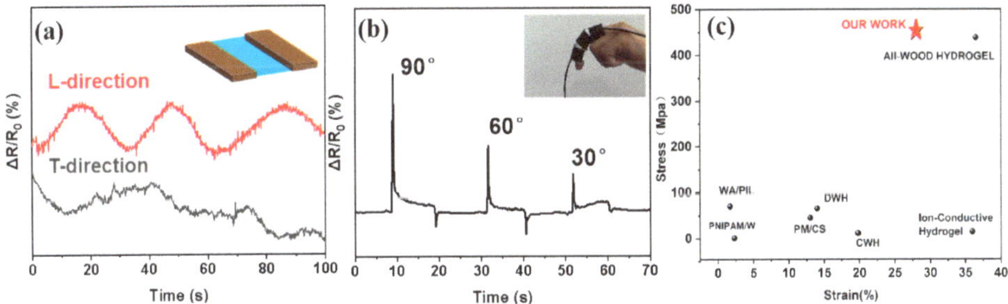

Figure 10. (**a**) Changes in the relative resistance values in the E-SWE L- and T-directions; (**b**) E-SWE detection of finger bending (90°~0°); (**c**) comparison of mechanical properties between SWE and other wood-based elastic materials.

The E-SWE has potential as a sensor for the real-time monitoring of human motions due to its good electrical conductivity, elasticity, and flexibility. As illustrated in Figure 10b, the E-SWE was sensitive enough to monitor the finger flexion at different angles ranging from 90° to ~0°. This suggests that the E-SWE has promising applications in wearable electronic devices. Moreover, in our work, its mechanical properties were superior to the majority of artificial elastomers reported in recent years (Table 2) [24,26,29,39–42]. These include wood-based stretchable elastomers prepared by integrating polymers such as poly acrylamide (PAM), poly(N-isopropylacrylamide) (PNIPAM), and polyvinyl alcohol (PVA) into a delignified wood. It is noteworthy that the all-wood hydrogels and ion-conductive hydrogels also exhibited excellent mechanical properties; however, they inevitably utilize chemical cross-linking in their preparation. In contrast, the SWE formed by physical cross-linking also exhibited mechanical properties that were nearly identical.

Table 2. Comparison of the SWE mechanical performance.

Name	Tensile Strength (MPa)	Strain (%)	Ref.
SWE	28	450	-
WA/PIL	1.7	70	[26]
All-wood hydrogels	36.5	438	[29]
Ion-conductive hydrogels	36	14.5	[39]
DWH	14	65	[40]
PM/CS	13	45	[41]
WCH	19.8	11.3	[24]
PNIPAM/WOOD	2.3	1.1	[42]

4. Conclusions

In conclusion, we synthesized a wood-based elastomeric material with strong mechanical properties by simple physical freeze–thawing without using any chemical cross-linking agent. The prepared SWE exhibited high mechanical strength and high flexibility due to the aligned cellulose as the rigid backbone and PVA as the crosslinking network. The SWE maintained the anisotropic structure of wood and its fiber spacing was significantly increased due to moisture action. The tensile strength in the SWE L-direction after freeze thaw cycles was 25.68 MPa and the strain was 463%, which were 5 and 2 times higher than those in the T-direction, respectively. In addition, we obtained an E-SWE with sensing properties by immersing it in a salt solution. The E-SWE could detect finger bending in the human body, demonstrating potential applications in wearable electronic devices. Compared to traditional elastomers, the SWE retains the anisotropic structure of wood and has excellent mechanical properties, which makes it more flexible and versatile in its application. Its preparation process is simple, greener, and more economical, and it is a new type of biomass elastomer that has great potential in the field of flexible wearable devices and smart response materials.

Author Contributions: Conceptualization, Y.Z., Y.L. and J.S.; Methodology, J.L.; Validation, J.S.; Formal analysis, Y.Z.; Writing—review and editing, Y.Z. and J.S. All authors have read and agreed to the published version of the manuscript.

Funding: This research was funded by the National Key Research Development Program of China (2023YFD2201405) and the Qing Lan Project to Jiangsu Province (2022).

Data Availability Statement: All data needed to evaluate the conclusions in the paper are presented in the paper. Additional data related to this paper may be requested from the authors.

Conflicts of Interest: The authors declare no conflicts of interest.

References

1. Dakneviciute, V.; Milasiene, D.; Ukvalbergiene, K. Tensile Strength, Elasticity and Cracking Character of Softwood Tissues. *Medžiagotyra* **2015**, *21*, 260–264.
2. Ramage, M.H.; Burridge, H.; Busse-Wicher, M.; Fereday, G.; Reynolds, T.; Shah, D.U.; Wu, G.; Yu, L.; Fleming, P.; Densley-Tingle, D.; et al. The wood from the trees: The use of timber in construction. *Renew. Sustain. Energy Rev.* **2017**, *68*, 333–359. [CrossRef]
3. Shi, X.; Luo, J.; Luo, J.; Li, X.; Han, K.; Li, D.; Cao, X.; Wang, Z.L. Flexible Wood-Based Triboelectric Self-Powered Smart Home System. *ACS Nano* **2022**, *16*, 3341–3350. [CrossRef] [PubMed]
4. Liu, Y.; Lu, C.; Bian, S.; Hu, K.; Zheng, K.; Sun, Q. Reversible photo-responsive smart wood with resistant to extreme weather. *Mater. Sci.* **2022**, *57*, 3337–3347. [CrossRef]
5. Ugolev, B.N. Wood as a natural smart material. *Wood Sci. Technol.* **2014**, *48*, 553–568. [CrossRef]
6. Takahashi, A.; Yamamoto, N.; Ooka, Y.; Toyohiro, T. Tensile Examination and Strength Evaluation of Latewood in Japanese Cedar. *Materials* **2022**, *15*, 2347. [CrossRef] [PubMed]
7. Longo, R.; Laux, D.; Pagano, S.; Delaunay, T.; Le Clézio, E.; Arnould, O. Elastic characterization of wood by Resonant Ultrasound Spectroscopy (RUS): A comprehensive study. *Wood Sci. Technol.* **2018**, *52*, 383–402. [CrossRef]
8. Chen, S.; Wu, Z.; Chu, C.; Ni, Y.; Neisiany, R.E.; You, Z. Biodegradable Elastomers and Gels for Elastic Electronics. *Adv. Sci.* **2022**, *9*, 2105146. [CrossRef] [PubMed]
9. Wan, Y.; Li, X.C.; Yuan, H.; Liu, D.; Lai, W.Y. Self-Healing Elastic Electronics: Materials Design, Mechanisms, and Applications. *Adv. Funct. Mater.* **2024**, *early view.* [CrossRef]

0. Walley, S.M.; Rogers, S.J. Is Wood a Material? Taking the Size Effect Seriously. *Materials* **2022**, *15*, 5403. [CrossRef]
1. Keying, L.; Dong, W.; Lanying, L.; Feng, F.U. Research Progress in Multi-scale Interface Structure and Mechanical Properties of Wood. *Trans. China Pulp Pap.* **2021**, *36*, 88–94. [CrossRef]
2. Shumin, Y.; Xing'E, L.; Lili, S.; Jianfeng, M.A.; Genlin, T.; Zehui, J. The Characteristics and Representation Methods of Lignin for Bamboo. *Mater. Rep.* **2020**, *34*, 7177–7182. [CrossRef]
3. Börcsök, Z.; Pásztory, Z. The role of lignin in wood working processes using elevated temperatures: An abbreviated literature survey. *Eur. J. Wood Wood Prod.* **2021**, *79*, 511–526. [CrossRef]
4. Galos, J.; Das, R.; Sutcliffe, M.P.; Mouritz, A.P. Review of balsa core sandwich composite structures. *Mater. Des.* **2022**, *221*, 111013. [CrossRef]
5. Zhang, H.; Zhu, P.; Wang, Z.; Ma, F.; Ji, H.; Li, X. Combined effect of rays and vessels to achieve high strength and toughness in balsa wood. *Mater. Lett.* **2023**, *352*, 135137. [CrossRef]
6. Gao, R.; Huang, Y.; Gan, W.; Xiao, S.; Gao, Y.; Fang, B.; Zhang, X.; Lyu, B.; Huang, R.; Li, J.; et al. Superhydrophobic elastomer with leaf-spring microstructure made from natural wood without any modification chemicals. *Chem. Eng. J.* **2022**, *442*, 136338. [CrossRef]
7. Wang, Z.; Lin, S.; Li, X.; Zou, H.; Zhuo, B.; Ti, P.; Yuan, Q. Optimization and absorption performance of wood sponge. *J. Mater. Sci.* **2021**, *56*, 8479–8496. [CrossRef]
8. Song, D.; Zeng, M.; Min, P.; Jia, X.; Gao, F.; Yu, Z.; Li, X. Electrically conductive and highly compressible anisotropic MXene-wood sponges for multifunctional and integrated wearable devices. *J. Mater. Sci. Technol.* **2023**, *144*, 102–110. [CrossRef]
9. Kuang, X.; Arıcan, M.O.; Zhou, T.; Zhao, X.; Zhang, Y.S. Functional Tough Hydrogels: Design, Processing, and Biomedical Applications. *Acc. Mater. Res.* **2023**, *4*, 101–114. [CrossRef]
0. Li, C.; Wang, C.; Keplinger, C.; Zuo, J.; Jin, L.; Sun, Y.; Zheng, P.; Cao, Y.; Lissel, F.; Linder, C.; et al. A highly stretchable autonomous self-healing elastomer. *Nat. Chem.* **2016**, *8*, 618–624. [CrossRef]
1. Steck, J.; Kim, J.; Yang, J.; Hassan, S.; Suo, Z. Topological adhesion. I. Rapid and strong topohesives. *Extrem. Mech. Lett.* **2020**, *39*, 100803. [CrossRef]
2. Ajdary, R.; Tardy, B.L.; Mattos, B.D.; Bai, L.; Rojas, O.J. Plant Nanomaterials and Inspiration from Nature: Water Interactions and Hierarchically Structured Hydrogels. *Adv. Mater.* **2021**, *33*, e2001085. [CrossRef]
3. Yongyue, Z.; Jiangtao, S.; Zongying, F.; Yun, L. Research Progress of Cellulose Self-Healing Hydrogels. *Sci. Silvae Sin.* **2024**, *60*, 128–138. [CrossRef]
4. Dong, Y.; Pan, N.; Zhu, M.; Tang, M.; Wu, Y.; You, Z.; Zhou, X.; Chen, M. An anti-swelling, strong and flexible wood-based composite hydrogel as strain sensor. *Ind. Crops Prod.* **2022**, *187*, 115491. [CrossRef]
5. Wang, S.; Li, K.; Zhou, Q. High strength and low swelling composite hydrogels from gelatin and delignified wood. *Sci. Rep.* **2020**, *10*, 17842. [CrossRef]
6. Shen, X.; Nie, K.; Zheng, L.; Wang, Z.; Wang, Z.; Li, S.; Jin, C.; Sun, Q. Muscle-inspired capacitive tactile sensors with superior sensitivity in an ultra-wide stress range. *J. Mater. Chem. C Mater. Opt. Electron. Devices* **2020**, *8*, 5913–5922. [CrossRef]
7. Adelnia, H.; Ensandoost, R.; Shebbrin Moonshi, S.; Gavgani, J.N.; Vasafi, E.I.; Ta, H.T. Freeze/thawed polyvinyl alcohol hydrogels: Present, past and future. *Eur. Polym. J.* **2022**, *164*, 110974. [CrossRef]
8. Holloway, J.L.; Lowman, A.M.; Palmese, G.R. The role of crystallization and phase separation in the formation of physically cross-linked PVA hydrogels. *Soft Matter* **2013**, *9*, 826–833. [CrossRef]
9. Yan, G.; He, S.; Chen, G.; Ma, S.; Zeng, A.; Chen, B.; Yang, S.; Tang, X.; Sun, Y.; Xu, F.; et al. Highly Flexible and Broad-Range Mechanically Tunable All-Wood Hydrogels with Nanoscale Channels via the Hofmeister Effect for Human Motion Monitoring. *Nano-Micro Lett.* **2022**, *14*, 84. [CrossRef]
0. Chen, C.; Wang, Y.; Zhou, T.; Wan, Z.; Yang, Q.; Xu, Z.; Li, D.; Jin, Y. Toward Strong and Tough Wood-Based Hydrogels for Sensors. *Biomacromolecules* **2021**, *22*, 5204–5213. [CrossRef]
1. Borrega, M.; Ahvenainen, P.; Serimaa, R.; Gibson, L. Composition and structure of balsa (*Ochroma pyramidale*) wood. *Wood Sci. Technol.* **2015**, *49*, 403–420. [CrossRef]
2. Jung, W.; Savithri, D.; Sharma-Shivappa, R.; Kolar, P. Changes in Lignin Chemistry of Switchgrass due to Delignification by Sodium Hydroxide Pretreatment. *Energies* **2018**, *11*, 376. [CrossRef]
3. Tarvo, V.; Lehtimaa, T.; Kuitunen, S.; Alopaeus, V.; Vuorinen, T.; Aittamaa, J. A Model for Chlorine Dioxide Delignification of Chemical Pulp. *J. Wood Chem. Technol.* **2010**, *30*, 230–268. [CrossRef]
4. Kubo, S.; Kadla, J.F. The Formation of Strong Intermolecular Interactions in Immiscible Blends of Poly(vinyl alcohol) (PVA) and Lignin. *Biomacromolecules* **2003**, *4*, 561–567. [CrossRef] [PubMed]
5. Bian, H.; Wei, L.; Lin, C.; Ma, Q.; Dai, H.; Zhu, J.Y. Lignin-Containing Cellulose Nanofibril-Reinforced Polyvinyl Alcohol Hydrogels. *ACS Sustain. Chem. Eng.* **2018**, *6*, 4821–4828. [CrossRef]
6. Ram, F.; Garemark, J.; Li, Y.; Pettersson, T.; Berglund, L.A. Functionalized Wood Veneers as Vibration Sensors: Exploring Wood Piezoelectricity and Hierarchical Structure Effects. *ACS Nano* **2022**, *16*, 15805–15813. [CrossRef] [PubMed]
7. Chen, P.; Li, Y.; Nishiyama, Y.; Pingali, S.V.; O Neill, H.M.; Zhang, Q.; Berglund, L.A. Small Angle Neutron Scattering Shows Nanoscale PMMA Distribution in Transparent Wood Biocomposites. *Nano Lett.* **2021**, *21*, 2883–2890. [CrossRef] [PubMed]
8. Wu, Y.; Tang, Q.; Yang, F.; Xu, L.; Wang, X.; Zhang, J. Mechanical and thermal properties of rice straw cellulose nanofibrils-enhanced polyvinyl alcohol films using freezing-and-thawing cycle method. *Cellulose* **2019**, *26*, 3193–3204. [CrossRef]

39. Kong, W.; Wang, C.; Jia, C.; Kuang, Y.; Pastel, G.; Chen, C.; Chen, G.; He, S.; Huang, H.; Zhang, J.; et al. Muscle-Inspired Highly Anisotropic, Strong, Ion-Conductive Hydrogels. *Adv. Mater.* **2018**, *30*, 1801934. [CrossRef]
40. Zhang, R.; Wu, C.; Yang, W.; Yao, C.; Jing, Y.; Yu, N.; Su, S.; Mahmud, S.; Zhang, X.; Zhu, J. Design of delignified wood-based high performance composite hydrogel electrolyte with double crosslinking of sodium alginate and PAM for flexible supercapacitors. *Ind. Crops Prod.* **2024**, *210*, 118187. [CrossRef]
41. Wang, Z.; Zhou, Z.; Wang, S.; Yao, X.; Han, X.; Cao, W.; Pu, J. An anti-freezing and strong wood-derived hydrogel for high performance electronic skin and wearable sensing. *Compos. Part B Eng.* **2022**, *239*, 109954. [CrossRef]
42. Chen, L.; Wei, X.; Wang, F.; Jian, S.; Yang, W.; Ma, C.; Duan, G.; Jiang, S. In-situ polymerization for mechanical strong composite actuators based on anisotropic wood and thermoresponsive polymer. *Chin. Chem. Lett.* **2022**, *33*, 2635–2638. [CrossRef]

forests

Article

Evaluation of Deterioration Degree of Archaeological Wood from Luoyang Canal No. 1 Ancient Ship

Weiwei Yang [1,2,3], Wanrong Ma [1,2,3] and Xinyou Liu [1,2,3,*]

1   Co-Innovation Center of Efficient Processing and Utilization of Forest Resources, Nanjing Forestry University, Nanjing 210037, China; yww2000922@njfu.edu.cn (W.Y.); mawan@njfu.edu.cn (W.M.)
2   College of Furnishing and Industrial Design, Nanjing Forestry University, Str. Longpan No.159, Nanjing 210037, China
3   Advanced Analysis and Testing Center, Nanjing Forestry University, Str. Longpan No.159, Nanjing 210037, China
*   Correspondence: liu.xinyou@njfu.edu.cn; Tel.: +86-25-8542-7408

Abstract: This study provides a detailed investigation of archaeological wood samples from the Luoyang Canal No. 1 site, focusing on wood species identification, physical properties, mechanical property analyses, and morphological examination. The identified wood species, belonging to the *Ulmus* genus, exhibited a 43% decline in compressive strength in waterlogged environments. Further, the wood exhibited increased moisture content, higher porosity, reduced basic density, and elevated shrinkage rates, indicating a mild level of degradation. X-ray diffraction was employed for the observation of cellulose structure, and Fourier transform infrared spectroscopy (FT-IR) demonstrated significant removal of cellulose and hemicellulose components. These findings emphasize the importance of understanding wood degradation mechanisms to evaluate structural integrity and durability in guiding the development of effective preservation strategies for archaeological wood artifacts. Continued research and conservation are crucial to deepen our knowledge of wood deterioration processes and enhance the implementation of preservation techniques.

Keywords: waterlogged wood; wood identification; physical properties; mechanical properties; XRD; FT-IR

Citation: Yang, W.; Ma, W.; Liu, X. Evaluation of Deterioration Degree of Archaeological Wood from Luoyang Canal No. 1 Ancient Ship. *Forests* 2024, 15, 963. https://doi.org/10.3390/f15060963

Academic Editor: Antonios Papadopoulos

Received: 30 April 2024
Revised: 18 May 2024
Accepted: 29 May 2024
Published: 31 May 2024

1. Introduction

Wood is an important natural material and was one of the earliest substances utilized by humans [1–3]. Numerous wooden objects discovered during archaeological excavations embody the cultural legacy of past generations, possessing substantial scientific, historical, and artistic importance [4]. Burial environments are usually wet or aquatic, which results in all of the pores of wood remains, including vessels, fibers, and micro capillaries, being entirely filled with water [5,6]. These archaeological woods can be extremely fragile due to various deteriorating factors, such as microbial attacks [7–11], temperature, humidity [7,9], oxygen, and chemical pollutants [7,11]. Hence, their preservation is imperative for maintaining the integrity of wooden artifacts [12,13]. Since the mode of wood degradation exhibits significant variability across different archaeological locations [14,15], a comprehensive understanding of the extent of deterioration in archaeological wood is essential for implementing scientific, accurate, and rigorous restoration and preservation plans [16].

Luoyang No. 1 is a wooden merchant ship that was discovered in September 2013 at the bottom of an ancient canal. Carbon-14 analyses of excavated porcelain, iron products, wooden artefacts, and soil confirmed that the shipwreck occurred during the early Qing Dynasty (1644–1902). The ship was uncovered after seven months of excavation and was transferred to the Luoyang Museum for further conservation [17].

This study focuses on the archaeological wood of the Luoyang Canal No. 1 ancient ship, beginning with wood species identification. Using recent healthy wood of identical species as a control group, the assessment of deterioration involved the evaluation of physical characteristics such as maximum moisture content, fundamental density, porosity, and rate of shrinkage. Furthermore, mechanical properties evaluation and chemical characterization techniques were utilized to assess wood degradation. The use of scanning electron microscopy (SEM) allowed for the observation of wood fiber structure and analysis of degradation across its various components. The findings of this research are expected to offer valuable insights for the restoration and conservation of the Luoyang Canal No. ancient ship, while also providing a benchmark for evaluating the extent of degradation.

2. Materials and Methods

2.1. Materials

The wood samples gathered from the archaeological site of the Luoyang Canal No. ancient ship were procured and relocated to the museum for further analysis. The sample consisted of more than 20 pieces of wood, varying in length from 12 to 26 cm, characterized by their lightweight, frayed edges, and dry surfaces coated with sediment. Subsequent examination revealed that these specimens were classified as hardwood, exhibiting ring porous characteristics. According to the excavation staff, the wooden specimens were probably part of the keel of the ancient ship discovered at the Luoyang Canal site. The wood underwent a preservation procedure at the excavation location, which included treatment with a 4% boron-based WP-1 water solution and a PEG composite solution consisting of 7% urea, 21% dimethyl urea, and 10% PEG 4000 [17]. The concentration of the PEG composite solution was incrementally raised until the saturation point of the wood was achieved, with a peak concentration of 40%. Further information on the ancient ship can be found in a prior study [17]. To assess the extent of deterioration in archaeological wood, recent healthy wood from the same species was selected for comparative analysis.

2.2. Microscopic Identification

The methodology employed in this study for determining the species of the archaeological wood primarily adhered to the sectioning protocol delineated in ISO 13061-(2014) [18]. Firstly, the archaeological wood was cut into 10 small blocks with dimensions of 5 mm $\times$ 5 mm $\times$ 10 mm (radial $\times$ tangential $\times$ longitudinal), which were subsequently boiled in distilled water until they sank. After cooling, the blocks were sliced using a Leica manual rotary microtome (Histo Core Biocutr, Ernst Leitz Company, Witzler, Hesse, Germany) to produce consistent thin sections (10–20 µm) from tangential, radial, and cross sectional surfaces. These sections were then treated with a 1.0% safranin solution for 30 min followed by two rounds of dehydration using 50%, 70%, and 95% ethanol. After the treatment, the samples were extracted with forceps and mounted on glass slides. Subsequently, the prepared archaeological wood specimens were examined for their microstructure using an Axio Scope A1 Zeiss polarizing microscope (Carl Zeiss Company, Jena, Germany) to identify the species, considering the macroscopic features of the wood blocks.

2.3. Physical Properties

The evaluation of physical properties of archaeological wood entailed a comparison of different parameters between the archaeological wood (AW) and recent healthy wood (RHW) of the same species. Each set of AW and RHW samples were prepared as 20 $\times$ 20 $\times$ 20 mm (radial $\times$ tangential $\times$ longitudinal) small blocks, with 10 blocks in each group. Surface moisture was removed with a towel every 24 h, and the blocks were promptly weighed (accurately to within 0.001 g). The data were recorded until the mass variance between successive measurements fell below 0.2%, signifying the point at which the wood reached saturation. The specimens were dried to a constant weight in an oven at 103 $\pm$ 2 °C, following ISO 13061-1 (2014) [18].

The maximum moisture content (MMC) and basic density (Db) can be calculated as follows [19,20]:

$$\text{MMC} = \frac{m - m_0}{m_0} \times 100\% \tag{1}$$

$$Db = \frac{m_0}{a_{max} \times b_{max} \times l_{max}} = \frac{m_0}{v_{max}} \tag{2}$$

where, m and m_0 are the masses of water-saturated and dry wood (g); and a_{max}, b_{max}, and l_{max} represent radial, tangential, and longitudinal dimensions of the waterlogged wood, respectively (cm).

Porosity (P), the maximum shrinkage rate of wood in three characteristic directions (ShD), and the total volume (ShV) were also calculated according to the formulae [21]:

$$P = (1 - \frac{m_0}{\rho_{ws} \times a_0 \times b_0 \times c_0}) \times 100\% \tag{3}$$

$$ShV = \frac{V_m - V_0}{V_m} \times 100\% \tag{4}$$

$$ShD = \frac{l_a - l_b}{l_a} \times 100\% \tag{5}$$

where a_0, b_0, and c_0 represent the radial, tangential, and longitudinal dimensions of the dried specimens (in cm), respectively, ρ_{ws} means the density of wood substance (g/cm^3), which is generally 1.54 according to Ugolev (1986) [22], and l_a and l_b are the lengths of the waterlogged and dry wood (in cm).

The results of these physical properties were then examined in SAS (version 9.4, SAS Institute, Cary, NC, USA) to conduct a comparative study of different wood types through *t*-tests at a significance level of $p = 0.05$.

2.4. Mechanical Properties and Morphological Characteristics

Specimens of AW and RHW with dimensions of $20 \times 20 \times 30$ mm (radial $\times$ tangential $\times$ longitudinal) were prepared for the assessment of mechanical properties, with 15 specimens assigned to each group. A destructive loading test was performed with a Shimadzu universal testing instrument AGS-X (Shimadzu Corporation, Kyoto, Japan) following ISO 13061-17 (2017) [23]. The specimens were placed at the center of the spherical movable support of the testing machine, loaded at a uniform speed, and broken within 1 to 5 min. The mean forces of five samples were recorded, and tensile strength was calculated by the formula:

$$\sigma_0 = \frac{P_{max}}{bt} \tag{6}$$

where σ_0 represents the longitudinal compressive strength of the dried specimens (MPa), P_{max} is the maximum destructive load (N), and b and t represent the width and thickness of the wood specimens (in mm), respectively.

The microstructure of the AW and RHW specimens was analyzed using a scanning electron microscope (SEM) Hitachi S-3400N II (Hitachi Ltd., Tokyo, Japan). To enhance the conductivity of the wood samples and obtain clearer images, a gold palladium SEM annular sputtering target 2″ ID $\times$ 3″ OD $\times$ 0.1 mm Anatech was applied to the surface of the specimens.

2.5. Chemical Properties

For the quantification of holocellulose and acid-insoluble lignin in wood, 500 g each of air-dried AW and RHW samples were ground to 40–60 mesh. Holocellulose and acid-insoluble lignin contents were determined using the TAPPI T204 cm-97 [24] and TAPPI T22 om-02 standards [25].

The crystal structures of the archaeological wood and recent healthy wood were investigated by X-ray diffractometer. Powdered samples that went through a 200 mesh were examined using Ultima IV X-ray diffractometry (Rigaku Corporation, Tokyo, Japan). The diffracted intensity of Cu Ka radiation (under a condition of 40 KV and 30 mA) was measured in a 2θ range between $5°$ and $60°$ and at a scan speed of $2°/\text{min}$. The crystallinity index (C_{rl}) was calculated by comparing the intensities of the crystalline and amorphous peaks [26]:

$$C_{rl} = (I_{002} - I_{am})/I_{002} \times 100\% \tag{7}$$

where C_{rl} represents the crystallinity of the wood cellulose (%), I_{002} is the maximum intensity of the diffraction peak of the (002) crystal plane near $2\theta = 22°$, and I_{am} stands for the intensity of the diffraction peak in the amorphous region near $2\theta = 18°$.

To examine the chemical deterioration of the archaeological wood, powdered samples from oven-dried wood were processed into 200 mesh KBr pellets for FT-IR analysis. The infrared spectra of the AW and RHW wood specimens were recorded in KBr tablets on a Bruker ALPHA FT-IR spectrometer (Bruker Corporation, Billerica, MA, USA), with a resolution of $0.01\ \text{cm}^{-1}$ using the 4000–$500\ \text{cm}^{-1}$ spectral range. For spectral processing, the OriginPro 2021 software (OriginLab Corporation, Northampton, MA, USA) was used. Five recordings were performed for each analyzed sample, and the average spectrum obtained was used for the evaluation.

3. Results and Discussion

3.1. Microscopic Identification

In wood identification, both the macroscopic characteristics and microscopic structures of wood provide important information [27]. In this research, after removing impurities from the surface of the selected samples, the cross-section clearly illustrated vessels and distinct growth rings. The drastic change in vessel size from earlywood to latewood indicates that the archaeological wood was a ring porous hardwood. Additionally, the large ray cells on both radial and tangential sections are prominently visible.

The microscopic image depicted in Figure 1a illustrates that in the earlywood zone of the cross-section, the vessels are elliptical in shape. Most individual vessel pores in the earlywood area are arranged continuously, forming groups of 2–3 pores aligned radially. The latewood vessel pores are small to medium in size, typically forming clusters of 4–10 vessel cells, with some arranged in a wave-like tangential pattern. Certain small vessels exhibit localized overlapping spiral thickenings on their walls. These vessels feature single perforations, either circular or oval, arranged alternately with bordered pits. Axial parenchyma cells are sparsely connected to the vessels, creating tangential bands in the earlywood region. The end walls of parenchyma cells show distinct or indistinct thickening, devoid of crystals but containing gum. Wood fibers exhibit varying thicknesses in their walls, with bordered pits present. Rays are non-storied, typically comprising 2–6 rows and measuring 8–20 cells in height. Ray tissue consist of multiseriate rays, occasionally with uniseriate rays, characterized by circular, oval and polygonal outlines of ray cells. These cells contain gum, possess thickened end walls and lack crystals. Based on the macroscopic structure and microscopic features [28,29], in conjunction with the distribution of forest resources and tree species at the excavation site, it has been concluded that the archaeological timber originated from *Ulmus parvifolia Jacq.*

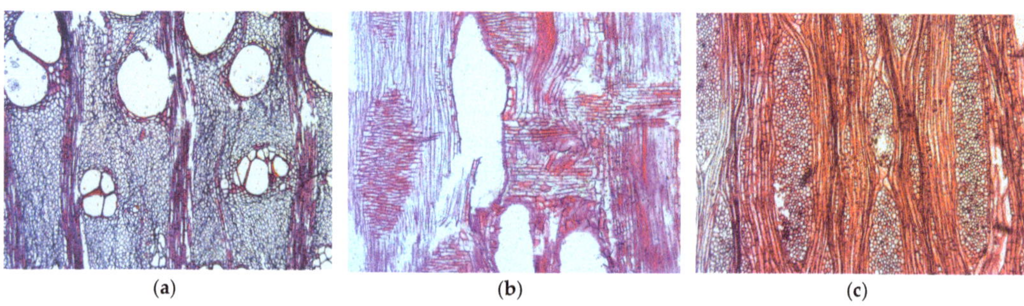

(a)   (b)   (c)

Figure 1. Microscopic images of archaeological wood in cross section (**a**), radial section (**b**) and tangential section (**c**).

3.2. Physical Characterization

Typically, waterlogged archaeological wood subjected to microbial attack undergoes varying degrees of cellulose and hemicellulose degradation [30,31]. The hydrophilic small molecules produced during deterioration increase the water absorption capacity of the wood, which explains the much higher MMC and lower density of AW samples compared to the RHW samples. Table 1 presents the maximum moisture content (MMC), basic density (Db), porosity (P), and shrinkage rate (Sh) of AW and RHW specimens for physical characterization. The maximum moisture content of 218.66% in AW is 2.13 times of the 102.61% observed in RHW. Referring to De Jong's criteria [32,33], archaeological wood with MMC ranging from 135% to 225% should be classified as the first grade of deterioration, which means the AW samples employed in this study are mildly degraded. The comparison reveals that the AW samples exhibit a fundamental density of 0.396 g/cm^3 and a porosity of 68.80%, while the RHW samples demonstrate values of 0.765 g/cm^3 and 43.89%, respectively. These results suggest a reduction in basic density and porosity resulting from degradation, which considerably weakens the mechanical strength of AW cell walls [34]. The statistical analysis ($p < 0.001$) further validates notable differences in MMC, basic density, and porosity between AW and RHW samples.

Table 1. Mean values of maximum moisture content (%), basic density (g/cm^3), porosity (%), and shrinkage rate (%) of archaeological wood (AW) and recent healthy wood (RHW). *p*-value less than 0.001 indicates a statistically significant result.

Wood Type → Parameters ↓		Archaeological Wood	Recent Healthy Wood	*p*-Value
Maximum moisture content (%)		218.66 ± 7.96 [1]	102.61 ± 8.51	<0.001 [2]
Basic density (g/cm^3)		0.396 ± 0.015	0.765 ± 0.016	<0.001
Porosity(%)		68.80 ± 1.18	43.89 ± 1.26	<0.001
Shrinkage rate (%)	Tangential	8.67 ± 0.43	5.84 ± 0.46	<0.001
	Radial	8.46 ± 0.56	3.80 ± 0.60	<0.001
	Longitudinal	2.23 ± 0.97	1.16 ± 0.91	0.063
	Volume	17.46 ± 0.92	11.44 ± 0.99	<0.001

Note: [1] Arithmetic mean of ten values ± standard error; [2] Wood type was used as a factor, *p* value for the significance analysis via *t*-test was at the 0.05 level.

Affected by prolonged deterioration, archaeological wood exhibits weakened cell walls due to the tension resulting from water evaporation and subsequent cell wall contraction [35–38]. Therefore, shrinkage rates are critical indicators of wood degradation. For the archaeological wood in the current study, the shrinkage rates in the tangential, radial, and longitudinal directions were measured at 8.67%, 8.46%, and 2.23%, respectively, higher than typical values. These results are in line with the volumetric shrinkage displayed, where the AW samples are 1.48 to 2.23 times higher than those of the RHW samples. It is important to stress that these multiples are significantly lower compared to the levels of severely degraded wood documented in prior research [39–41]. Consequently, the archaeological wood analyzed in this research should be considered as mildly degraded.

3.3. Mechanical Properties and Morphological Characteristics

Mechanical property assessments play a crucial role in determining the structural integrity and long-term durability of archaeological wood [42]. As illustrated in Figure 2, the compressive strength of the archaeological wood (AW) samples was recorded at 36.12 MPa, a notable decrease compared to the 62.91 MPa found in the modern healthy reference wood (RHW). This decline can be primarily attributed to the deterioration of wood cell walls and the breakdown of the fiber structure [9]. Additionally, the increased presence of pores in the archaeological wood diminishes its ability to withstand pressure,

resulting in compromised mechanical properties that render it more susceptible to damag
when subjected to external forces.

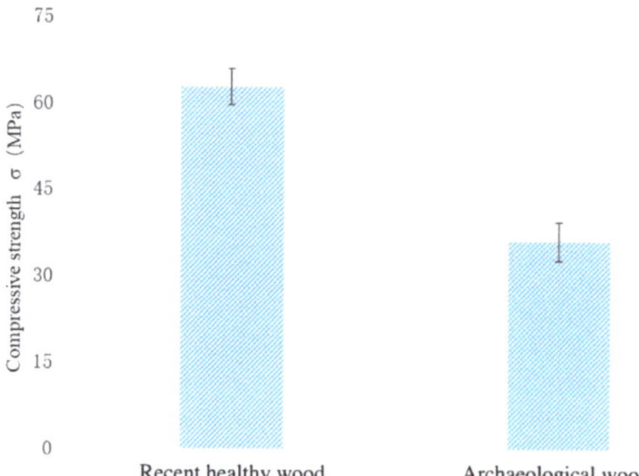

Figure 2. Comparison between the compressive strength (MPa) of recent healthy wood and archaec
logical wood samples.

According to the scanning electron microscope images (Figure 3), the growth ring
and vessels are clearly characterized on the cross section. Despite the accumulation o
wood fragments within vessels, the cell lumens of the archaeological wood samples retai
a relatively intact microscopic morphology. This suggests that the internal tissues hav
maintained a degree of mechanical strength sufficient to resist collapse during the dry
ing process. The structural characteristics of the wood rays in the radial section remai
discernible (the arrows), regardless of a noticeable relaxation in the arrangement of th
wood fiber cells. The image of the tangential section reveals clearly identifiable fiber cell
and rays, indicating that the level of deterioration in this wood sample has not reached
critically severe stage.

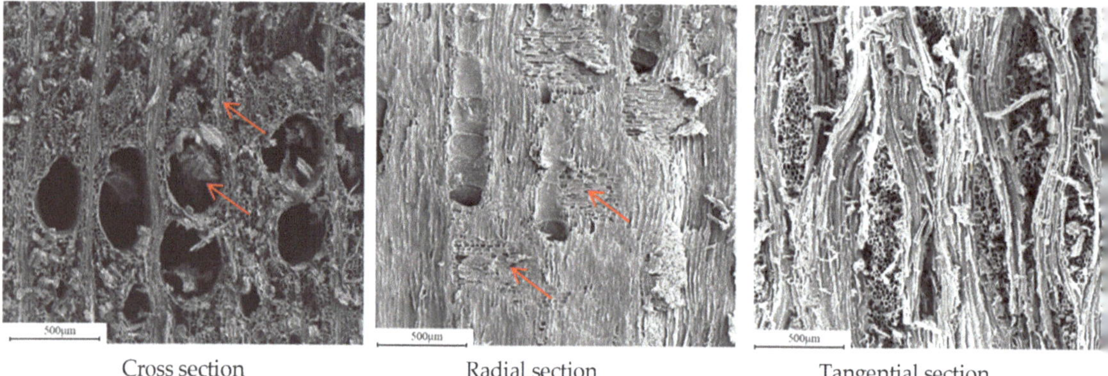

Cross section   Radial section   Tangential section

Figure 3. Scanning electron microscope (SEM) images of archaeological wood samples i
different sections.

3.4. Chemical Properties

The changes in the physical and mechanical properties of archaeological wood ar
attributed to alterations in its chemical composition. Table 2 presents the content of holoce
lulose and acid-insoluble lignin in AW and RHW. For archaeological wood, holocellulos
accounts for 48.09% of the composition, which is only 61.78% of that present in the RHV

samples. This finding is consistent with previous studies [43–46], underscoring the importance of holocellulose degradation and the inherent chemical and biological resistance of lignin in the decay mechanisms of archaeological wood.

Table 2. The content of holocellulose and acid-insoluble lignin in AW and RHW.

Chemical Components	Archaeological Wood	Recent Healthy Wood
Holocellulose (%)	48.09 ± 2.96	77.83 ± 3.51
Acid-insoluble lignin (%)	39.41 ± 2.15	23.12 ± 1.16

Diffractometric analysis has been employed to evaluate the crystallinity index (CI) of cellulose, which is the sole crystalline constituent in wood, given that hemicellulose and lignin are completely amorphous.

Figure 4 displays the X-ray diffraction patterns of both the AW and RHW, demonstrating essential similarities in their profiles. In the recent wood spectra, the characteristics peaks at $2\theta = 15.2°, 22.2°, 24°$, and $30°$ can be observed, corresponding to crystallographic planes of cellulose [47]. The XRD spectrum of AW presents additional diffraction peaks at $2\theta = 21°, 26.7°$, and $27.8°$, which can be ascribed to crystalline compounds due to waterlogging. Notably, the diffraction intensity of the RHW is marginally higher than that of the AW, suggesting a decline in cellulose content in the AW samples used for analysis. According to the formula outlined in Section 2.5, the cellulose crystallinity of archaeological wood was determined to be 38.21%, which is slightly lower than the 42.46% found in recent healthy wood. These results indicate that the cellulose in the AW had become degraded to some extent [48,49].

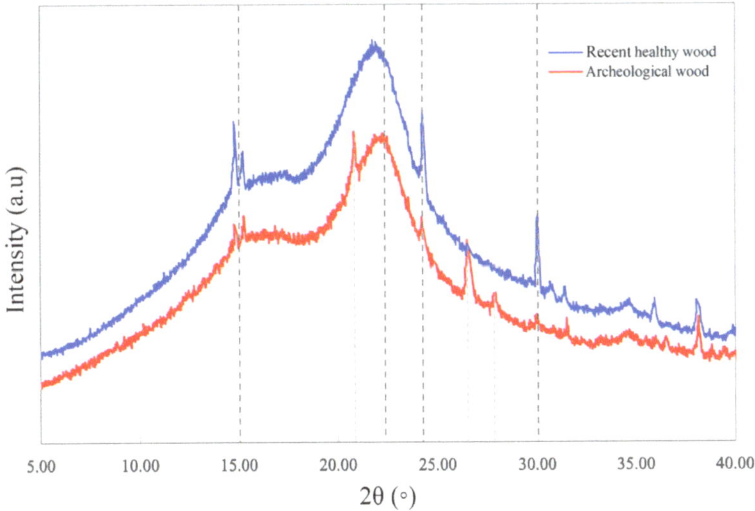

Figure 4. X-RD spectra of archaeological wood and recent healthy wood.

Fourier transform infrared (FT-IR) spectroscopy is an essential technique for assessing the degradation or transformation of wood components, as changes in specific characteristic peaks, including their emergence, disappearance, enhancement, or weakening, indicate alterations in the wood's chemical composition [50–53]. Figure 5 represents the FT-IR spectra of the AW and RHW, revealing a fundamental similarity between them. Prominent peaks at 3400 cm^{-1}, 2925 cm^{-1}, 2855 cm^{-1}, and 1025 cm^{-1} are visualized [54], corresponding to the stretching vibration of hydroxyl (-OH), C-H asymmetrical stretching, C-H symmetrical stretching, and -C-O-C- stretching vibration, respectively.

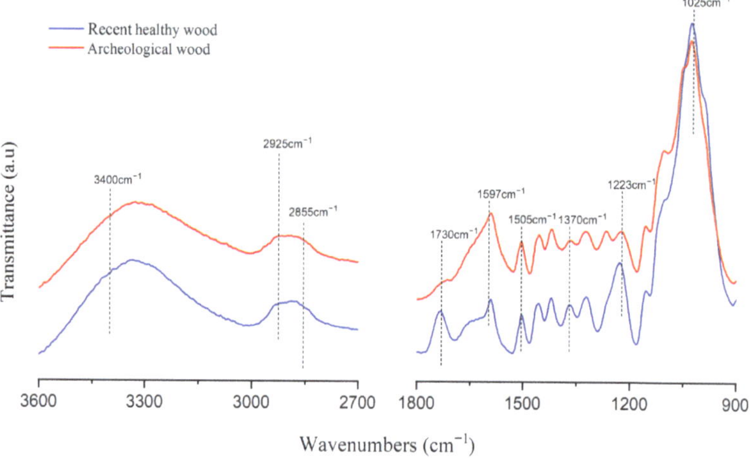

Figure 5. FT-IR spectra of archaeological wood and recent healthy wood.

Significant changes are observed in the spectral comparison between AW and RHW in the range of 1800–1200 cm^{-1}. One of the most striking differences is seen at the peak 1730 cm^{-1}, associated with the C=O stretching of acetyl or carboxylic acid [55]. This peak is nearly absent in the AW spectra, indicating a notable deterioration of cellulose and hemicellulose within the archaeological samples. However, the two characteristic peaks associated with lignin at 1597 cm^{-1} and 1505 cm^{-1} in AW remain intact and exhibit slightly higher intensities compared to those in RHW [56]. This feature agrees with the higher relative content of acid-insoluble lignin presented in AW as outlined in Table 2. Furthermore, slight weakening of the peaks at 1370 cm^{-1} and 1223 cm^{-1} in the archaeological wood suggests cellulose degradation [57]. Chemical property analyses reveal that the primary mode of degradation in archaeological wood involves the breakdown of cellulose and hemicellulose. Specifically, cellulose undergoes considerable degradation, whereas lignin exhibits relative stability with a slight increase in its proportion.

4. Conclusions

The comprehensive analysis of the physical, mechanical, morphological, and chemical properties of archaeological wood samples, identified as belonging to the *Ulmus parvifolia Jacq*, has provided valuable insights into the degradation processes and characteristics of the wood. The identification of the tree species underlines the historical significance of *Ulmus* wood as a crucial material for shipbuilding during the Qing Dynasty. Examination of parameters such as maximum moisture content, basic density, and porosity confirmed mild degradation levels in the wood from the Luoyang Canal No.1 ancient ship. Additionally, detailed chemical property analyses revealed that the degradation of the archaeological wood primarily stemmed from the breakdown of cellulose and hemicellulose, while lignin demonstrated a relatively stable nature. These results not only deepen our comprehension of the wood's condition, but also provide critical values for the consequent restoration and preservation. They stress the urgent need for conservation strategies to protect this historical artifact for future generations.

Author Contributions: Conceptualization, W.Y., X.L. and W.M.; methodology, W.Y. and X.L.; software, W.M. and X.L.; validation, W.Y., X.L. and W.M.; formal analysis, W.Y., X.L. and W.M.; investigation, W.Y., X.L. and W.M.; data curation, W.Y., X.L. and W.M.; writing, original draft preparation, W.Y., X.L. and W.M.; writing, review and editing, W.Y. and X.L.; visualization, W.M.; supervision, X.L. All authors have read and agreed to the published version of the manuscript.

Funding: This work was funded by the Nanjing Forestry University Foundation for Basic Research, Grant No. 163104127 and the Priority Academic Program Development (PAPD) of Jiangsu Province, China.

Data Availability Statement: The raw data supporting the conclusions of this article will be made available by the authors on request.

Acknowledgments: The authors would like to thank the College of Furnishings and Industrial Design, Nanjing Forestry University for providing the experimental conditions.

Conflicts of Interest: The authors declare no conflicts of interest.

References

1. Sun, G.R.; He, Y.R.; Wu, Z.H. Effects of thermal treatment on the dimensional stability and chemical constituents of new and aged camphorwood. *BioResources* **2022**, *17*, 4186–4195. [CrossRef]
2. Feng, X.; Sheng, Y.; Ge, X.; Wu, Z.; Huang, Q. Evaluation of The Properties of Hybrid Yellow Poplar (Liriodendron Sino-Americanum): A Comparison Study with Yellow Poplar (*Liriodendron tulipifera*). *Maderas-Cienc. Y Tecnol.* **2021**, *23*. [CrossRef]
3. Feng, X.; Chen, J.; Yu, S.; Wu, Z.; Huang, Q. Mild Hydrothermal Modification of Beech Wood (*Zelkova schneideriana* Hand-Mzt): Its Physical, Structural, and Mechanical Properties. *Eur. J. Wood Wood Prod.* **2022**, *80*, 933–945. [CrossRef]
4. Liu, X.; Xu, X.; Tu, X.; Ma, W.; Huang, H.; Varodi, A.M. Characteristics of Ancient Ship Wood from Taicang of the Yuan Dynasty. *Materials* **2023**, *16*, 104. [CrossRef]
5. Grattan, D.W. Waterlogged Wood. *Conserv. Mar. Archaeol. Objects* **1987**, *1*, 55–67.
6. Björdal, C.G. Microbial Degradation of Waterlogged Archaeological Wood. *J. Cult. Herit.* **2012**, *13*, 118–122. [CrossRef]
7. Broda, M.; Hill, C.A. Conservation of Waterlogged Wood—Past, Present and Future Perspectives. *Forests* **2021**, *12*, 1193. [CrossRef]
8. Majka, J.; Zborowska, M.; Fejfer, M.; Waliszewska, B.; Olek, W. Dimensional Stability and Hygroscopic Properties of PEG Treated Irregularly Degraded Waterlogged Scots Pine Wood. *J. Cult. Herit.* **2018**, *31*, 133–140. [CrossRef]
9. Bjurhager, I.; Halonen, H.; Lindfors, E.L.; Iversen, T.; Almkvist, G.; Gamstedt, E.K.; Berglund, L.A. State of Degradation in Archaeological Oak from the 17th Century Vasa Ship: Substantial Strength Loss Correlates with Reduction in (Holo) Cellulose Molecular Weight. *Biomacromolecules* **2012**, *13*, 2521–2527. [CrossRef]
10. Cao, H.; Gao, X.; Chen, J.; Xi, G.; Yin, Y.; Guo, J. Changes in Moisture Characteristics of Waterlogged Archaeological Wood Owing to Microbial Degradation. *Forests* **2022**, *14*, 9. [CrossRef]
11. Li, R.; Guo, J.; Macchioni, N.; Pizzo, B.; Xi, G.; Tian, X.; Chen, J.; Sun, J.; Jiang, X.; Cao, J.; et al. Characterisation of Waterlogged Archaeological Wood from Nanhai No. 1 Shipwreck by Multidisciplinary Diagnostic Methods. *J. Cult. Herit.* **2022**, *56*, 25–35. [CrossRef]
12. Muyzer, G.; Stams, A.J.M. The Ecology and Biotechnology of Sulphate-Reducing Bacteria. *Nat. Rev. Microbiol.* **2008**, *6*, 441–454. [CrossRef] [PubMed]
13. Kilic, N.; Kiliç, A.G. Analysis of Waterlogged Woods: Example of Yenikapi Shipwreck. *Art-Sanat Derg.* **2018**, *9*, 1–11.
14. Blanchette, R.A.; Nilsson, T.; Daniel, G.; Abad, A.R. Biological Degradation of Wood. In *Archaeological Wood: Properties, Chemistry, and Preservation*; Rowell, R.M., Barbour, R.J., Eds.; American Chemical Society: Washington, DC, USA, 1990; pp. 141–174.
15. Giachi, G.; Capretti, C.; Macchioni, N.; Pizzo, B.; Donato, I.D. A Methodological Approach in the Evaluation of the Efficacy of Treatments for the Dimensional Stabilisation of Waterlogged Archaeological Wood. *J. Cult. Herit.* **2010**, *11*, 91–101. [CrossRef]
16. Broda, M.; Mazela, B.; Krolikowska-Pataraja, K.; Siuda, J. The State of Degradation of Waterlogged Wood from Different Environments. In *Annals of Warsaw University of Life Sciences-SGGW. Forestry and Wood Technology*; Warsaw University of Life Sciences: Warsaw, Poland, 2015.
17. Liu, X.; Ma, W.; Tu, X.; Huang, H.; Varodi, A.M. Study on the Wood Characteristics of the Chinese Ancient Ship Luoyang I. *Materials* **2023**, *6*, 1145. [CrossRef] [PubMed]
18. *ISO 13061–1:2014*; Physical and Mechanical Properties of Wood—Test Methods for Small Clear Wood Specimens—Part 1: Determination of Moisture Content for Physical and Mechanical Tests. International Organization for Standardization: Geneva, Switzerland, 2014.
19. Hoffmann, P.; Jones, M.A. Structure and Degradation Process for Waterlogged Archaeological Wood. In *Archaeological Wood: Properties, Chemistry, and Preservation*; Rowell, R.M., Barbour, R.J., Eds.; American Chemical Society: Washington, DC, USA, 1990; pp. 35–65.
20. Jensen, P.; Gregory, D.J. Selected Physical Parameters to Characterize the State of Preservation of Archaeological Wood: A Practical Guide for Their Determination. *J. Archaeol. Sci.* **2006**, *33*, 551–559. [CrossRef]
21. Varivodina, I.; Kosichenko, N.; Varivodin, V.; Sedliačik, J. Interconnections Among the Rate of Growth, Porosity, and Wood Water Absorption. *Wood Res.* **2010**, *55*, 59–66.
22. Ugolev, B.N. *Wood Science and Bases of Forest Commodity Science: Textbook for Higher Educational Institutions of Forestry Engineering*; MGUL: Moscow, Russia, 1986; p. 351.
23. *ISO 13061–17:2017*; Physical and Mechanical Properties of Wood—Test Methods for Small Clear Wood Specimens—Part 17: Determination of Ultimate Stress in Compression Parallel to Grain. International Organization for Standardization: Geneva, Switzerland, 2017.

24. Tappi, T. *204 cm-97, Solvent Extractives of Wood and Pulp*; Technical Association of the Pulp and Paper Industry (TAPPI): Atlanta, GA, USA, 2007.

25. Tappi, T. *222 cm-02, Acid-Insoluble Lignin in Wood and Pulp*; Technical Association of the Pulp and Paper Industry (TAPPI): Atlanta, GA, USA, 1997.

26. Segal, L.; Creely, J.J.; Martin, A.E., Jr.; Conrad, C.M. An Empirical Method for Estimating the Degree of Crystallinity of Native Cellulose Using the X-ray Diffractometer. *Text. Res. J.* **1959**, *29*, 786–794. [CrossRef]

27. Steele, J.H.; Ifju, G.; Johnson, J.A. Quantitative Characterization of Wood Microstructure. *J. Microsc.* **1976**, *107*, 297–311. [CrossRef]

28. Cheng, J.; Yang, J.; Liu, P. *Chinese Wood Flora*; China Forestry Publishing House: Beijing, China, 1992.

29. Wang, C.; Yu, C.; Liu, M.; Peter, B. Formation and Influencing Factors of Dew in Sparse Elm Woods and Grassland in a Semi-Arid area. *Acta Ecol. Sin.* **2017**, *37*, 125–132. [CrossRef]

30. Liu, X.; Zhu, L.; Tu, X.; Zhang, C.; Huang, H.; Varodi, A.M. Characteristics of Ancient Shipwreck Wood from Huaguang Jiao No. 1 after Desalination. *Materials* **2023**, *16*, 510. [CrossRef] [PubMed]

31. Gao, J.; Li, J.; Qiu, J.; Guo, M. Degradation assessment of waterlogged wood at Haimenkou site. *Frat. Ed Integrità Strutt.* **2014**, *8*, 495–501.

32. Jong, J. Conservation Techniques for Old Archaeological Wood from Shipwrecks Found in the Netherlands. In *Biodeterioration Investigation Techniques*; Walters, A.H., Ed.; Applied Science: London, UK, 1977; pp. 295–338.

33. Kennedy, A.; Pennington, E.R. Conservation of Chemically Degraded Waterlogged Wood with Sugars. *Stud. Conserv.* **2014**, *59*, 194–201. [CrossRef]

34. Babiński, L.; Izdebska-Mucha, D.; Waliszewska, B. Evaluation of the State of Preservation of Waterlogged Archaeological Wood Based on Its Physical Properties: Basic Density vs. Wood Substance Density. *J. Archaeol. Sci.* **2014**, *46*, 372–383. [CrossRef]

35. Zhou, T.; Liu, H. Research Progress of Wood Cell Wall Modification and Functional Improvement: A Review. *Materials* **2022**, *15*, 1598. [CrossRef] [PubMed]

36. Yin, Q.; Liu, H. Drying Stress and Strain of Wood: A Review. *Appl. Sci.* **2021**, *11*, 5023. [CrossRef]

37. Zhang, X.; Wang, X.; Zhao, M.; Liu, H. Study on free shrinkage and rheological characteristics of *Eucalyptus urophylla* × *E. grandis* during conventional drying. *Wood Mater. Sci. Eng.* **2024**, 2311827. [CrossRef]

38. Zheng, J.; Xu, X.; Yang, L. Dewatering characteristics of Juglans mandshurica wood using supercritical carbon dioxide: A comparison with conventional drying. *Dry. Technol.* **2024**, *42*, 2324931. [CrossRef]

39. Lionetto, F.; Quarta, G.; Cataldi, A.; Auriemma, R.; Calcagnile, L.; Frigione, M. Characterization and Dating of Waterlogged Woods from an Ancient Harbor in Italy. *J. Cult. Herit.* **2014**, *15*, 213–217. [CrossRef]

40. Han, Y.; Du, J.; Huang, X.; Ma, K.; Wang, Y.; Guo, P.; Pan, J. Chemical Properties and Microbial Analysis of Waterlogged Archaeological Wood from the Nanhai No. 1 Shipwreck. *Forests* **2021**, *12*, 587. [CrossRef]

41. Guo, J.; Xiao, L.; Han, L.; Wu, H.; Yang, T.; Wu, S.; Yin, Y. Deterioration of the Cell Wall in Waterlogged Wooden Archaeological Artifacts, 2400 Years Old. *IAWA J.* **2019**, *40*, 820–844. [CrossRef]

42. Emara, M.; Barris, C.; Baena, M.; Torres, L.; Barros, J. Bond behavior of NSM CFRP laminates in concrete under sustained loading. *Constr. Build. Mater.* **2018**, *177*, 237–246. [CrossRef]

43. Pecoraro, E.; Pizzo, B.; Alves, A.; Macchioni, N.; Rodrigues, J.C. Measuring the Chemical Composition of Waterlogged Decayed Wood by Near Infrared Spectroscopy. *Microchem. J.* **2015**, *122*, 176–188. [CrossRef]

44. Passialis, C.N. Physico-Chemical Characteristics of Waterlogged Archaeological Wood. *Holzforschung* **1997**, *51*, 111–113. [CrossRef]

45. Xia, Y.; Chen, T.Y.; Wen, J.L.; Zhao, Y.-L.; Qiu, J.; Sun, R.-C. Multi-Analysis of Chemical Transformations of Lignin Macromolecules from Waterlogged Archaeological Wood. *Int. J. Biol. Macromol.* **2018**, *109*, 407–416. [CrossRef] [PubMed]

46. Colombini, M.P.; Lucejko, J.J.; Modugno, F.; Orlandi, M.; Tolppa, E.L.; Zoia, L. A multi-analytical study of degradation of lignin in archaeological waterlogged wood. *Talanta* **2009**, *80*, 61–70. [CrossRef] [PubMed]

47. Toba, K.; Yamamoto, H.; Yoshida, M. Crystallization of Cellulose Microfibrils in Wood Cell Wall by Repeated Dry-and-Wet Treatment, Using X-Ray Diffraction Technique. *Cellulose* **2013**, *20*, 633–643. [CrossRef]

48. High, K.E.; Penkman, K.E. A Review of Analytical Methods for Assessing Preservation in Waterlogged Archaeological Wood and Their Application in Practice. *Heritage Sci.* **2020**, *8*, 83. [CrossRef]

49. Han, L.; Tian, X.; Keplinger, T.; Zhou, H.; Li, R.; Svedström, K.; Burgert, I.; Yin, Y.; Guo, J. Even Visually Intact Cell Walls in Waterlogged Archaeological Wood Are Chemically Deteriorated and Mechanically Fragile: A Case of a 170 Year-Old Shipwreck. *Molecules* **2020**, *25*, 1113. [CrossRef]

50. Liu, Y.; Liu, H.; Shen, Z. Nanocellulose Based Filtration Membrane in Industrial Waste Water Treatment: A Review. *Materials* **2021**, *14*, 5398. [CrossRef]

51. Antonelli, F.; Galotta, G.; Sidoti, G.; Zikeli, F.; Nisi, R.; Davidde Petriaggi, B.; Romagnoli, M. Cellulose and lignin nano-scale consolidants for waterlogged archaeological wood. *Front. Chem.* **2020**, *8*, 32. [CrossRef] [PubMed]

52. Liu, X.; Tu, X.; Ma, W.; Zhang, C.; Huang, H.; Varodi, A.M. Consolidation and Dehydration of Waterlogged Archaeological Wood from Site Huaguangjiao No.1. *Forests* **2022**, *13*, 1919. [CrossRef]

53. Christensen, M.; Frosch, M.; Jensen, P.; Schnell, U.; Shashoua, Y.; Nielsen, O.F. Waterlogged Archaeological Wood—Chemical Changes by Conservation and Degradation. *J. Raman Spectrosc.* **2006**, *37*, 1171–1178. [CrossRef]

4. Zhu, X.; Kaal, J.; Traoré, M.; Kuang, Y. Characterization of modern and waterlogged archaeological cypress (*Glyptostrobus pensilis*) wood: An analytical pyrolysis (Py-GC-MS and THM-GC-MS) and infrared spectroscopy (FTIR-ATR) study of within tree (radial) and decay-induced compositional variations. *J. Anal. Appl. Pyrolysis* **2024**, *177*, 106347.

5. Han, L.; Guo, J.; Tian, X.; Jiang, X.; Yin, Y. Evaluation of PEG and sugars consolidated fragile waterlogged archaeological wood using nanoindentation and ATR-FTIR imaging. *Int. Biodeterior. Biodegrad.* **2022**, *170*, 105390. [CrossRef]

6. Jiang, J.; Li, J.; Gao, Q. Effect of flame retardant treatment on dimensional stability and thermal degradation of wood. *Constr. Build. Mater.* **2015**, *75*, 74–81. [CrossRef]

7. Popescu, M.C.; Jones, D.; Krzisnik, D.; Humar, M. Determination of the effectiveness of a combined thermal/chemical wood modification by the use of FTIR spectroscopy and chemometric methods. *J. Mol. Struct.* **2020**, *1200*, 127–133. [CrossRef]

Article

The Gradient Variation of Location Distribution, Cross-Section Area, and Mechanical Properties of Moso Bamboo Vascular Bundles along the Radial Direction

Hongbo Li [1,2,*], Qipeng Zhu [1], Pengchen Lu [1], Xi Chen [1] and Yu Xian [3]

[1] College of Agricultural Engineering, Shanxi Agricultural University, Jinzhong 030801, China; s20212062@stu.sxau.edu.cn (Q.Z.); s20222074@stu.sxau.edu.cn (P.L.); 20233033@stu.sxau.edu.cn (X.C.)
[2] Dryland Farm Machinery Key Technology and Equipment Key Laboratory of Shanxi Province, Jinzhong 030801, China
[3] College of Forestry, Shanxi Agricultural University, Jinzhong 030801, China; xianyu@sxau.edu.cn
[*] Correspondence: lihb@sxau.edu.cn; Tel.: +86-135-9310-0742

Abstract: Bamboo is a typical natural fiber-reinforced composite with excellent mechanical properties, which are determined by its special micro-structure. As the reinforcing phase, the vascular bundles play a central role in the control of the mechanical properties of bamboo macro-structure. To find the exact gradient variation of the mechanical properties of these continuously distributed vascular bundles within the bamboo culm, 4-year-old Moso bamboo was chosen to investigate the variation of locate-distribution, cross-section area, and mechanical properties of single vascular bundles along the longitudinal and radial directions with respect to their location from the base, middle, and top sections of bamboo culm, respectively. It shows that the spatial distribution of vascular bundles along the column is distributed exponentially from the inside to the outside of the culm. The cross-section area of the vascular bundles decreased exponentially from the inside to the outside along the radial direction. All the vascular bundles were then carefully separated from bamboo strips and tested via the tensile tests. Test results show that the longitudinal tensile strengths of vascular bundles ranged from 180.44 to 774.10 MPa, and the longitudinal Young's modulus ranged from 9.00 to 44.76 GPa. The tensile strength of vascular bundles at the outer side was three times higher than that of the inner side, while Young's modulus at the outer side was three to four times higher than that of the inner side. For all three height positions, the strengths and Young's modulus of vascular bundles are all exponentially increased from the inner side to the outer side along the radial direction. This work will provide a basis for the highly processed product's application of bamboo resources and a reference for further study on the trans-scale analysis of the mechanical properties of bamboo.

Keywords: Moso bamboo; vascular bundle; functionally graded material; mechanical properties

Citation: Li, H.; Zhu, Q.; Lu, P.; Chen, X.; Xian, Y. The Gradient Variation of Location Distribution, Cross-Section Area, and Mechanical Properties of Moso Bamboo Vascular Bundles along the Radial Direction. *Forests* **2024**, *15*, 1023. https://doi.org/10.3390/f15061023

Academic Editor: Alain Cloutier

Received: 16 May 2024
Revised: 5 June 2024
Accepted: 7 June 2024
Published: 13 June 2024

1. Introduction

As renewable resources and grows throughout the world's tropical and subtropical regions, bamboo has a long and well-established tradition as a traditional material used in various aspects of life and production, such as household materials, furniture, crafts, paper, and textiles. Compared with the rich resources of bamboo, the available wood supply is decreasing quickly. To relieve the supply and demand stresses in wood resources, public attention has turned to bamboo as an alternative raw material for wood products. At the same time, plastic waste pollution is another serious environmental pollution crisis on Earth. To reduce plastic pollution and address climate change, researchers have tried many methods to find candidates to replace plastic products. Bamboo, recognized as a natural and renewable biomass material [1], emerges as an ideal alternative to plastic. An initiative of "bamboo as a substitute for plastic" was proposed in 2022 [2].

Among the typical natural biological composite, bamboo has one of the most favorable combinations of low density and high mechanical strength [3]. Its specific properties are comparable with conventional materials such as low-carbon steel and glass-reinforced plastics [3,4], which made it one of the best raw materials in Southeast Asia for the construction of scaffolding, packing boards, and furniture panels. As a biological material, bamboo exhibits many levels of hierarchical structures, from subcells, cells, and tissue to macroscopic plant bodies [5–7]. Its exceptional mechanical properties are determined by its unique macroscopic (bamboo culm), mesoscale (vascular bundles), microscopic (fibers), and nanoscopic (microfibril) structures [8–11].

Along the bamboo culm cross-section, it can be observed that the vascular bundles are distributed non-uniformly through the culm thickness. Most of the vascular bundles are concentrated densely near the exterior periphery while distributed sparsely near the inner zone. With the vascular bundles' volume density, type, and size spatially varying continuously, the bamboo's macroscopic structure possesses a continuously graded mechanical property. So, the bamboo structure is viewed as a natural functional graded composite material [11–14].

To investigate the distribution regularity of bamboo's mechanical properties along the culm thickness, bamboo slices were split and tested via tensile tests. For tensile strength and Young's modulus of bamboo culm, test results showed that although both of them were increased continuously from the inner side to the outer side, the gradient regularity was not consistent in the same way [15–18]. For Moso bamboo, Amada et al. [15,16] got the strength and Young's modulus increased parabolically, and Li and Shen [17] regarded Young's modulus increased cubically, while Nogata and Takahashi [18] found the strength and Young's modulus increased exponentially with radial distance.

Bamboo is a typical natural fiber-reinforced composite. Its gradient mechanical properties are determined by its special micro-structure. As the reinforcing phase, the vascular bundles play a central role in the appearance of bamboo culm with its number, shape, size, content, and mechanical properties [19–21].

Research showed that the distribution of vascular bundles volume fraction along the radial direction may follow different forms, namely linear [17], quadratic [13], or exponential curve [22]. Using an image binarization method based on the K-means clustering algorithm, Xu et al. [14,23] found that the vascular bundles' volume fraction of Moso bamboo [*Phyllostachys edulis*] decreased exponentially, the length-to-width ratio of vascular bundle decreased quadratically, and the width of vascular bundle increased linearly along the radial direction from the outer side to the inner side.

The bamboo vascular bundle is hard but tiny, with its diameter ranging between 100 and 500 μm. The strong adhesive between rigid vascular bundles and their surrounding parenchyma matrix made it very difficult to directly separate intact vascular bundles from bamboo slices with enough length. To obtain a better and more comprehensive understanding of the mechanical behavior of bamboo vascular bundles, many isolation techniques, such as mechanical [4,24,25], chemical [17,26], or biological [27] treatment methods, were used to extract vascular bundles from bamboo culm. Osorio et al. [4] developed a novel mechanical extraction and obtained strength values. Young's modulus of vascular bundles of the bamboo species *Guadua angustifolia* were 800 MPa and 43 GPa, respectively. Wang and Shao [24] extracted vascular bundles of Moso bamboo by handwork and obtained the average tensile strength and Young's modulus with 523.2 MPa and 22.3 GPa, respectively. The mechanical process can separate bamboo vascular bundles effectively. However, the extracted vascular bundles can be damaged due to the strong rigidity of the bamboo culm.

Compared to the mechanical method, the chemical isolation technique could produce more intact vascular bundles [28]. For example, Li and Shen [17] separated vascular bundles of Moso bamboo from three height positions with an alkali treatment method and tested the tensile strength varied from 495.2 to 916.2 MPa, and Young's modulus varied from 17.0 to 35.9 GPa. Also, they pointed out that both the tensile strength and Young's modulus were

linearly increased from the inner zone to the outer surface. Recently, Li et al. [26] reported
a top-down strategy by using a two-step process involving chemical delignification and
air-drying to extract high-performance vascular bundles with an average tensile strength of
1.90 ± 0.32 GPa, Young's modulus of 91.3 ± 29.7 GPa, and toughness of 25.4 ± 4.5 MJ m^{-3},
which exceed those of the mechanically extracted vascular bundles. This work provides a
good reference method for successfully extracting scalable long and undamaged vascular
bundles from tough bamboo material.

The structure of a bamboo culm in a transverse section is characterized by numerous
vascular bundles embedded in the parenchymatous ground tissue. Through anatomical
investigations, it can be found that there are thousands of vascular bundles within a bamboo
culm cross-section. Previous studies have conducted mechanical tests on the extracted
vascular bundles. However, these tests were limited to part of vascular bundles. Only
a few studies have examined the mechanical properties variation of vascular bundles
with respect to their layer location within a bamboo culm [17,25]. To characterize bamboo
mechanical properties and graded variation within the bamboo culm, it is necessary to
identify the relationship between the variation of mechanical properties and the exact
location of vascular bundles.

This investigation aims to find the exact gradient variation of the mechanical properties of these continuously distributed vascular bundles within the bamboo culm. This
work will provide a basis for the highly processed product's application of bamboo resources and a reference for further study on the trans-scale analysis of the mechanical
properties of bamboo.

2. Materials and Methods

2.1. Origin and Sampling of Bamboo Culms

The 4-year-old Moso bamboo *Phyllostachys edulis* was collected from a plantation
located in Suqian City, Jiangsu Province, China. Where the bamboo with a breast diameter
of 7–9 cm and height of 6–7 m was selected. Every culm was cut into four long segments in
the longitudinal direction. These segments were kept for about six months for seasoning.

Three internodes from heights of 1, 3, and 5 m were cut and named as the base, middle,
and top sections, respectively. Along the longitudinal direction, each internode was cut
into two segments, namely the short part (about 3 cm length) and the long part (about
10 cm length). The short part (Figure 1) was used to observe the location distribution and
measure the cross-section area of vascular bundles along the radial direction of the bamboo
culm wall. At the same time, the longer part was used to extract vascular bundles and test
the mechanical properties.

Figure 1. Bamboo rings from different height positions.

2.2. The Vascular Bundles' Precise Location and Area Calculation

The short bamboo ring was divided into 8 equal parts along the clockwise direction,
as shown in Figure 2. Within each part, two represented columns of vascular bundles,
arranged from the inner to the outer side, were chosen for the vascular bundles' precise

localization and observation. Consequently, there were a total of 16 columns of vascular bundles that needed to be observed in this experiment.

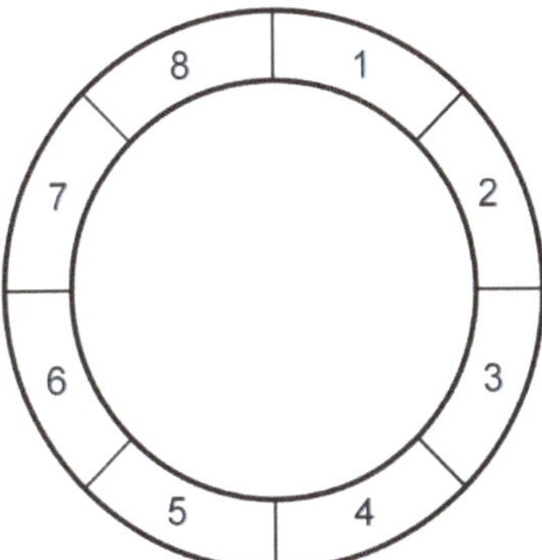

Figure 2. Bamboo ring division.

With the changes in height, the number of vascular bundles decreases steadily from the base to the top. In order to determine the variability of the location of vascular bundles, the location of the scatted vascular bundles in the section plane was described by row and column for comparison. From different height sections, the vascular bundles were divided into 15 rows at the base part, 13 rows at the middle part, and 12 rows at the top part. To determine the location of each vascular bundle along a column, a coordinate system was further established, where the innermost position of the vascular bundle is designated as the origin of coordinates, and the tangential and radial direction are named as the x-axis and R-axis, respectively (Figure 3a). A non-dimensional radius, $\bar{r}$, defined by the distance from the inner surface divided by the culm thickness t, was described as Equation (1):

$$\bar{r} = \frac{(r - r_0)}{t} \tag{1}$$

where r is the radial position of the vascular bundle in bamboo culm, r_0 is the inner surface, t is the culm thickness, $\bar{r} = 0$ corresponds to the inner surface, and $\bar{r} = 1$ to the outer surface.

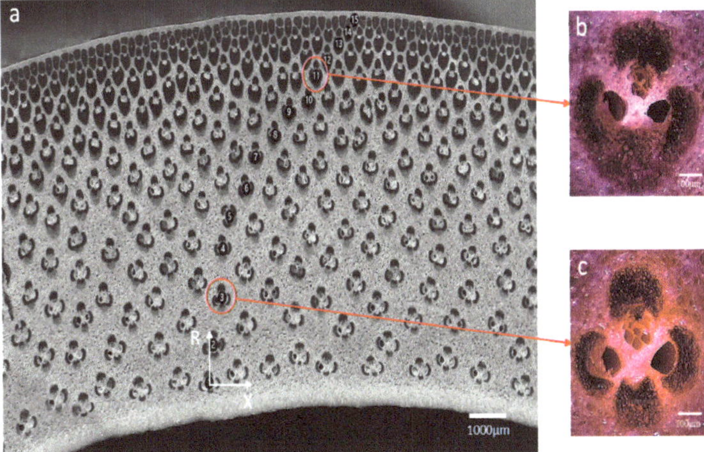

Figure 3. Vascular bundles in bamboo culm: (**a**) location of vascular bundles along a column, number 1–15 means vascular bundle from the inner to the outer periphery of the culm wall. (**b**) semi-open type, and (**c**) open type.

The short bamboo segments in Figure 1 were polished. The targeted vascular bundle photograph was captured using a stereomicroscope. The cross-sectional area of the vascular bundles was measured using the image processing software Digimizer 5.44. The software's built-in image processing tools were employed to capture the outline of the vascular bundle cross-sections, ensuring a distinct differentiation between the vascular bundles and the parenchymatous ground tissue. Points were manually plotted along the boundary of the vascular bundle to measure the cross-sectional area (Figure 4).

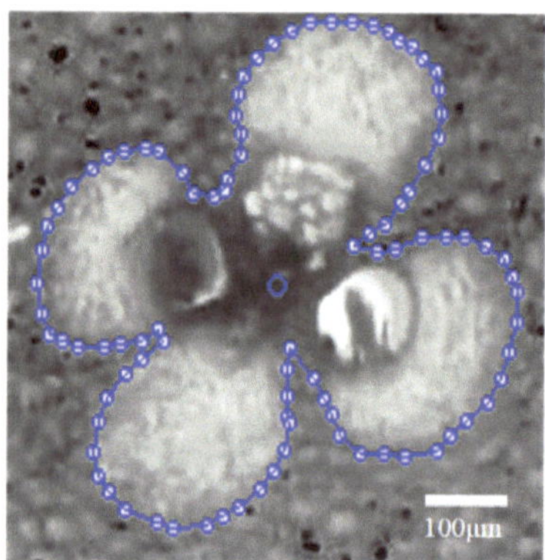

Figure 4. Vascular bundle cross-section area measurement.

2.3. Sample Preparation and Tensile Test

The longer parts of bamboo segments taken from the base, middle, and top internodes were divided into eight equal bamboo strips along the circumferential direction. Following the vascular bundle separation method from Li et al. [26], the prepared bamboo strips were soaked in peroxy formic acid solution (synthesized from a mixture of 30% hydrogen peroxide (Tianjin Damao chemical reagent factory, Tainjin, China) and formic acid (Tianjin Fengboat Chemical reagent Technology Co., Ltd. Tainjin, China) at a mole-to-mole ratio of 1:1, 1% sulfuric acid (Tianjin Fengboat Chemical reagent Technology Co., Ltd. Tainjin, China) was added as a catalyst) for 10 h at a constant temperature of 50 °C, as shown in Figure 5a. Subsequently, the strips were soaked in a 0.5% NaOH (Tianjin Damao chemical reagent factory, Tainjin, China) solution for 10 min and washed with distilled water. In this way, the softened bamboo strip samples can be obtained. Since the treatment was applied to the whole bamboo strip but not thin slices, it can ensure the location of vascular bundles is fixed in the strip. Under a stereomicroscope, parenchymatous ground tissue surrounding the vascular bundles was carefully eliminated with a sharp blade. Finally, all of the vascular bundles from the same column were picked out from the bamboo strip (Figure 5b).

Two strengthening plates, each 0.3 mm thick and 10 mm long, were bonded at both ends of the specimens to protect the specimens from clamping damage (Figure 5c). The gauge length of the specimen was set at 40 mm. A single vascular bundle tensile test was performed longitudinally using a universal tensile test machine (INSTRON5544, Instron, MA, USA) with a 2000 N sensor at a 2 mm min^{-1} displacement rate (Figure 5d). Once the vascular bundle was broken, the fracture morphology was observed by a scanning electron microscope (SEM, JSM-7500F, JEOL, TYO, Japan).

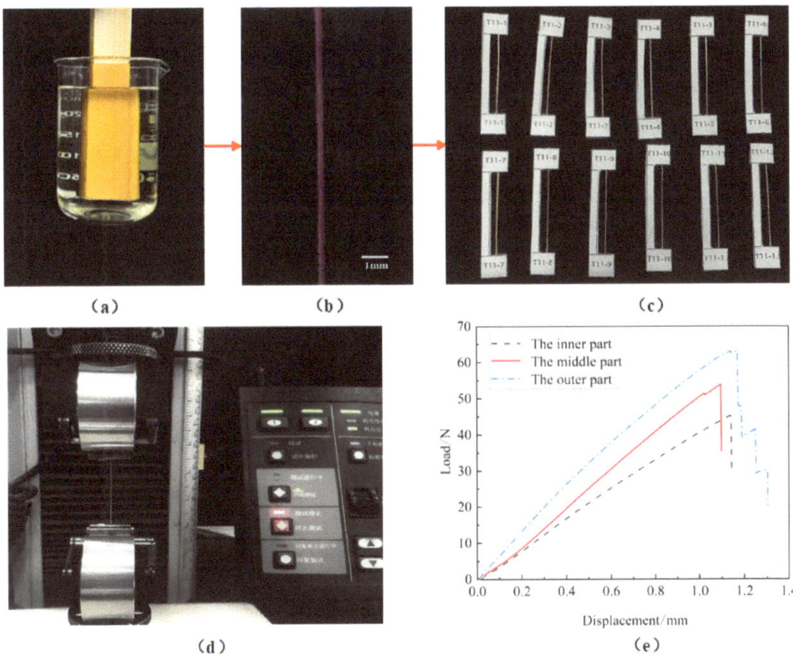

Figure 5. Separation process and tensile test of vascular bundles: (**a**) bamboo strips delignification, (**b**) the isolated vascular bundle, (**c**) preparation of vascular bundle specimen, (**d**) tensile test of single vascular bundle, and (**e**) load–displacement curves of vascular bundle specimen in tension.

During the test, the computer collected data through a force sensor, and the material testing software Bluehill 2.17 plotted the load–displacement curve of the specimen (Figure 5e). Parameters such as Young's modulus, tensile strength, and maximum tensile force of the vascular bundle were obtained from the test results. From the load–displacement curve, the linear segment within the initial elastic deformation segment was selected to calculate Young's modulus of the vascular bundle, and Young's modulus was calculated according to Equation (2). The tensile strength of the vascular bundles was measured as Equation (3):

$$E = \frac{Fl}{\Delta l A} \tag{2}$$

where E is Young's modulus, GPa; F is the tensile force, N; l is the specimen scale distance, mm; Δl is the specimen deformation, mm; and A is the cross-section area of the vascular bundle, mm^2.

$$\sigma_b = \frac{F_{max}}{A} \tag{3}$$

where σ_b is the tensile strength, MPa, and F_{max} is the maximum tensile force, N.

2.4. Data Processing Methods

Statistical analysis software SAS 9.2 was utilized to perform an analysis of variance (ANOVA) on the experimental data. Before performing test F, the normal distribution and homogeneity of variances tests were performed using the graphical method and Levene's test, respectively. The Duncan multiple range test is one kind of the popular multiple comparison procedures in SAS. So, it was applied for mean comparisons to analyze the differences in cross-sectional area and mechanical properties of vascular bundles at various radial positions.

3. Results

3.1. Vascular Bundles Position Distribution within the Bamboo Culm

The distribution pattern of many plants, including succulent leaves, pine cone fruit, and sunflower seeds, follows a golden spiral curve [29]. The distribution of vascular bundles in bamboo culm also exhibits a similar pattern (Figure 3). With the positioning method and coordinate system in Section 2.2 the spatial distribution of vascular bundles within the bamboo culm is shown in Figure 6. The fitted curve along the radial direction can be expressed in exponential form (Equation (4)), and the equations are obtained for the base part, middle part, and top part.

$$\bar{r} = a + be^{cx} \tag{4}$$

where the coefficients in the three height locations are the following:

The base part: $a = 0.92, \quad b = -0.84, \quad c = -0.53$

The middle part: $a = 1.08, \quad b = -1.06, \quad c = -0.38$

The top part: $a = 1.04, \quad b = -0.98, \quad c = -0.56$

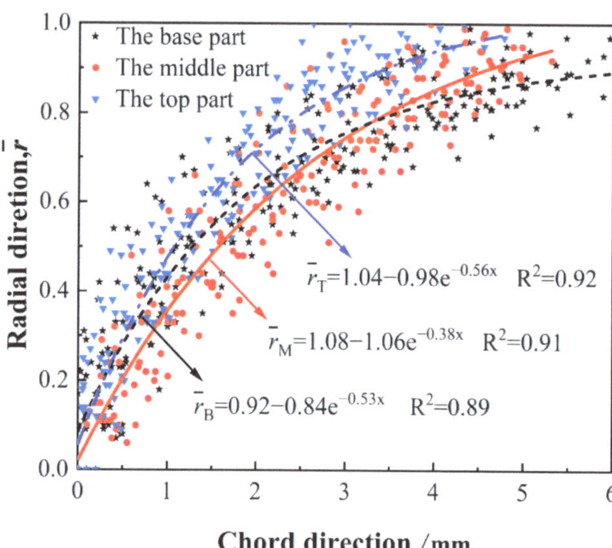

Figure 6. Locate distribution of vascular bundles in bamboo culm.

3.2. The Radial Variation of the Cross-Section Area of Vascular Bundles

The cross-sectional area of the vascular bundles at different positions is shown in Table 1. It can be seen that the range of cross-sectional areas of the vascular bundles within the bamboo culm varies from 0.061 to 0.194 mm². The results indicate that the variation in the cross-sectional area of the vascular bundles within the bamboo culm wall is highly significant along the radial direction ($p < 0.01$) but not with height ($p > 0.05$), as shown in Table 2.

Table 1. Test results of vascular bundles.

No.	Cross-Section Area/mm^2			Tensile Strength/MPa			Young's Modulus/GPa		
	Base	Middle	Top	Base	Middle	Top	Base	Middle	Top
1	0.178 ± 0.02 AB	0.171 ± 0.02 BC	0.185 ± 0.02 AB	180.44 ± 41.20 A	215.16 ± 61.07 A	270.78 ± 46.72 A	9.00 ± 1.92 A	11.25 ± 2.18 AB	13.16 ± 2.48 A
2	0.186 ± 0.01 A	0.194 ± 0.01 A	0.191 ± 0.01 A	185.34 ± 58.23 A	226.80 ± 67.30 A	276.56 ± 56.71 A	9.77 ± 2.10 A	10.32 ± 1.46 A	13.29 ± 1.27 A
3	0.185 ± 0.01 A	0.177 ± 0.01 B	0.187 ± 0.01 AB	190.43 ± 58.36 A	226.06 ± 55.71 A	284.02 ± 49.05 A	9.62 ± 1.63 A	10.86 ± 1.56 AB	14.07 ± 1.22 AB
4	0.184 ± 0.01 AB	0.176 ± 0.01 BC	0.179 ± 0.01 B	203.43 ± 65.12 AB	258.78 ± 66.78 AB	283.33 ± 43.63 A	10.64 ± 1.91 AB	11.89 ± 2.04 AB	14.37 ± 1.39 AB
5	0.181 ± 0.01 AB	0.177 ± 0.02 B	0.179 ± 0.01 B	224.44 ± 66.18 ABC	265.47 ± 55.97 AB	286.13 ± 55.59 A	11.2 ± 2.22 AB	12.46 ± 1.92 AB	14.77 ± 1.63 AB
6	0.179 ± 0.01 AB	0.171 ± 0.01 BC	0.167 ± 0.02 C	249.32 ± 89.82 ABC	335.73 ± 87.54 BC	315.96 ± 80.86 A	11.81 ± 2.70 AB	13.86 ± 2.50 AB	17.78 ± 2.41 BC
7	0.175 ± 0.02 AB	0.165 ± 0.01 C	0.169 ± 0.02 C	276.41 ± 83.75 CD	333.33 ± 60.01 BC	337.06 ± 45.70 AB	13.88 ± 3.50 BC	15.06 ± 2.88 BC	19.11 ± 1.57 CD
8	0.173 ± 0.01 B	0.153 ± 0.01 D	0.154 ± 0.02 D	289.47 ± 103.84 DE	391.66 ± 66.23 CD	407.95 ± 110.73 BC	14.93 ± 3.50 C	18.15 ± 3.78 CD	22.01 ± 3.17 D
9	0.161 ± 0.01 C	0.147 ± 0.01 D	0.135 ± 0.01 E	301.24 ± 69.44 DE	427.33 ± 93.83 DE	467.57 ± 128.37 CD	15.27 ± 4.00 C	20.61 ± 3.87 DE	25.97 ± 4.89 E
10	0.144 ± 0.02 D	0.131 ± 0.01 E	0.121 ± 0.01 F	357.55 ± 89.37 EF	485.88 ± 80.21 E	518.79 ± 169.93 D	20.15 ± 2.87 D	23.76 ± 2.71 E	29.76 ± 5.99 E
11	0.129 ± 0.02 E	0.114 ± 0.01 F	0.104 ± 0.02 G	395.41 ± 103.69 FG	581.84 ± 128.50 F	615.31 ± 182.61 E	21.72 ± 4.43 DE	28.64 ± 5.49 F	35.42 ± 9.68 F
12	0.119 ± 0.01 E	0.096 ± 0.02 G	0.074 ± 0.02 H	438.15 ± 94.46 G	625.88 ± 192.34 F	721.38 ± 215.75 F	24.78 ± 6.59 EF	33.32 ± 8.34 G	44.76 ± 13.68 G
13	0.105 ± 0.01 F	0.061 ± 0.01 H	-	457.56 ± 85.60 G	774.10 ± 246.60 G	-	27.18 ± 4.02 F	44.37 ± 14.25 H	-
14	0.095 ± 0.01 G	-	-	557.71 ± 203.73 H	-	-	32.76 ± 9.15 G	-	-
15	0.072 ± 0.02 H	-	-	602.86 ± 215.62 H	-	-	35.72 ± 11.14 G	-	-

Note: $p < 0.01$ means highly significant, $0.01 \leq p < 0.05$ means significant, and $p > 0.05$ means not significant. Different capital letters A, B, C, D, E, F, G, and H indicate significant differences for radial position at 0.05 level according to Duncan's multiple range test.

Table 2. Analysis of variation of vascular bundle cross-sectional area and tensile test results.

Source of Variation	*p*-Values		
	Cross-Section Area	Tensile Strength	Young's Modulus
Radial	<0.001	<0.001	<0.001
Height	0.4267	<0.001	<0.001

Note: $p < 0.01$ means highly significant, $0.01 \leq p < 0.05$ means significant, and $p > 0.05$ means not significant according to test F.

Figure 7 shows the radial variation of the cross-sectional area of a single vascular bundle along the column. The cross-sectional area of single vascular bundles decreases slightly from the inner side to the middle part, with the maximum area near the inner zone decreasing rapidly from the middle part to the outer zone. Overall, the radial variation pattern of the vascular bundle cross-section area can be expressed as Equation (5):

$$A = 0.19 - 8.17e^{5.27\bar{r}} \tag{5}$$

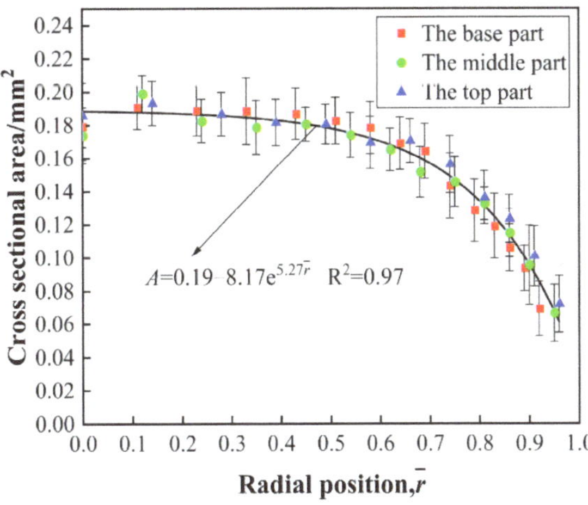

Figure 7. The radial variation of the cross-section area of vascular bundles.

Vascular bundle cross-sectional area variation with radial is strongly related to morphology. There are two types of morphology for vascular bundles in Moso bamboo: semi-open type and open type [25]. The vascular bundles near the inner side of the bamboo culm show an open-type morphology, as shown in Figure 3c. They consist of four independent fiber sheaths, fully differentiated xylem and phloem fractions, which made them nearly the same size. From the middle to the periphery, the vascular bundles show a semi-open-type morphology, with two fiber sheaths gradually diminishing, as shown in Figure 3b. The size of differentiated vessels and sieve tubes also decreases radially, resulting in a rapid reduction in the cross-section area of the vascular bundles.

3.3. Gradient Variation of Vascular Bundle Tensile Properties

The mechanical properties of air-dried Moso bamboo vascular bundles were obtained with the tensile tests, and the test results are shown in Table 1. It shows that the tensile strength of vascular bundles in the base internode, middle internode, and top internode ranged from 180.64 to 602.86 MPa, 215.16 to 774.10 MPa, and 270.78 to 721.38 MPa, respectively. Young's modulus of vascular bundles in the base internode, middle internode, and top internode ranged from 9 to 35.72 GPa, 11.25 to 44.37 GPa, and 13.16 to 44.76 GPa respectively. Comparing the test results, it can be found that the tensile strength of vascular bundles at the outer side is three times higher than that of the inner side, while Young's modulus at the outer side is nearly three to four times higher than that of the inner side.

3.3.1. Gradient Variation of Tensile Strength

The tensile strength of vascular bundles varies significantly with radial and heights ($p < 0.01$), as shown in Table 2. The change rule of vascular bundle tensile strength with radial at different internodes is shown in Figure 8. It can be seen that the tensile strength of vascular bundles of all three height internodes follow similar trends along the radial direction: the tensile strength of vascular bundles near the inner side of bamboo culm changes slowly, while those from the middle to the outer side increases rapidly and reached to the maximum at the outer periphery. The gradient variation of tensile strength of all three heights can be expressed exponentially as Equation (6):

$$\sigma_b(\bar{r}) = \sigma_{b0} + Be^{k_1\bar{r}} \tag{6}$$

where the coefficients in the three height locations are the following:

The base part: $\sigma_{b0} = 183.24, \quad B = 4.19, \quad k_1 = 4.99$

The middle part: $\sigma_{b0} = 215.27, \quad B = 7.63, \quad k_1 = 4.49$

The top part: $\sigma_{b0} = 267.47, \quad B = 1.82, \quad k_1 = 5.75$

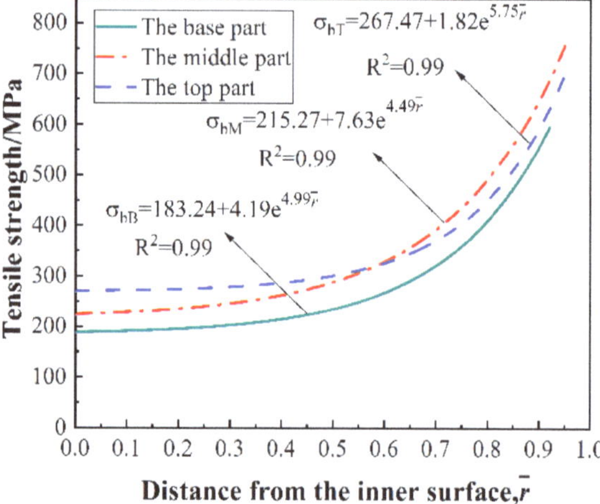

Figure 8. The radial variation of tensile strength of vascular bundles.

A comparison of the tensile strength of vascular bundles at different height positions reveals that both the top and middle parts are stronger than the culm base. In the top and middle of the culm investigated along the culm wall, although the strength in the inner wall of the middle part is smaller than that of the top part, the strength of the middle part increases much more than the top part which made the outer wall of the middle part stronger than that of the top part. With the fitted curves, it can be found that the strength of vascular bundles in the middle and top internodes meet equally at $\bar{r} = 0.58$. Li [7] obtained a similar result to our work; the strength of vascular bundles in the middle and top of the Moso bamboo culm reached an equal level at $\bar{r} = 0.62$. A very interesting observation from these test results is that the ratio of the whole culm wall thickness to the distances from the inner side to the strength equal point are all approximately near the golden ratio (1.618), and this phenomenon has been widely observed in plant structure [30,31].

3.3.2. Gradient Variation of Young's Modulus

The Young's modulus of vascular bundles varied significantly in different radial and height positions within the bamboo culm ($p < 0.01$), as shown in Table 2. Figure 9 shows the variation of Young's modulus of vascular bundles with radial in different culm heights.

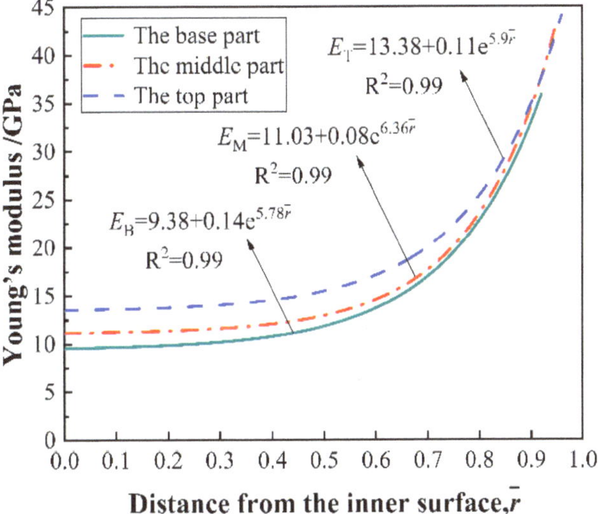

Figure 9. The radial variation of Young's modulus of vascular bundles.

Similar to the feature of tensile strength, the vascular bundles near the outer periphery are stiffer than those near the inner side. The Young's modulus of vascular bundles increases gradually from the inner side to the outer side at different culm heights. The general trend of vascular bundle stiffness as a function of radius is the same for all three positions, which follows an exponential way as in Equation (7):

$$E(\bar{r}) = E_0 + Ce^{k_2\bar{r}} \tag{7}$$

where the coefficients in the three height locations are the following:

The base part: $E_0 = 9.38$, $C = 0.14$, $k_2 = 5.78$

The middle part: $E_0 = 11.03$, $C = 0.08$, $k_2 = 6.36$

$$\text{The top part: } E_0 = 13.38,\ C = 0.11,\ k_2 = 5.9$$

For the vascular bundles at different height locations, it can be seen from Figure 9 that Young's modulus is increased from the base to the top of the culm. Along the thickness direction, due to the different increase rate k_2 in Equation (7), the trends of Young's modulus of vascular bundles at different height culm are different. Finally, Young's modulus approach has nearly the same value at the periphery ($\bar{r} = 1$): 54.71 GPa for the base part, 57.29 GPa for the middle part, and 53.53 GPa for the top part.

3.4. Fracture Mode

The tensile load–displacement curves of bamboo vascular bundles at different wall positions are shown in Figure 5e. Although all of the stress–strain behavior under loading shows approximately linear elastic deformation before reaching the fracture stress, they are broken with different fracture properties. Owing to the variation of shape, size, and concentration of vascular bundles across the internode wall, the vascular bundles are broken into different fracture modes, namely the brittle type and broom-like type.

Figure 10 shows photographs of the vascular bundle samples after tensile testing. From this, it can be found that there are different broken properties in the inner (Figure 10a–c) and outer (Figure 10d–f) culm walls.

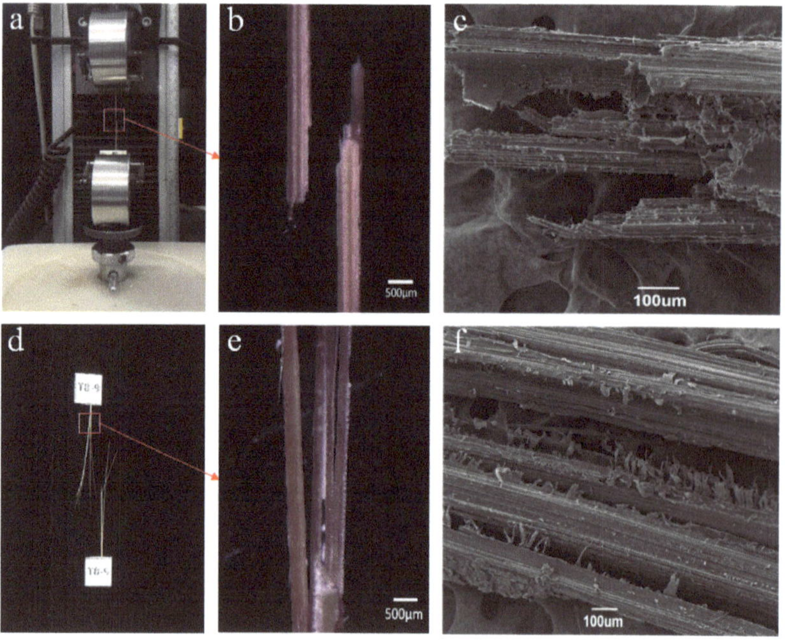

Figure 10. The fracture modes of vascular bundles: (**a–c**) fracture modes of inner vascular bundles and (**d–f**) fracture modes of outer vascular bundles.

Vascular bundles can be viewed as composite materials made up of sclerenchyma fiber and multi-porous matrix phases. The vascular bundles near the inner side belong to the open-type morphology with a supporting tissue of four sclerenchyma fibers on the sides. With the reason that the area of central multi-porous matrix phases is strikingly larger, but the strength is much weaker than the sclerenchyma sheaths, which made the vascular bundles broken with a brittle type (Figure 10b,c).

Different from the inner vascular bundles, vascular bundles from the middle to the outer side of the bamboo culm show a semi-open-type morphology with the lateral and inner sclerenchyma fiber linked together. The area of multi-porous matrix phases is extremely small and trends to zero on the outer periphery. During the tensile process, the stress–displacement curve shows a linear segment until it reaches the fracture load. The cracks first appear in the weak areas between parenchyma cells and fiber bundles, causing the fiber bundles to gradually separate (Figure 10d–f), with the separated fiber bundles having a smaller area and continuing to sustain the load. Under larger loads, the separated fiber bundles begin to fracture gradually according to the size of their area until the vascular bundle specimen fails as a whole [32]. The fracture shows a staggered decrease in load and made the vascular bundles finally broken with a broom-like type (Figure 10d).

4. Discussion

To investigate the mechanical properties of single vascular bundles in bamboo culm walls, the splitting method was often used to divide the culm wall into several thin slices; then, the single vascular bundles were extracted from thin slices by mechanical or chemical methods. This splitting method has disadvantages and advantages. The advantage is it is easy to extract some contact single vascular bundles. The disadvantage is that it can only obtain some vascular bundles but not all vascular bundles in the culm wall.

Because the extracted vascular bundles from thin slices were measured by their rough but not precise location in the culm wall, it is difficult to find the exact gradient variation of the mechanical properties of these continuously distributed vascular bundles along the bamboo culm wall. Li and Shen [17] divided Moso bamboo strips into six slices to extract single vascular bundles and found that both the vascular bundles' strength and Young's modulus were linearly increased from the inner side to the outer side. Shang et al. [25] divided the vascular bundles into 14 layers to extract single vascular bundles and found that the strength and Young's modulus of the vascular bundles all gradually increased from the inner side to the outer side. From this, we can find that the more layers the strips were peeled, the more exact gradient variation of the mechanical properties of vascular bundles can be obtained.

In this paper, whole long and undamaged vascular bundles were successfully extracted from bamboo culm walls by the chemical delignification methods. This method ensured that the position of the vascular bundles remained unchanged during the extraction process. The test results obtained in this paper are more accurate than previous research in the gradient variation of the mechanical properties of vascular bundles along the culm wall.

The results of Figure 7 indicate that the vascular bundle cross-sectional area exhibited significant variation with radial but not with height. Along the height direction, although the area of vascular bundles was somewhat larger in the base internodes according to absolute position (Table 1), the variation of the area was not obvious according to the relative position, $\bar{r}$. From the analysis results shown in Table 2, it can be found that $P = 0.4267$, which shows that the cross-sectional area of the vascular bundles was not statistically significant along the height direction. Many researchers have come to the same results [25,33].

Bamboo is a fast-growing, cheap, and green resource with superior mechanical and physical properties; it offers great potential as an alternative to wood, plastic, and other materials for structural and product applications. There are approximately 1500 species of bamboo worldwide. In this paper, the Moso bamboo was chosen as a typical representative species to investigate the exact gradient variation of the mechanical properties of the continuously distributed vascular bundles. This work can supply references for other typical bamboo species with different vascular bundle types. Furthermore, a better understanding of the optimal design in natural bamboo will be helpful for designing functionally graded materials with excellent mechanical properties.

5. Conclusions

(1) The radial distribution of vascular bundles in the bamboo culm follows an exponential pattern at different heights.

(2) The cross-section area of vascular bundles in the bamboo culm is larger on the inside and smaller on the outside and decreases exponentially from inside to outside along the radial direction.

(3) Tests showed that the tensile strength of Moso bamboo vascular bundles varied from 180.44 to 774.10 MPa, and Young's modulus varied from 9.00 to 44.76 GPa. Both the vascular bundles' strength and Young's modulus are increased with the exponential gradient variation from the inside to the outer periphery along the culm wall.

(4) The single vascular bundles from different bamboo culm wall positions are broken with different fracture modes. The vascular bundles near the inner side are broken in a brittle manner, while the vascular bundles near the outer side are broken in a broom-like manner.

(5) This work can supply references for other typical bamboo species with different vascular bundle types.

Author Contributions: Conceptualization and supervision, H.L. and Q.Z.; methodology, H.L.; investigation, X.C.; resources, P.L.; formal analysis, Y.X.; writing—original draft preparation, Q.Z; writing—review and editing, H.L.; project administration, H.L.; funding acquisition, H.L. All authors have read and agreed to the published version of the manuscript.

Funding: This research was funded by a Research Project Supported by the Shanxi Scholarship Council of China, grant number 2021-075, and supported by the Fundamental Research Program of Shanxi Province, grant number 202303021221090.

Data Availability Statement: The original contributions presented in the study are included in the article, further inquiries can be directed to the corresponding author.

Conflicts of Interest: The authors declare no conflicts of interest.

References

1. Xu, D.; He, S.; Leng, W.; Chen, Y.; Wu, Z. Replacing Plastic with Bamboo: A Review of the Properties and Green Applications of Bamboo-Fiber-Reinforced Polymer Composites. *Polymers* **2023**, *15*, 4276. [CrossRef] [PubMed]
2. Zhao, X.; Ye, H.; Chen, F.; Wang, G. Bamboo as a Substitute for Plastic: Research on the Application Performance and Influencing Mechanism of Bamboo Buttons. *J. Clean. Prod.* **2024**, *446*, 141297. [CrossRef]
3. Low, I.M.; Che, Z.Y.; Latella, B.A. Mapping the Structure, Composition and Mechanical Properties of Bamboo. *J. Mater. Res.* **2006**, *21*, 1969–1976. [CrossRef]
4. Osorio, L.; Trujillo, E.; Van Vuure, A.W.; Verpoest, I. Morphological Aspects and Mechanical Properties of Single Bamboo Fibers and Flexural Characterization of Bamboo/Epoxy Composites. *J. Reinf. Plast. Compos.* **2011**, *30*, 396–408. [CrossRef]
5. Liu, C.; Liu, M.; Liu, W.; Li, Z.; Xu, F. Interlaminar Fracture Property of Moso Bamboo Strips Influenced by Fiber Distributions in Bamboo Internode and Node. *Compos. Struct.* **2022**, *294*, 115777. [CrossRef]
6. Wegst, U.G.K.; Bai, H.; Saiz, E.; Tomsia, A.P.; Ritchie, R.O. Bioinspired Structural Materials. *Nat. Mater.* **2015**, *14*, 23–36. [CrossRef] [PubMed]
7. Li, H.B. Trans-Scale Analysis of the Mechanical Properties of Moso Bamboo. Ph.D. Thesis, Xi'an Jiaotong University, Xian, China, 2012.
8. Wang, D.; Lin, L.; Fu, F. Fracture Mechanisms of Moso Bamboo (*Phyllostachys pubescens*) under Longitudinal Tensile Loading. *Ind. Crops Prod.* **2020**, *153*, 112574. [CrossRef]
9. Wang, X.; Chen, X.; Huang, B.; Chen, L.; Fang, C.-H.; Ma, X.; Fei, B. Gradient Changes in Fiber Bundle Content and Mechanical Properties Lead to Asymmetric Bending of Bamboo. *Constr. Build. Mater.* **2023**, *395*, 132328. [CrossRef]
10. Gao, X. Structural and Mechanical Properties of Bamboo Fiber Bundle and Fiber/Bundle Reinforced Composites: A Review. *J. Mater. Res. Technol.* **2022**, *19*, 1162–1190. [CrossRef]
11. Han, S.; Xu, H.; Chen, F.; Wang, G. Construction Relationship between a Functionally Graded Structure of Bamboo and Its Strength and Toughness: Underlying Mechanisms. *Constr. Build. Mater.* **2023**, *379*, 131241. [CrossRef]
12. Sun, H.; Li, H.; Dauletbek, A.; Lorenzo, R.; Corbi, I.; Corbi, O.; Ashraf, M. Review on Materials and Structures Inspired by Bamboo. *Constr. Build. Mater.* **2022**, *325*, 126656. [CrossRef]
13. Ghavami, K.; Rodrigues, C.S.; Paciornik, S. Bamboo: Functionally Graded Composite Material. *Asian J. Civ. Eng. Build. Hous.* **2003**, *4*, 1–10.
14. Xu, H.; Zhang, Y.; Wang, J.; Li, J.; Zhong, T.; Ma, X.; Wang, H. A Universal Transfer-Learning-Based Detection Model for Characterizing Vascular Bundles in *Phyllostachys*. *Ind. Crops Prod.* **2022**, *180*, 114705. [CrossRef]
15. Amada, S.; Ichikawa, Y.; Munekata, T.; Nagase, Y.; Shimizu, H. Fiber Texture and Mechanical Graded Structure of Bamboo. *Compos. Part B Eng.* **1997**, *28*, 13–20. [CrossRef]

5. Amada, S.; Untao, S. Fracture Properties of Bamboo. *Compos. Part B Eng.* **2001**, *32*, 451–459. [CrossRef]
7. Li, H.; Shen, S. The Mechanical Properties of Bamboo and Vascular Bundles. *J. Mater. Res.* **2011**, *26*, 2749–2756. [CrossRef]
8. Nogata, F.; Takahashi, H. Intelligent Functionally Graded Material: Bamboo. *Compos. Eng.* **1995**, *5*, 743–751. [CrossRef]
9. Shao, Z.-P.; Fang, C.-H.; Huang, S.-X.; Tian, G.-L. Tensile Properties of Moso Bamboo (*Phyllostachys pubescens*) and Its Components with Respect to Its Fiber-Reinforced Composite Structure. *Wood Sci. Technol.* **2010**, *44*, 655–666. [CrossRef]
10. Chen, K.; Long, L.C. Analysis of the Effects of Fiber Gradient Distribution on the Mechanical Properties of Moso Bamboo. *Appl. Mech. Mater.* **2014**, *590*, 13–18. [CrossRef]
11. Zhang, X.; Li, J.; Yu, Z.; Yu, Y.; Wang, H. Compressive Failure Mechanism and Buckling Analysis of the Graded Hierarchical Bamboo Structure. *J. Mater. Sci.* **2017**, *52*, 6999–7007. [CrossRef]
12. Dixon, P.G.; Gibson, L.J. The Structure and Mechanics of Moso Bamboo Material. *J. R. Soc. Interface* **2014**, *11*, 20140321. [CrossRef] [PubMed]
13. Xu, H.; Li, J.; Ma, X.; Yi, W.; Wang, H. Intelligent Analysis Technology of Bamboo Structure. Part II: The Variability of Radial Distribution of Fiber Volume Fraction. *Ind. Crops Prod.* **2021**, *174*, 114164. [CrossRef]
14. Wang, F.; Shao, Z. Study on the Variation Law of Bamboo Fibers' Tensile Properties and the Organization Structure on the Radial Direction of Bamboo Stem. *Ind. Crops Prod.* **2020**, *152*, 112521. [CrossRef]
15. Shang, L.; Liu, X.; Jiang, Z.; Tian, G.; Yang, S. Variation in Tensile Properties of Single Vascular Bundles in Moso Bamboo. *For. Prod. J.* **2021**, *71*, 246–251. [CrossRef]
16. Li, Z.; Chen, C.; Xie, H.; Yao, Y.; Zhang, X.; Brozena, A.; Li, J.; Ding, Y.; Zhao, X.; Hong, M.; et al. Sustainable High-Strength Macrofibres Extracted from Natural Bamboo. *Nat. Sustain.* **2022**, *5*, 235–244. [CrossRef]
17. Fu, J.; Zhang, X.; Yu, C.; Guebite, G.M.; Cavaco-Paulo, A. Bioprocessing of Bamboo Materials. *Fibres Text. East. Eur.* **2012**, *20*, 13–19.
18. Li, J.; Lian, C.; Wu, J.; Zhong, T.; Zou, Y.; Chen, H. Morphology, Chemical Composition and Thermal Stability of Bamboo Parenchyma Cells and Fibers Isolated by Different Methods. *Cellulose* **2023**, *30*, 2007–2021. [CrossRef]
19. Minarova, N. The Fibonacci Sequence: Nature's Little Secret. *CRIS-Bull. Cent. Res. Interdiscip. Study* **2014**, *2014*, 7–17. [CrossRef]
20. Zeng, L.; Wang, G. Modeling Golden Section in Plants. *Prog. Nat. Sci.* **2009**, *19*, 255–260. [CrossRef]
21. Sun, Z.; Cui, T.; Zhu, Y.; Zhang, W.; Shi, S.; Tang, S.; Du, Z.; Liu, C.; Cui, R.; Chen, H.; et al. The Mechanical Principles behind the Golden Ratio Distribution of Veins in Plant Leaves. *Sci. Rep.* **2018**, *8*, 13859. [CrossRef]
22. Li, Z.; Chen, C.; Mi, R.; Gan, W.; Dai, J.; Jiao, M.; Xie, H.; Yao, Y.; Xiao, S.; Hu, L. A Strong, Tough, and Scalable Structural Material from Fast-Growing Bamboo. *Adv. Mater.* **2020**, *32*, 1906308. [CrossRef] [PubMed]
23. Tsuyama, T.; Hamai, K.; Kijidani, Y.; Sugiyama, J. Quantitative Morphological Transformation of Vascular Bundles in the Culm of Moso Bamboo (*Phyllostachys pubescens*). *PLoS ONE* **2023**, *18*, e0290732. [CrossRef] [PubMed]

Article

Consolidation and Dehydration Effects of Mildly Degraded Wood from Luoyang Canal No. 1 Ancient Ship

Weiwei Yang [1,2,3], Wanrong Ma [1,2,3], Xinyou Liu [1,2,3] and Wei Wang [1,2,3,*]

1   Co-Innovation Center of Efficient Processing and Utilization of Forest Resources, Nanjing Forestry University, Nanjing 210037, China; yww2000922@njfu.edu.cn (W.Y.); mawan@njfu.edu.cn (W.M.); liu.xinyou@njfu.edu.cn (X.L.)
2   College of Furnishing and Industrial Design, Nanjing Forestry University, Str. Longpan No.159, Nanjing 210037, China
3   Advanced Analysis and Testing Center, Nanjing Forestry University, Str. Longpan No.159, Nanjing 210037, China
*   Correspondence: wangwei1219@njfu.edu.cn; Tel.: +86-25-8542-7408

Abstract: To ensure the conservation of waterlogged archaeological wood, sustainable, safe, and effective methods must be implemented, with consolidation and dehydration being crucial for long-term preservation to maintain dimensional stability and structural integrity. This study compared the permeability of 45% methyltrimethoxysilane (MTMS) and 45% trehalose solutions to evaluate the dimensional changes, hygroscopicity, and mechanical properties of treated wood. Since the collected samples (from an ancient ship, Luoyang Canal No. 1) were mildly degraded, the drying method had a slight impact on the properties of archaeological wood. Consolidated with trehalose and MTMS agents, the longitudinal compressive strength of the waterlogged wood's cell walls increased by 66.8% and 23.5%, respectively. Trehalose proved to be more advantageous in filling pores and reducing overall shrinkage, while MTMS significantly reduced the hygroscopicity and surface hydrophilicity of the wood substance. Overall, the MTMS treatment has a smaller effect on the appearance of samples, making it more suitable for the consolidation of mildly degraded waterlogged archaeological wood.

Keywords: waterlogged archaeological wood; mechanical properties; chemical properties; dimensional stability; methyltrimethoxysilane; trehalose; morphological characteristics

Citation: Yang, W.; Ma, W.; Liu, X.; Wang, W. Consolidation and Dehydration Effects of Mildly Degraded Wood from Luoyang Canal No. 1 Ancient Ship. *Forests* **2024**, *15*, 1089. https://doi.org/10.3390/f15071089

Academic Editor: Jesús Julio Camarero

Received: 5 June 2024
Revised: 15 June 2024
Accepted: 21 June 2024
Published: 23 June 2024

1. Introduction

As a renewable natural polymer, wood has been extensively used in numerous human activities throughout history owing to its superior properties, abundant availability, and ease of processing. A significant portion of archaeological wood was waterlogged or wet when excavated [1,2]. Microbial degradation weakens cell walls by decomposing cellulose and hemicellulose polymers [3–5]. Although this deterioration may not be visually apparent, wood becomes highly susceptible to irreversible shrinkage and cracking upon drying [6,7]. Consequently, consolidants that reinforce cell walls and prevent collapse by filling pores and micropores are essential for long-term preservation.

Early attempts to conserve waterlogged wood involved the use of oils, waxes, and alum ($KAl(SO_4)_2 \cdot 12H_2O$), but these failed to provide adequate reinforcement [8,9]. Currently, polyethylene glycol (PEG) is the most common consolidant, and it has been widely used in the cases of the Vasa warship [10–12], Mary Rose [13], Bremen Cog [14], and other waterlogged wooden artifacts due to its non-toxicity, cost-effectiveness, and ability to enhance dimensional stability. However, PEG-treated wood is more sensitive to heat, metal ions, salts, and microbial degradation [9,15]. The decomposition products of PEG, such as formic acid, can further chemically degrade the wood, posing a threat to its long-term preservation [16].

The conservation of waterlogged archaeological wood (WAW) necessitates a safe and sustainable approach, with the consolidation process being reversible and free from inducing subsequent degradation [17]. Non-reducing sugars, such as trehalose, have shown promise as consolidants for WAW due to their stability and antioxidant capabilities [18–21] and can markedly enhance the dimensional stability and bending performance of waterlogged archeological wood. Trehalose, with its low molecular mass, offers advantages over PEG in terms of its ability to improve mechanical properties and its conservation effectiveness [22]. Therefore, investigating the application of trehalose as a consolidant for WAW may offer a promising alternative to conventional PEG treatments.

It has been proven by many researchers that by using organosilicons, according to the type of functional group within the silane molecule, the consequent chemical modification of waterlogged wood can enhance decay resistance as well as dimensional stability [23–25]. A study by Broada et al. [26] revealed that methyltrimethoxysilane (MTMS) can effectively encrust and cover the microstructure of WAW. Recent studies have confirmed that MTMS is also satisfactory in reducing the hygroscopicity of treated wood [27,28]. Since MTMS molecules are able to penetrate and form a uniform coating on the cell wall surface, the pore area of WAW can be filled effectively.

The dehydration process also plays a crucial role in stabilizing the wood and reducing the risk of shrinkage and microbial degradation post-excavation [29]. Compared to sound wood, multiple deterioration factors weaken the structure of WAW, making the cell walls more susceptible to collapse due to capillary forces and the high surface tension of evaporating water [30]. Specialized drying should therefore take place under strictly controlled conditions. Vacuum freeze-drying and long-lasting air-drying are the most common approaches for waterlogged archaeological wood dehydration [31]. The application of these methods varies depending on the anatomical characteristics of the wood, the degree of degradation, and the size of the object to be treated. For instance, slow air-drying requires careful monitoring of the temperature and humidity to ensure gradual and uniform dehydration without causing damage to the wood. Vacuum freeze-drying is particularly suitable for delicate or highly degraded wood that may be prone to collapse during traditional drying methods [18]. This approach proved to be effective in removing moisture from wood while minimizing the risks of structural destruction, shrinkage, and cracking, but it is less applicable for large pieces [32]. Moreover, supercritical CO_2 fluid dehydration as a method of preservation has also been investigated, which utilizes CO_2's excellent solubility and heat transfer properties to dissolve moisture from WAW.

The purpose of this research was to determine the influences of trehalose and MTMS consolidation on the mechanical properties, dimensional stability, and hygroscopicity of WAW. *Ulmus* samples from the mildly degraded ancient ship of Luoyang Canal No. 1 were employed [33] using various drying methods. Statistical analyses were also involved in evaluating the effectiveness of both the consolidation and dehydration processes. The findings presented in this paper will serve as important references for future generations, enabling such historical treasures to be properly preserved.

2. Materials and Methods

2.1. Materials

Waterlogged elm (*Ulmus parvifolia* Jacq.) wrecks from the Luoyang Canal No. 1 ancient ship (Figure 1a) [33] were cut into 90 pieces measuring $30 \times 30 \times 10$ mm (longitudinal direction $\times$ tangential direction $\times$ radial direction) for experiments. The objects were classified as mildly degraded with an average maximum moisture content of $212.03 \pm 28.33\%$ and a basic density of 0.396 ± 0.015 g/cm^3 [34].

(a)   (b)

Figure 1. Samples from the ancient ship "Luoyang Canal No. 1" (**a**) the excavated ship planks and (**b**) samples submerged in consolidants. Note: The Chinese word "海藻糖" in the (**b**) means trehalose.

Methyltrimethoxysilane (MTMS) and trehalose were used as consolidation agents for waterlogged wood treatment, which were provided by Macklin Ltd., Shanghai, China.

2.2. Methods

2.2.1. Consolidation Process

The specimens were initially immersed in a 70% ethanol solution for one week and subsequently transferred to a 96% ethanol solution for another week. This process led to a decrease in the moisture content ranging from 78% to 109%, as indicated by the alteration in the ethanol concentration. Subsequently, the samples were randomly divided into three groups: the first group remained untreated as a control group, the second was submerged in a 45% solution of D-trehalose consolidation for 14 days, and the third group was immersed in a 45% solution of MTMS for the same period of time (Figure 1b).

2.2.2. Dehydration and Air Conditioning

Each group of the aforementioned samples were dehydrated under the low-temperature; high-humidity (LT-HH); vacuum-freezing (VF), and supercritical CO_2(SC) drying methods, respectively. In LT-HH air-drying, the specimens were conditioned at 45 °C with 70% relative humidity until they were absolutely dry (the mass difference between two consecutive measurements was less than 0.2%). Vacuum freeze-drying and supercritical CO_2 fluid dehydration processes were carried out following the methods established in previous studies [35–37].

To examine the hygroscopicity of treated/untreated wood, all specimens were re-conditioned at 20 °C and 65%, from which the hygroscopic equilibrium moisture content was determined.

2.2.3. Mechanical Properties and Morphological Characteristics

The destructive loading test was performed with a Shimadzu universal testing instrument, AGS-X (Shimadzu Corporation, Kyoto, Japan), following ISO 13061-17 (2017) [38]. The specimens were placed at the center of the spherical movable support of the testing machine, loaded at a uniform speed, and broken within 1.0 to 5.0 min. The mean forces of 5 samples were recorded, and compressive strength was calculated using the following equation:

$$\sigma_0 = \frac{P_{max}}{bt} \tag{1}$$

where σ_0 represents the longitudinal compressive strength of dried specimens (MPa), P_{max} is the maximum destructive load (N), and b and t represent the width and thickness of the wood specimens (mm), respectively.

To observe the morphology of wood cells, specimens were imaged in the Hitachi S-3400N II Scanning Electron Microscope (SEM) (Hitachi Ltd., Tokyo, Japan).

2.2.4. Dimensional Stability

Physical parameters, including the shrinkage/swelling rate and weight percent gain (WPG), were measured. The effectiveness of the consolidation treatment was calculated as the weight percent gain (WPG), which was determined by the differences in the dry weight of the sample before (W_0) and after impregnation (W_1), using the following formula [39]:

$$\text{WPG} = \frac{W_0 - W_1}{W_0} \times 100\% \tag{2}$$

Linear wood shrinkage/swelling in tangential, radial, and longitudinal directions was calculated using the following equation [16]:

$$\beta = \frac{l_0 - l_1}{l_0} \times 100\% \tag{3}$$

where β represents the linear wood shrinkage/swelling (%), and l_0 and l_1 represent the initial and final lengths of waterlogged/dried wood samples (cm), respectively.

The volumetric shrinkage/swelling rate can be calculated as follows [40]:

$$\gamma = \frac{V_0 - V_1}{V_0} \times 100\% \tag{4}$$

where γ is the volumetric shrinkage/swelling (%), and V_0 and V_1 are the initial and final volumes of waterlogged/dried wood samples (cm^3), respectively.

The results of these physical characteristics were then examined by SAS (version 9.4, SAS Institute, Cary, NC, USA) to investigate their statistical significance.

2.2.5. Surface Hydrophilicity

The hydrophilicity of WAW specimens before and after the treatments was evaluated using a Drop Shape Analyzer (Model: DSA100S, Kruss, Germany). Distilled water droplets were used to record the state of the water droplet on the tangential section of the dehydrated wood surfaces, and the contact angle (θ) was then calculated through a water droplet image analysis.

2.2.6. Chemical Properties Analysis Using FT-IR

To investigate the chemical changes in WAW after reinforcement, samples from the control group and the consolidated group were ground, sieved through an 80-mesh sieve, and prepared as KBr pellets. The prepared samples were subjected to FT-IR analysis using a standard FTIR spectrometer (Tensor 27, Bruker, Germany), with 32 scans conducted at a resolution of 4 cm^{-1} over the 700 to 4000 cm^{-1} wavenumber range.

3. Results and Discussion

3.1. Mechanical Properties and Morphological Characteristics

The compressive strength (MPa) results of different wood samples vary significantly among the groups (Figure 2). The untreated archaeological wood (control) exhibited a compressive strength of 41.70 MPa, while the recent sound wood displayed a higher strength of 60.46 MPa. The wood treated with trehalose (Tre) demonstrated the highest compressive strength at 69.54 MPa, indicating a significant enhancement in strength due to the trehalose treatment. In contrast, the wood treated with Trimethoxymethylsilane (MTMS) showed a lower compressive strength of 51.49 MPa compared to the trehalose-treated and healthy wood samples [41].

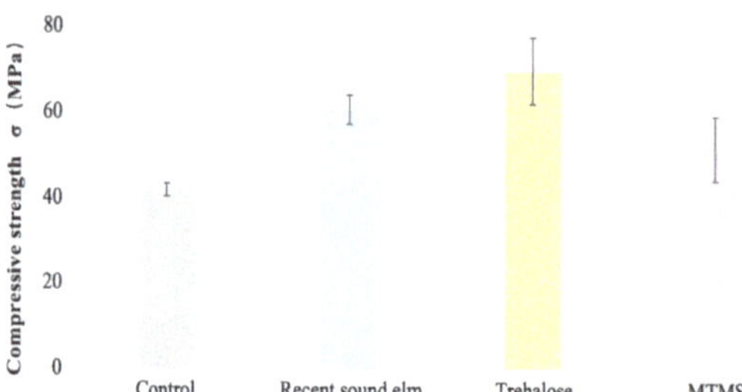

Figure 2. Compressive strength (MPa) of recent sound elm, untreated archaeological wood (control) and samples treated with trehalose and MTMS.

Figure 3 illustrates the visual effects of the MTMS and trehalose treatments on the waterlogged wood samples compared to the untreated samples. The MTMS-treated wood shows a noticeable change in color, with the treated samples appearing significantly darker and more uniform than the untreated ones. This suggests that MTMS treatment impacts the surface coloration, likely due to the formation of a siloxane network within the wood structure. In contrast, the trehalose-treated wood exhibits a lighter and more even coloration, indicating that trehalose treatment does not significantly darken the wood's surface, possibly due to its different interaction with the wood structure, forming hydrogen bonds rather than a siloxane network. Both treatments result in a more uniform surface texture compared to the untreated samples, implying improved structural consolidation. These visual differences underscore that while both MTMS and trehalose effectively consolidate and stabilize waterlogged wood, they do so in ways that differently impact the wood's surface appearance and internal structure, aligning with previous findings about their unique advantages in preserving archaeological wood.

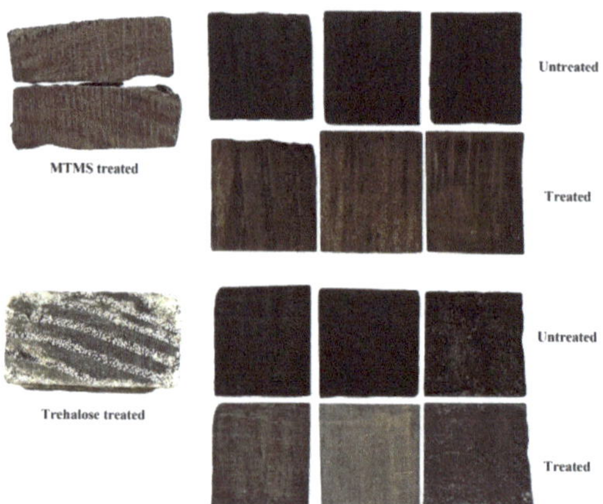

Figure 3. Cross-section and radial-section photographs of wood samples after MTMS and trehalose reinforcement.

A SEM analysis of the treated and untreated samples (Figure 4) confirmed these mechanical properties. Both the MTMS and trehalose solvents effectively penetrated the cell walls to form a thick coating on their surfaces and filled the cell lumina. Comparative observations of SEM images in the tangential section revealed that trehalose deposition filled the cell wall pits more efficiently than MTMS.

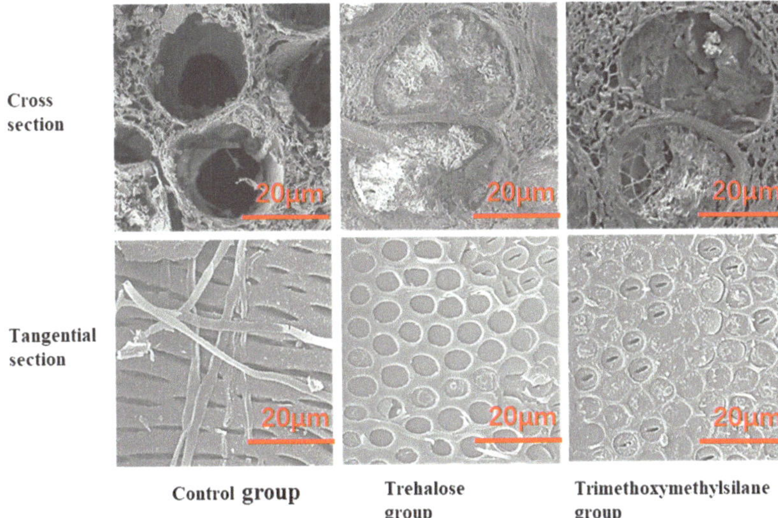

Figure 4. Scanning electron microscope (SEM) images of archaeological wood samples in the cross section and tangential section.

3.2. Evaluation of Dimensional Stability

The results of the weight percent gain (WPG) and shrinkage rate of the untreated and consolidated archaeological elm dried under various conditions are presented in Tables 1 and 2.

Table 1. Average weight percent gain (WPG) and shrinkage of waterlogged archaeological wood samples unmodified and modified with trehalose and MTMS, respectively.

Consolidation Treatment	WPG (%)	Shrinkage (%)			
		Tangential	Radial	Longitudinal	Volume
Control (untreated)	-	6.61	5.18	1.46	13.8
Trehalose	92.19	2.49	1.88	0.97	5.46
MTMS	44.63	3.65	3.82	0.68	8.33
p-value	<0.001	<0.001	<0.001	0.012	<0.001

Table 2. Average weight percent gain (WPG) and shrinkage of waterlogged archaeological wood samples that were low-temperature, high-humidity dried (LT-HH-D), vacuum freeze-dried (VF-D), and supercritical CO_2 fluid-dried (SC-D), respectively.

Dehydration Method	WPG (%)	Shrinkage (%)			
		Tangential	Radial	Longitudinal	Volume
LT-HH-D	80.97	3.08	3.22	0.72	7.24
VF-D	69.48	4.14	4.04	0.66	9.06
SC-D	54.78	3.52	2.30	1.36	7.34
p-value	0.042	0.496	0.005	0.004	0.088

It is evident that the retention of both consolidants corresponds to wood shrinkage and weight change regardless of the drying method employed. This observation agrees with the SEM images, which show that the vessels and wood fiber cells were sufficiently filled. Compared with the untreated wood, the trehalose and MTMS interventions reduced the shrinkage rate by approximately 50% or greater either in the three anatomical directions or in the total volume of specimens. Similar results were observed by Broda et al. and Kennedy et al. [42–47], where physical measurements revealed that the decreased porosity may contribute to less shrinkage and swelling in archaeological wood [27].

The reduced porosity in waterlogged wood helps preserve the structural integrity of the wood cell walls, which, in turn, reduces volume loss and directly minimizes overall dimensional changes [48–50]. This effect was particularly evident in the case of the trehalose-treated samples (Table 1), where the agent most substantially limited wood shrinkage and caused the highest WPG.

The relatively higher density of trehalose (378.3 g/mol) may contribute to the WPG value being twice as much as that observed with the MTMS treatment (136.22 g/mol). Different chemical interactions between the two consolidants and the wood also have an impact. Trehalose can form hydrogen bonds with the hydroxyl groups in the cellulose of the wood, leading to more effective penetration into the wood material. MTMS, on the other hand, primarily forms a siloxane network on the surface, which might not contribute as significantly to weight gain [42,43]. For comparison, with severely degraded wood, the WPG values could reach approximately 200% after consolidation [46–52], whereas this rate is obviously lower in Table 1, primarily because the specimens used were only mildly degraded [33].

Moreover, p-values less than 0.001 based on the SAS analysis indicate that the consolidation process has a significant effect on the WPG and shrinkage rate. It is noticeable that the drying conditions do not present strong relevance with the dimensional stability of the archaeological wood samples in this study.

3.3. Hygroscopicity of Treated Wood

To estimate the effects of the treatments with respect to moisture absorption, the archeological wood samples were conditioned at 20 °C with 65%RH until reaching EMC [53,54].

Table 3 presents the hygroscopicity of the WAW samples. Due to the decay of cellulosic fraction [55,56], the most pronounced swelling was observed in the unreinforced group, where the volumetric swelling reached up to 7.21%, while 4.99% and 2.06% swelling were measured in the tangential and radial directions, respectively.

Table 3. The average swelling rate (%), absorption equilibrium moisture content (%), and surface contact angle (°) of the treated and untreated WAW specimens.

Specimens	Swelling (%)		EMC at 65%RH (%)	Contact Angle (°)
Control	Tangential	4.99	12.68 ± 0.90	
	Radial	2.06		
	Longitudinal	0.30		θ = 36.4°
	Volume	7.21		
Trehalose (Tre)	Tangential	2.01	11.18 ± 0.57	
	Radial	1.82		
	Longitudinal	0.31		θ = 65.4°
	Volume	4.08		
Trimethoxymethylsilane (MTMS)	Tangential	2.75	8.01 ± 0.46	
	Radial	1.11		
	Longitudinal	0.28		
	Volume	4.09		θ = 101.9°

MTMS and trehalose proved to be efficient in decreasing moisture absorption. The swelling rate demonstrated in Table 3 suggests that there is no significant difference between consolidated specimens, whereas the hydrophilicity of wood is variable after air humidity conditioning. Despite the fact that the hygroscopicity of wood was permanently reduced at a higher relative humidity after the initial desorption, the trehalose treated samples presented an equilibrium moisture content of 11.18%, indicating that trehalose is less capable of reducing the hygroscopicity of consolidated wood. For comparison, this value presented a 4.6% decrease after MTMS reinforcement, which implies that the absorption behavior was diminished. The results are in line with the desorption and adsorption isotherms depicted with the GAB model, as Tahira et al. [21] and Majka et al. [57] reported in their previous studies.

The surface hydrophobicity of the untreated samples (control) and those treated with MTMS and trehalose is reflected by the contact angle values (Table 3). It is clearly revealed that the modification of the initial wetting behavior occurred on waterlogged elm surfaces after reinforcement. The mean water contact angle on the untreated sample surface is 36.4°, which is typical of hydrophilic wood (<65°). The largest contact angle θ was observed in the MTMS-treated samples, suggesting that MTMS conservation improved the hydrophobicity of waterlogged archaeological wood [58,59].

3.4. Chemical Properties

To confirm the presence of disaccharides and organosilicons inside the treated wood samples, FT-IR analyses were performed, as shown in Figure 5.

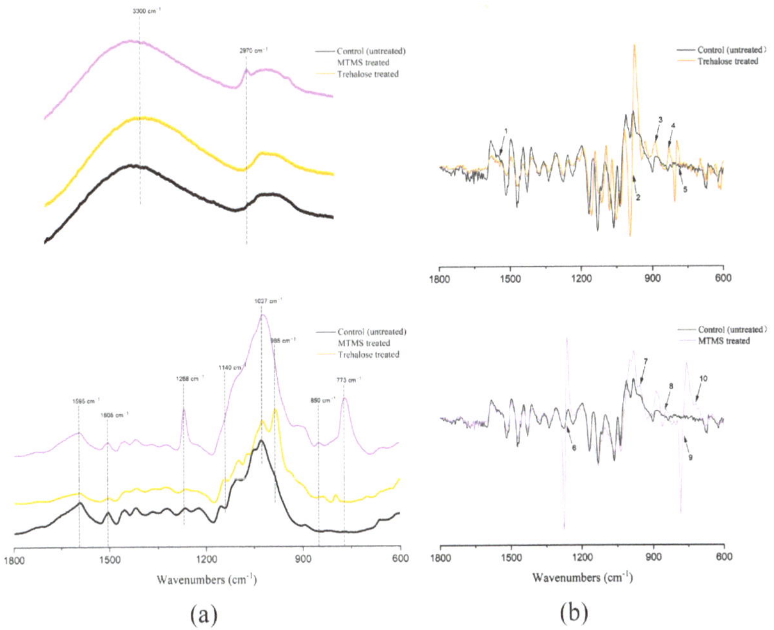

Figure 5. FT-IR spectra (**a**) and their derivatives (**b**) of untreated, trehalose-treated, and MTMS-treated WAW.

Compared to the untreated spectrum, the spectra of the samples treated with MTMS and trehalose appear to decrease in intensity during the band from 3600 to 3100 cm^{-1}, indicating that the hydroxyl groups assigned to this region are less available due to their reaction with consolidants. The reduction in the accessible hydroxyl groups was also evidenced by the aforementioned hydroscopicity experiments, revealing that the wood treatment caused reductions in the equilibrium moisture content and sorption hysteresis in comparison with the untreated wood [27,42]. The wood samples treated with MTMS show the presence of a strong absorption band at 2970 cm^{-1} assigned to the symmetric stretching vibration of the C-H bonds in CH$_3$ groups in wood and silanes.

Greater differences can be observed in the fingerprint region (Figure 5a). In order to identify the interactions between the consolidate solvent and the wood substrate, derivatives were plotted to record the slope variation in the FT-IR spectra (Figure 5b). The MTMS and trehalose spectra present lower intensities for the bands between 1595 and 1505 cm^{-1} (1), suggesting the alteration in C=C stretching assigned to lignin [60,61]. For the MTMS-treated spectrum, the distinguishable strong absorption at 1268 cm^{-1} (6) may be attributed to the Si-C stretching vibration in Si-CH$_3$. The same behavior show bands from 1140, 1027, 850 (8), and 773 (9) cm^{-1} in the silane spectrum, which demonstrate the symmetric and asymmetric stretching vibrations of Si-C, Si-O-C, Si-C-Si, and C-O, as well as the -Si-C rocking in -SiCH$_3$ [62–70].

It is thought that the main difference in mechanism between organosilicons and sugar treatment lies in their modes of interaction with the wood structure [43]. Owing to the reaction with hydroxyl groups, MTMS is able to form siloxy bands (Si-O-C) and potentially create a spatial network that binds wood polymers together, which would explain the appearance of the bands near 850 (8) and 715 (10) cm^{-1} in the derivative spectrum (Figure 5b). The analyzed spectrum of the samples treated with trehalose at 890 (3) and 835 (2) cm^{-1} clearly indicate their presence in the wood. Han et al. interpreted the band at 995 cm^{-1} as a characteristic absorption of trehalose due to unique molecular vibrations involving their structural groups [71]. The feature is also shown in Figure 5b, as arrow 3 points out.

4. Discussion

The results indicate that trehalose treatment significantly enhances the compressive strength of archaeological wood, surpassing the strengths of both untreated and MTMS-treated wood. The SEM analysis supports these findings, showing that trehalose more effectively fills cell wall pits compared to MTMS, thereby contributing to the higher compressive strength. This suggests that trehalose is a more effective consolidant for improving the mechanical properties of degraded wood. The weight percent gain and shrinkage data highlight the effectiveness of trehalose and MTMS in stabilizing archaeological wood. The substantial reduction in shrinkage rates and the significant weight gains, particularly with trehalose, indicate a strong interaction between the consolidants and the wood structure. Trehalose's higher molecular weight and its ability to form hydrogen bonds with cellulose hydroxyl groups likely contribute to its superior performance in limiting shrinkage and enhancing dimensional stability compared to MTMS. The hygroscopicity results show that both the trehalose and MTMS treatments reduce moisture absorption in archaeological wood, with MTMS being slightly more effective in reducing the equilibrium moisture content. However, the trehalose-treated samples still exhibit considerable improvements in hygroscopicity and surface hydrophobicity, as evidenced by the contact angle measurements. This indicates that while MTMS might offer better moisture resistance, trehalose also significantly enhances wood's resistance to moisture absorption. The FT-IR analysis reveals that both the trehalose and MTMS treatments result in a decrease in the availability of hydroxyl groups, supporting the observed reduction in hygroscopicity. The specific absorption bands in the MTMS-treated wood spectrum suggest the formation of a siloxane network, while the trehalose-treated wood spectrum indicates the presence of trehalose molecules. These chemical interactions highlight the different mechanisms by which trehalose and MTMS consolidate and stabilize the wood structure, with trehalose forming hydrogen bonds and MTMS creating a siloxane network. Overall, the findings demonstrate that trehalose is highly effective in improving the mechanical properties, dimensional stability, and moisture resistance of archaeological wood, making it a superior consolidant compared to MTMS for the preservation of mildly-degraded waterlogged wood.

5. Conclusions

Trehalose and MTMS treatments significantly enhanced the mechanical properties of waterlogged wood cell walls, with MTMS increasing the strength by 1.23 times and trehalose increasing the strength by 1.67 times. These treatments also improved the dimensional stability of the specimens, with MTMS achieving a 39.6% improvement and trehalose achieving an even more impressive 60.4% enhancement. MTMS proved to be a versatile stabilizing agent, particularly effective in reducing the hygroscopicity of the waterlogged *Ulmus* samples. Despite causing a slightly darker surface coloration, trehalose can also be considered satisfactory in strengthening the mechanical properties and dimensional stability of archaeological wood. The FT-IR and derivative spectra analyses provided further evidence that both treatments contribute to the consolidation of the wood structure by bulking the cell walls or filling the lumina. These chemical interactions underscore the efficacy of both trehalose and MTMS in preserving and stabilizing waterlogged archaeological wood, each offering unique advantages.

Author Contributions: Conceptualization, W.Y., X.L. and W.W.; methodology, W.Y. and X.L.; software, W.Y. and W.M.; validation, W.Y., W.W. and X.L.; formal analysis, W.Y., W.M. and X.L.; investigation, W.Y., W.W. and X.L.; data curation, W.Y., W.W. and X.L.; writing—original draft preparation, W.Y., X.L. and W.W.; writing—review and editing, W.W. and X.L.; visualization, X.L.; supervision, W.W. All authors have read and agreed to the published version of the manuscript.

Funding: This work was funded by the Nanjing Forestry University Foundation for Basic Research, Grant No. 163104127, and the Priority Academic Program Development (PAPD) of Jiangsu Province, China.

Data Availability Statement: The raw data supporting the conclusions of this article will be made available by the authors upon request.

Acknowledgments: The authors would like to thank the College of Furnishings and Industrial Design, Nanjing Forestry University for providing the experimental conditions.

Conflicts of Interest: The authors declare no conflicts of interest.

References

1. Singh, A.P.; Kim, Y.S.; Chavan, R.R. Advances in Understanding Microbial Deterioration of Buried and Waterlogged Archaeological Woods: A Review. *Forests* **2022**, *13*, 394. [CrossRef]
2. Welling, J.; Schwarz, T.; Bauch, J. Biological, Chemical and Technological Characteristics of Waterlogged Archaeological Piles (Quercus Petraea (Matt.)Liebl.) of a Medieval Bridge Foundation in Bavaria. *Eur. J. Wood Prod.* **2018**, *76*, 1173–1186. [CrossRef]
3. Weimer, P.J. Degradation of Cellulose and Hemicellulose by Ruminal Microorganisms. *Microorganisms* **2022**, *10*, 2345. [CrossRef]
4. Reese, E.T. Degradation of Polymeric Carbohydrates by Microbial Enzymes. In *The Structure, Biosynthesis, and Degradation of Wood*; Loewus, F.A., Runeckles, V.C., Eds.; Springer: Boston, MA, USA, 1977; pp. 311–367.
5. Christensen, M.; Frosch, M.; Jensen, P.; Schnell, U.; Shashoua, Y.; Nielsen, O.F. Waterlogged Archaeological Wood—Chemical Changes by Conservation and Degradation. *J. Raman Spectrosc.* **2006**, *37*, 1171–1178. [CrossRef]
6. Walsh-Korb, Z.; Avérous, L. Recent Developments in the Conservation of Materials Properties of Historical Wood. *Prog. Mater. Sci.* **2019**, *102*, 167–221. [CrossRef]
7. Blanchette, R.A. A Review of Microbial Deterioration Found in Archaeological Wood from Different Environments. *Int. Biodeterior. Biodegrad.* **2000**, *46*, 189–204. [CrossRef]
8. Braovac, S.; McQueen, C.M.A.; Sahlstedt, M.; Kutzke, H.; Lucejko, J.J.; Klokkernes, T. Navigating Conservation Strategies: Linking Material Research on Alum-Treated Wood from the Oseberg Collection to Conservation Decisions. *Herit. Sci.* **2018**, *6*, 77. [CrossRef]
9. Zoia, L.; Tamburini, D.; Orlandi, M.; Lucejko, J.J.; Salanti, A.; Tolppa, E.-L.; Modugno, F.; Colombini, M.P. Chemical Characterisation of the Whole Plant Cell Wall of Archaeological Wood: An Integrated Approach. *Anal. Bioanal. Chem.* **2017**, *409*, 4233–4245. [CrossRef]
10. Seborg, R.M.; Inverarity, R.B. Preservation of Old, Waterlogged Wood by Treatment with Polyethylene Glycol. *Science* **1962**, *136*, 649–650. [CrossRef]

11. Hocker, E.; Almkvist, G.; Sahlstedt, M. The Vasa Experience with Polyethylene Glycol: A Conservator's Perspective. *J. Cu* *Herit.* **2012**, *13*, S175–S182. [CrossRef]
12. Wagner, L.; Almkvist, G.; Bader, T.K.; Bjurhager, I.; Rautkari, L.; Gamstedt, E.K. The Influence of Chemical Degradatic and Polyethylene Glycol on Moisture-Dependent Cell Wall Properties of Archeological Wooden Objects: A Case Study the Vasa Shipwreck. *Wood Sci. Technol.* **2016**, *50*, 1103–1123. [CrossRef]
13. Fors, Y.; Sandström, M. Sulfur and Iron in Shipwrecks Cause Conservation Concerns. *Chem. Soc. Rev.* **2006**, *35*, 399–41 [CrossRef] [PubMed]
14. Hoffmann, P.; Singh, A.; Kim, Y.S.; Wi, S.G.; Kim, I.-J.; Schmitt, U. The Bremen Cog of 1380—An Electron Microscopic Stud of Its Degraded Wood before and after Stabilization with PEG. *Holzforschung* **2004**, *58*, 211–218. [CrossRef]
15. Glastrup, J.; Shashoua, Y.; Egsgaard, H.; Mortensen, M.N. Degradation of PEG in the Warship Vasa. *Macromol. Symp.* **200** *238*, 22–29. [CrossRef]
16. Broda, M.; Mazela, B.; Radka, K. Methyltrimethoxysilane as a Stabilising Agent for Archaeological Waterlogged Woo Differing in the Degree of Degradation. *J. Cult. Herit.* **2019**, *35*, 129–139. [CrossRef]
17. Bugani, S.; Cloetens, P.; Colombini, M.; Giachi, G.; Janssens, K.; Modugno, F.; Morselli, L.; Van de Casteele, E. Evaluatio of Conservation Treatments for Archaeological Waterlogged Wooden Artefacts. In Proceedings of the 9th Internation; Conference on NDT of Art, Jerusalem, Israel, 25 May 2008; pp. 25–30.
18. Jones, S.P.; Slater, N.K.; Jones, M.; Ward, K.; Smith, A.D. Investigating the Processes Necessary for Satisfactory Freez Drying of Waterlogged Archaeological Wood. *J. Archaeol. Sci.* **2009**, *36*, 2177–2183. [CrossRef]
19. Giachi, G.; Capretti, C.; Macchioni, N.; Pizzo, B.; Donato, I.D. A Methodological Approach in the Evaluation of the Efficac of Treatments for the Dimensional Stabilisation of Waterlogged Archaeological Wood. *J. Cult. Herit.* **2010**, *11*, 91–10 [CrossRef]
20. Gallina, M.E.; Sassi, P.; Paolantoni, M.; Morresi, A.; Cataliotti, R.S. Vibrational Analysis of Molecular Interactions in Aqueou Glucose Solutions. Temperature and Concentration Effects. *J. Phys. Chem. B* **2006**, *110*, 8856–8864. [CrossRef] [PubMed]
21. Tahira, A.; Howard, W.; Pennington, E.R.; Kennedy, A. Mechanical Strength Studies on Degraded Waterlogged Woo Treated with Sugars. *Stud. Conserv.* **2017**, *62*, 223–228. [CrossRef]
22. Imazu, S.; Morgos, A. Conservation of Waterlogged Wood Using Sugar Alcohol and Comparison the Effectiveness of Lac titol, Sucrose and PEG 4000 Treatment. In Proceedings of the 6th ICOM Group on Wet Organic Archaeological Material Conference, York, UK, 9–13 September 1997; pp. 235–254.
23. Mai, C.; Militz, H. Modification of Wood with Silicon Compounds. Treatment Systems Based on Organic Silicon Compounds? Review. *Wood Sci. Technol.* **2004**, *37*, 453–461. [CrossRef]
24. Donath, S.; Militz, H.; Mai, C. Creating Water-Repellent Effects on Wood by Treatment with Silanes. *Holzforschung* **2006**, *6* 40–46. [CrossRef]
25. Giudice, C.A.; Alfieri, P.V.; Canosa, G. Decay Resistance and Dimensional Stability of Araucaria Angustifolia Using Silo> anes Synthesized by Sol–Gel Process. *Int. Biodeterior. Biodegrad.* **2013**, *83*, 166–170. [CrossRef]
26. Broda, M.; Mazela, B. Application of Methyltrimethoxysilane to Increase Dimensional Stability of Waterlogged Wood. *Cult. Herit.* **2017**, *25*, 149–156. [CrossRef]
27. Broda, M.; Curling, S.F.; Spear, M.J.; Hill, C.A.S. Effect of Methyltrimethoxysilane Impregnation on the Cell Wall Porosit and Water Vapour Sorption of Archaeological Waterlogged Oak. *Wood Sci. Technol.* **2019**, *53*, 703–726. [CrossRef]
28. Broda, M.; Plaza, N.Z. Durability of Model Degraded Wood Treated with Organosilicon Compounds against Fungal Deca) *Int. Biodeterior. Biodegrad.* **2023**, *178*, 105562. [CrossRef]
29. Nguyen, T.D.; Sakakibara, K.; Imai, T.; Tsujii, Y.; Kohdzuma, Y.; Sugiyama, J. Shrinkage and swelling behavior of archaeolog ical waterlogged wood preserved with slightly crosslinked sodium polyacrylate. *J. Wood Sci.* **2018**, *64*, 294–300. [CrossRef
30. Fejfer, M.; Majka, J.; Zborowska, M. Dimensional Stability of Waterlogged Scots Pine Wood Treated with PEG and Driec Using an Alternative Approach. *Forests* **2020**, *11*, 1254. [CrossRef]
31. Liu, X.; Tu, X.; Ma, W.; Zhang, C.; Huang, H.; Varodi, A.M. Consolidation and Dehydration of Waterlogged Archaeologica Wood from Site Huaguangjiao No. 1. *Forests* **2022**, *13*, 1919. [CrossRef]
32. Blanchet, P.; Kaboorani, A.K.; Bustos, C. Understanding the Effects of Drying Methods on Wood Mechanical Properties a Ultra and Cellular Levels. *Wood Fiber Sci.* **2016**, *48*, 117–128.
33. Yang, W.; Ma, W.; Liu, X. Evaluation of Deterioration Degree of Archaeological Wood from Luoyang Canal No. 1 Ancien Ship. *Forests* **2024**, *15*, 963. [CrossRef]
34. Jong, J. Conservation Techniques for Old Archaeological Wood from Shipwrecks Found in The Netherlands. In *Biodeteriora tion Investigation Techniques*; Walters, A.H., Ed.; Applied Science: London, UK, 1977; pp. 295–338.
35. Jensen, P.; Jensen, J.B. Dynamic Model for Vacuum Freeze-Drying of Waterlogged Archaeological Wooden Artefacts. *J. Cult Herit.* **2006**, *7*, 156–165. [CrossRef]
36. Liu, H.; Xie, J.; Zhang, J. Moisture transfer and drying stress of eucalyptus wood during supercritical CO_2 (ScCO2) dewater ing and ScCO2 combined oven drying. *BioResources* **2022**, *17*, 5116–5128. [CrossRef]
37. Yang, L. Effect of Temperature and Pressure of Supercritical CO_2 on Dewatering, Shrinkage and Stresses of Eucalyptu: Wood. *Appl. Sci.* **2021**, *11*, 8730. [CrossRef]

28. ISO 13061-17:2017; Physical and Mechanical Properties of Wood—Test Methods for Small Clear Wood Specimens—Part 17: Determination of Ultimate Stress in Compression Parallel to Grain. International Organization for Standardization: Geneva, Switzerland, 2017.
29. Varivodina, I.; Kosichenko, N.; Varivodin, V.; Sedliačik, J. Interconnections Among the Rate of Growth, Porosity, and Wood Water Absorption. *Wood Res.* **2010**, *55*, 59–66.
30. ISO 13061-16:2017; Physical and Mechanical Properties of Wood—Test Methods for Small Clear Wood Specimens—Part 16: Determination of Volumetric Swelling. International Organization for Standardization: Geneva, Switzerland, 2017.
31. Broda, M.; Spear, M.J.; Curling, S.F.; Ormondroyd, G.A. The Viscoelastic Behaviour of Waterlogged Archaeological Wood Treated with Methyltrimethoxysilane. *Materials* **2021**, *14*, 5150. [CrossRef] [PubMed]
32. Broda, M.; Majka, J.; Olek, W.; Mazela, B. Dimensional Stability and Hygroscopic Properties of Waterlogged Archaeological Wood Treated with Alkoxysilanes. *Int. Biodeterior. Biodegrad.* **2018**, *133*, 34–41. [CrossRef]
33. Broda, M.; Dąbek, I.; Dutkiewicz, A.; Dutkiewicz, M.; Popescu, C.-M.; Mazela, B.; Maciejewski, H. Organosilicons of Different Molecular Size and Chemical Structure as Consolidants for Waterlogged Archaeological Wood—A New Reversible and Retreatable Method. *Sci. Rep.* **2020**, *10*, 2188. [CrossRef] [PubMed]
34. Morgós, A.; Imazu, S.; Ito, K. Sugar Conservation of Waterlogged Archaeological Finds in the Last 30 Years. In Proceedings of the 2015 Conservation and Digitalization Conference, Gdańsk, Poland, 19–22 May 2015; pp. 15–20.
35. Kennedy, A.; Pennington, E.R. Conservation of Chemically Degraded Waterlogged Wood with Sugars. *Stud. Conserv.* **2014**, *59*, 194–201. [CrossRef]
36. Nguyen, T.D.; Kohdzuma, Y.; Endo, R.; Sugiyama, J. Evaluation of Chemical Treatments on Dimensional Stabilization of Archaeological Waterlogged Hardwoods Obtained from the Thang Long Imperial Citadel Site, Vietnam. *J. Wood Sci.* **2018**, *64*, 436–443. [CrossRef]
37. Liu, L.; Zhang, L.; Zhang, B.; Hu, Y. A Comparative Study of Reinforcement Materials for Waterlogged Wood Relics in Laboratory. *J. Cult. Herit.* **2019**, *36*, 94–102. [CrossRef]
38. Eriksson, K.-E.L.; Blanchette, R.A.; Ander, P. *Microbial and Enzymatic Degradation of Wood and Wood Components*; Springer Science & Business Media: Berlin/Heidelberg, Germany, 2012.
39. Cao, H.; Gao, X.; Chen, J.; Xi, G.; Yin, Y.; Guo, J. Changes in Moisture Characteristics of Waterlogged Archaeological Wood Owing to Microbial Degradation. *Forests* **2022**, *14*, 9. [CrossRef]
40. Babiński, L.; Fabisiak, E.; Zborowska, M.; Michalska, D.; Prądzyński, W. Changes in Oak Wood Buried in Waterlogged Peat: Shrinkage as a Complementary Indicator of the Wood Degradation Rate. *Eur. J. Wood Prod.* **2019**, *77*, 691–703. [CrossRef]
41. Zhang, J.; Li, Y.; Ke, D.; Wang, C.; Pan, H.; Chen, K.; Zhang, H. Modified Lignin Nanoparticles as Potential Conservation Materials for Waterlogged Archaeological Wood. *ACS Appl. Nano Mater.* **2023**, *6*, 12351–12363. [CrossRef]
42. Endo, R.; Sugiyama, J. New Attempts with the Keratin-Metal/Magnesium Process for the Conservation of Archaeological Waterlogged Wood. *J. Cult. Herit.* **2022**, *54*, 53–58. [CrossRef]
43. Giachi, G.; Capretti, C.; Donato, I.D.; Macchioni, N.; Pizzo, B. New Trials in the Consolidation of Waterlogged Archaeological Wood with Different Acetone-Carried Products. *J. Archaeol. Sci.* **2011**, *38*, 2957–2967. [CrossRef]
44. Pecoraro, E.; Pizzo, B.; Salvini, A.; Macchioni, N. Dynamic Mechanical Analysis (DMA) at Room Temperature of Archaeological Wood Treated with Various Consolidants. *Holzforschung* **2019**, *73*, 757–772. [CrossRef]
45. Skinner, T.P.W.C. *Dimensional Stabilisation of Waterlogged Archaeological Wood: An Investigation of the Water Content of the Cell Wall of Waterlogged Archaeological Wood and Its Replacement with Water-Soluble Compounds*; University of London, University College London: London, UK, 2001.
46. Lindfors, E.-L.; Lindström, M.; Iversen, T. Polysaccharide Degradation in Waterlogged Oak Wood from the Ancient Warship Vasa. *Holzforschung* **2008**, *62*, 57–63. [CrossRef]
47. Majka, J.; Babiński, L.; Olek, W. Sorption Isotherms of Waterlogged Subfossil Scots Pine Wood Impregnated with a Lactitol and Trehalose Mixture. *Holzforschung* **2017**, *71*, 813–819. [CrossRef]
48. Zhou, Y.; Wang, K.; Hu, D. High Retreatability and Dimensional Stability of Polymer Grafted Waterlogged Archaeological Wood Achieved by ARGET ATRP. *Sci. Rep.* **2019**, *9*, 9879. [CrossRef]
49. Barkai, H.; Soumya, E.; Sadiki, M.; Mounyr, B.; Ibnsouda, K.S. Impact of Enzymatic Treatment on Wood Surface Free Energy: Contact Angle Analysis. *J. Adhes. Sci. Technol.* **2017**, *31*, 726–734. [CrossRef]
50. Broda, M.; Hill, C.A. Conservation of Waterlogged Wood—Past, Present and Future Perspectives. *Forests* **2021**, *12*, 1193. [CrossRef]
51. Colombini, M.P.; Lucejko, J.J.; Modugno, F.; Orlandi, M.; Zoia, L. A multi-analytical study of degradation of lignin in archaeological waterlogged wood. *Talanta* **2010**, *80*, 61–70. [CrossRef]
52. Popescu, M.C.; Froidevaux, J.; Navi, P.; Popescu, C.M. Structural modifications of Tilia cordata wood during heat treatment investigated by FT-IR and 2D IR correlation spectroscopy. *J. Mol. Struct.* **2013**, *1033*, 176–186. [CrossRef]
53. Al-Oweini, R.; El-Rassy, H. Synthesis and characterization by FTIR spectroscopy of silica aerogels prepared using several Si (OR) and R00 Si (OR0) 3 precursors. *J. Mol. Struct.* **2009**, *919*, 140–145. [CrossRef]
54. Benmouhoub, C.; Gauthier-Manuel, B.; Zegadi, A.; Robert, L. A Quantitative Fourier Transform Infrared Study of the Grafting of Aminosilane Layers on Lithium Niobate Surface. *Appl. Spectrosc.* **2017**, *71*, 1568–1577. [CrossRef] [PubMed]

65. Pasteur, G.A.; Schonhorn, H. Interaction of Silanes with Antimony Oxide to Facilitate Particulate Dispersion in Organic Media and to Enhance Flame Retardance. *Appl. Spectrosc.* **1975**, *29*, 512–517. [CrossRef]
66. Kavale, M.S.; Mahadik, D.B.; Parale, V.G.; Wagh, P.B.; Gupta, S.C.; Rao, A.V.; Barshilia, H.C. Optically transparent, super hydrophobic methyltrimethoxysilane based silica coatings without silylating reagent. *Appl. Surf. Sci.* **2011**, *258*, 158–16. [CrossRef]
67. Latthe, S.S.; Imai, H.; Ganesan, V.; Rao, A.V. Porous superhydrophobic silica films by sol–gel process. *Microporous Mes porous Mater.* **2010**, *130*, 115–121. [CrossRef]
68. Lin, J.; Chen, H.; Fei, T.; Zhang, J. Highly transparent superhydrophobic organic–inorganic nanocoating from the aggrega tion of silica nanoparticles. *Colloids Surf.* **2013**, *421*, 51–62. [CrossRef]
69. Robles, E.; Csóka, L.; Labidi, J. Effect of reaction conditions on the surface modification of cellulose nanofibrils with amino propyl triethoxysilane. *Coatings* **2018**, *8*, 139. [CrossRef]
70. Popescu, C.M.; Popescu, M.C.; Vasile, C. Characterization of fungal degraded lime wood by FT-IR and 2D IR correlation spectroscopy. *Microchem. J.* **2010**, *95*, 377–387. [CrossRef]
71. Han, L.; Guo, J.; Tian, X.; Jiang, X.; Yin, Y. Evaluation of PEG and Sugars Consolidated Fragile Waterlogged Archaeological Wood Using Nanoindentation and ATR-FTIR Imaging. *Int. Biodeterior. Biodegrad.* **2022**, *170*, 105390. [CrossRef]

Article

Optimizing the Preparation Process of Bamboo Scrimber with Bamboo Waste Bio-Oil Phenolic Resin Using Response Surface Methodology

Ying Li [1], Chunmiao Li [1], Xueyong Ren [1,*], Fuming Chen [2] and Linbi Chen [2]

1 National Forestry and Grassland Engineering Technology Center for Wood Resources Recycling, School of Materials Science and Technology, Beijing Forestry University, Beijing 100083, China; ly3230330@bjfu.edu.cn (Y.L.); lcm0408@bjfu.edu.cn (C.L.)
2 Institute of Biomaterials for Bamboo and Rattan Resources, International Centre for Bamboo and Rattan, Beijing 100102, China; cnfuming@icbr.ac.cn (F.C.); chenlinbi@icbr.ac.cn (L.C.)
* Correspondence: rxueyong@bjfu.edu.cn; Tel.: +86-010-62336907

Abstract: Bamboo scrimber is a new type of biomass fiber-based composite material with broad application. In this study, self-developed bio-oil phenolic resin (BPF) was used to prepare bamboo scrimber. The effects of hot-pressing temperature, hot-pressing time, and BPF resin solid content on the modulus of rupture (MOR) and modulus of elasticity (MOE) were systematically investigated through single-factor experiments and response surface methodology (RSM). According to the Box-Behnken design (BBD) experiment of the RSM, the effects of all three factors on MOR and MOE are significant. The effects of the main factors affecting the MOR and MOE decreased in the order of resin solid content, hot-pressing temperature, and hot-pressing time. Based on BBD, the optimal conditions for the preparation of bamboo scrimber were determined as follows: a hot-pressing temperature of 150 °C, a hot-pressing time of 27.5 min, and a resin solid content of 29%. Under these conditions, the MOR is 150.05 MPa and the MOE is 12,802 MPa, which are close to the theoretical values, indicating that the optimization results are credible. This study helps to promote the full utilization of bamboo components and provides a reference for the development of high-quality bamboo scrimber.

Keywords: bamboo scrimber; bio-oil phenolic resin; response surface methodology; regression model; MOR; MOE

Citation: Li, Y.; Li, C.; Ren, X.; Chen, F.; Chen, L. Optimizing the Preparation Process of Bamboo Scrimber with Bamboo Waste Bio-Oil Phenolic Resin Using Response Surface Methodology. *Forests* **2024**, *15*, 1173. https://doi.org/10.3390/f15071173

Academic Editor: Yongfeng Luo

Received: 29 May 2024
Revised: 21 June 2024
Accepted: 2 July 2024
Published: 5 July 2024

1. Introduction

As a green and biodegradable biomass material, bamboo is characterized by abundant resources, a short growth cycle (3–5 years to maturity), excellent performance, and having a wide range of applications in the fields of construction, home furnishing, papermaking, and textile [1–3]. Moreover, it can absorb a large number of carbon dioxide during the growth process and has a carbon sequestration capacity in the ecosystem, which gives it great potential to address climate change, energy conservation, and emission reduction [4–6]. In the context of low-carbon development and the "Bamboo as a Substitute for Plastic" initiative, the demand for bamboo products is gradually increasing, and the research on the high-value utilization of bamboo has also received widespread attention. At present, a large number of bamboo-based materials have been developed, including bamboo scrimber, plybamboo, and laminated bamboo [7–9].

Among them, bamboo scrimber is a bamboo-based composite material obtained from bundles by defibering, impregnating adhesive, oriented assembling, drying, and hot-pressing. In contrast to other composites, bamboo scrimber has a greater usage rate because it can be made from raw materials such as small-sized bamboo and waste [10]. Because it maintains the bamboo fibers' inherent orientation and traits, they have notable mechanical properties [11]. It also has a beautiful texture and is commonly applied in decorative materials, indoor and outdoor flooring, and engineering structural materials [12].

Phenolic (PF) resin is the most common adhesive for the production of bamboo scrimber, which can bring good mechanical properties and dimensional stability to bamboo boards. For example, with a resin content of 25 wt%, the water-absorbing width and thickness expansion of bamboo scrimber were 3.74% and 3.72%, respectively [13]. With a density of 1.0 g/cm^3 and a PF content of 16%, the moso bamboo (*Phyllostachys heterocycla*) scrimber had a modulus of rupture (MOR) of 271.05 MPa and a modulus of elasticity (MOE) of 23.7 GPa [14]. However, PFM resin also has drawbacks such as high brittleness and a long curing time [15]. At the same time, with the increasing global attention to sustainable development, people are committed to using green materials to replace phenol to synthesize PF resin. There are many natural polymers in nature that contain phenolic substances, including tannin [16], lignin [17], and biomass-derived products such as cardanol [18] and bio-oil [19].

Bio-oil can be obtained by pyrolysis of wood or bamboo processing wastes such as sawdust and small chips [20,21]. Bio-oil contains a high content of phenolic substances and exhibits good reactivity. It can be added to phenolic resin to increase the mechanical strength, toughness, and aging resistance of the resin [22]. Additionally, it can reduce the consumption of petrochemical resources and lower production costs [23]. A relevant study demonstrated that bamboo scrimber prepared using bio-oil phenolic resin (BPF) had good mechanical and anti-mildew properties. Among them, the MOR was 143 MPa and the MOE was 9269 MPa. According to tests on mildew resistance, the anti-mildew level increased from slight to high due to BPF resin [24].

During the preparation of bamboo scrimber, many factors have a strong influence on the quality and performance of products. For example, Lu et al. studied the effects of various hot-pressing temperatures on bamboo scrimber [25]. Ji et al. investigated the effects of density and PF resin adhesive solid content on the water resistance of bamboo scrimber [26]. However, few previous studies have systematically investigated the impact of multiple factors on bamboo scrimber. Finding the optimal process parameters based on the interactions of the factors can provide important guidance for the high-quality production of bamboo scrimber. The response surface methodology (RSM) is an accurate and effective statistical and mathematical tool for optimizing experimental processes [27]. The theoretical optimal process parameters also can be corrected and reconfirmed according to the experimental data, which can greatly improve work efficiency and save resources [28].

In this study, BPF resin was synthesized from bamboo waste bio-oil and was used to prepare bamboo scrimber. The effects of hot-pressing temperature, hot-pressing time, and the adhesive solid content of BPF resin on the properties of bamboo scrimber were evaluated by statistical modeling using RSM. According to the actual application scenario of bamboo scrimber, MOR and MOE were selected as the response values to evaluate the quality of the boards. High-performance bamboo scrimber was obtained through the optimization of process parameters. Based on this method, bamboo scrimber can achieve the full utilization of bamboo components. These results can offer a technical and theoretical foundation for the bamboo scrimber's quality control and preparation process.

2. Materials and Methods

2.1. Materials

Thirty-five-centimeter moso bamboo (*Phyllostachys heterocycla*) bundles were supplied by Fujian Youzhu Technology Co., Ltd., Yongan, China. A bio-oil phenolic resin with a solid content of 40% and a viscosity of 68 mPa·s was self-developed. Bamboo waste bio-oil, an acid liquid (pH 3.5), was obtained by quickly pyrolyzing moso bamboo waste in a fluidized bed at 550 °C in the Laboratory of Beijing Forestry University [24]. Phenol was produced by Shanghai McLean Biochemical Technology Co., Ltd., Shanghai, China. Sodium hydroxide was produced by Beijing Chemical Factory, Beijing, China. 37% formaldehyde was obtained by Shanghai Aladdin Biochemical Technology Co., Ltd., Shanghai, China. All reagents used were analytically pure.

2.2. Preparation of Bamboo Scrimber

Firstly, according to Table 1, the BPF resin was diluted into different solid contents, and the bamboo bundles were soaked in the adhesive at atmospheric pressure for 8 min, aged for 2 days, and dried naturally to a moisture content of 8%. Then, according to the set board density (1.05 g/cm^3), the board was assembled with a length × width × thickness of 35 cm × 35 cm × 5 mm, and the board was put into the hot press (QD, SWPM, Shanghai, China) for hot-pressing according to Table 1, and the hot-pressing pressure was set to 5.0 MPa. Finally, the pressed bamboo scrimber boards were kept at constant room conditions (20 °C and 65% RH) for 2 days and then trimmed, sanded, and sawn for samples (Figure 1).

Figure 1. Detailed preparation process of bamboo scrimber.

2.3. Experimental Design

2.3.1. Single-Factor Experiment Design

The effects of hot-pressing temperature, hot-pressing time, and resin solid content on the MOR and MOE of bamboo scrimber were studied by single-factor experiments to clarify the optimization level range of influencing factors. The number of MOR and MOE test samples was six, and the test results were averaged. The error bars were represented as a standard deviation to reflect the fluctuations in the data for each sample. According to the relevant studies [24,25], the levels of each factor are shown in Table 1.

Table 1. Factors and levels of single-factor experiment design.

Factors	Level
Hot-pressing temperature (°C)	120, 130, 140, 150, 160, 170
Hot-pressing time (min)	15, 20, 25, 30, 35, 40
Solid content (%)	15, 20, 25, 30, 35, 40

2.3.2. Multi-Factor Experiment Design

Based on a single factor experiment, the Box-Behnken design (BBD) scheme in RSM was used to design the central composite experiment. Three variables and their actual experimental scope and coding level are shown in Table 2. A total of 17 runs were made in random order. In the optimization level interval, the optimal process for the preparation of bamboo scrimber was found and three confirmatory tests were conducted. The test results were averaged, and there were six specimens for each performance test. The experiment results of the optimized samples were compared to match the predictions.

Table 2. Actual and coded experimental factors and corresponding levels in BBD.

Factors	Code	Level		
		Low (−1)	**Middle (0)**	**High (1)**
Hot-pressing temperature (°C)	A	120	145	170
Hot-pressing time (min)	B	20	27.5	35
Solid content (%)	C	15	25	35

The independent variables (x_1: hot-pressing temperature (°C), x_2: hot-pressing time (min), x_3: resin solid content (%)) were at three levels, exploring the impact of each factor on the response values in terms of MOE and MOR. The specific model of the quadratic multinomial regression equation was as follows: y represents the model prediction response value. The coefficient for constant regression was denoted as β_0, while the coefficient for linear regression were β_1, β_2, and β_3. The quadratic coefficients were β_{11}, β_{22}, and β_{33}, while the interaction coefficients were β_{12}, β_{13}, and β_{23}. These coefficients were determined in the second-order model using Design-Expert software V.8.0.6 (Stat-Ease, Inc, Minneapolis, MN, USA). An analysis of variance (ANOVA) was used to assess the model's statistical significance based on p values less than 0.05. The quality of the created model was assessed using the model regression R^2 and the lack of fit from the ANOVA [29].

$$y = \beta_0 + \beta_1 x_1 + \beta_2 x_2 + \beta_3 x_3 + \beta_{11} x_1^2 + \beta_{22} x_2^2 + \beta_{33} x_3^2 + \beta_{12} x_1 x_2 + \beta_{13} x_1 x_3 + \beta_{23} x_2 x_3 \quad (1)$$

2.4. Performance Testing

Before testing, all specimens were kept in a climate chamber (20 °C and 65% RH) until the mass was constant. Each group of experiments was replicated three times. The number of each performance test specimen was six, and the test results were averaged. There were 18 groups in the single-factor experiment, and a total of 324 samples were prepared. There were also 17 groups in the multi-factor experiment, and a total of 306 samples were prepared. There were 3 groups in the optimal process validation experiment, and a total of 108 samples were prepared. The standard deviation was used to describe the level of error in samples. The tests of MOR, MOE, and water-absorbing expansion were carried out in accordance with the Chinese national standard GB/T 17657-2022 [30]. The water-absorbing expansion tests were conducted in a water bath at (20 ± 1 °C) and pH (7 ± 1) using the indoor test method. Among them, a universal testing machine (MMW-50, NAIER, Jinan, China) was used to measure the MOE and MOR at a constant speed of 10 mm/min. The cross-sectional microstructure of bamboo scrimber was observed using scanning electron microscopy (Regulus8100, HITACHI, Tokyo, Japan), and samples were gold-plated in a vacuum for 5 min before observation using ion sputter (MC1000, HITACHI, Tokyo, Japan). The experimental conditions for electron microscopy were a vacuum environment and an excitation voltage of 3.0 kV. The sample size was 10 mm × 10 mm × 5 mm. The size of the samples for measuring MOR and MOE was 15 cm × 5 cm × 5 mm. The size of the samples for measuring thickness swelling rate (TSR) and width swelling rate (WSR) was 5 cm × 5 cm × 5 mm. The equations are shown in (2) and (3), respectively:

$$\text{TSR} = \frac{t_2 - t_1}{t_1} \, 100\% \quad (2)$$

$$\text{WSR} = \frac{d_2 - d_1}{d_1} \, 100\% \quad (3)$$

In the formula, t_1 is the initial thickness of the bamboo scrimber specimen in mm; t_2 is the final thickness of the bamboo scrimber specimen in mm. d_1 is the initial width of the bamboo scrimber specimen in mm; d_2 is the final width of the bamboo scrimber specimen in mm (Figure 2).

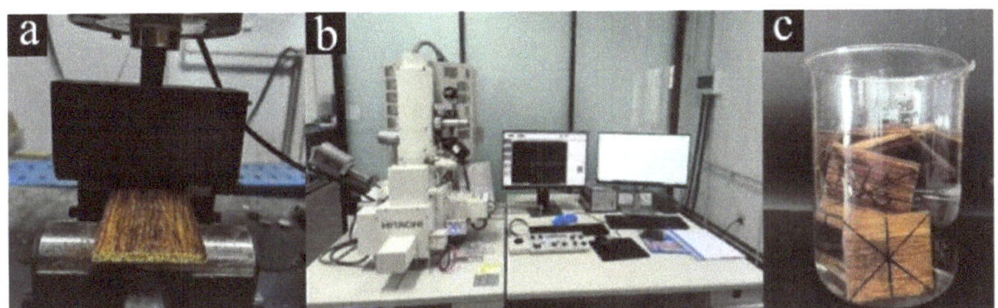

Figure 2. (**a**) MOR and MOE testing, (**b**) scanning electron microscope and (**c**) dimensional stability test.

3. Results and Discussion

3.1. Single-Factor Experiment

3.1.1. Effect of Hot-Pressing Temperature

The effect of hot-pressing temperature on MOR and MOE is shown in Figure 3. From Figure 3a,b, it can be observed that as the hot-pressing temperature increased from 120 °C to 140 °C, the increase in MOR of the bamboo scrimber was relatively small, with a maximum increase of 35.1%; while the increase in MOE was more significant, with a maximum increase of 84.8%. This indicates that within the temperature range of 120 °C to 170 °C, there is difference in the curing degree of the adhesive. The MOR and MOE of the bamboo scrimber gradually improved as the temperature increased from 120 °C to 140 °C. However, after reaching 140 °C, the mechanical properties gradually decreased. This may be due to the high temperature causing thermal degradation of the bamboo surface or curing of the adhesive [31].

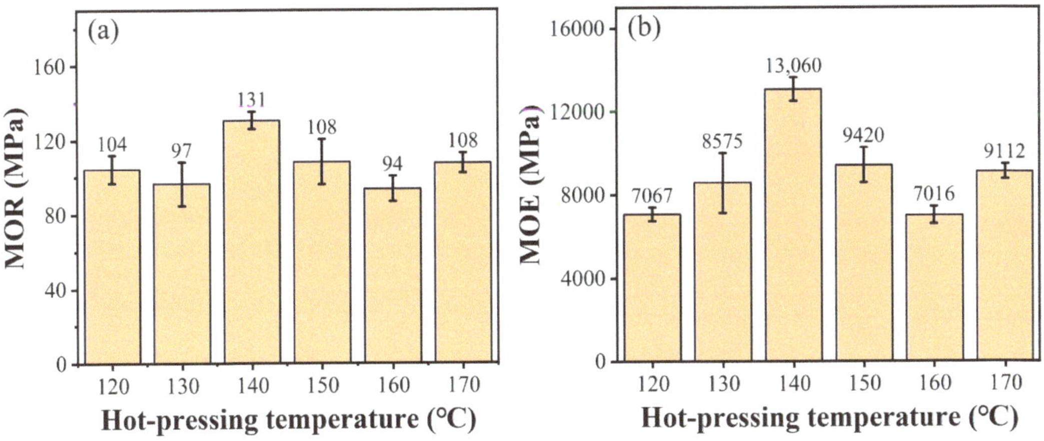

Figure 3. Effect of hot-pressing temperature on the (**a**) MOR and (**b**) MOE of bamboo scrimber.

3.1.2. Effect of Hot Pressing Time

The effect of hot-pressing time on MOR and MOE is shown in Figure 4. From Figure 4a,b, it can be seen that the influence of hot-pressing time on the MOR and MOE of bamboo scrimber was greater compared to the hot-pressing temperature. Within the certain experimental range, both performance indicators increased with the extension of hot-pressing time. When the hot-pressing time was 15 min, the MOR was 84 MPa and the MOE was 7247 MPa, both of which did not meet the minimum requirements of GB/T 40247-2021 (The MOE is 9000 MPa and the MOR is 80 MPa) [32], indicating that the hot-pressing time was too short and the resin had not fully cured. However, as the hot-pressing time increased, the curing of the PF resin in the bamboo scrimber surface formed a hard shell easily, which was not conducive to the conduction of pressure [33]. When the hot-pressing time was 40 min, although both performance indicators would meet the national standards, excessive curing could make the board brittle, leading to a decrease in MOR and MOE. From an economic perspective, this would reduce production efficiency and increase production costs. Therefore, while ensuring performance requirements, a shorter hot-pressing time should be chosen.

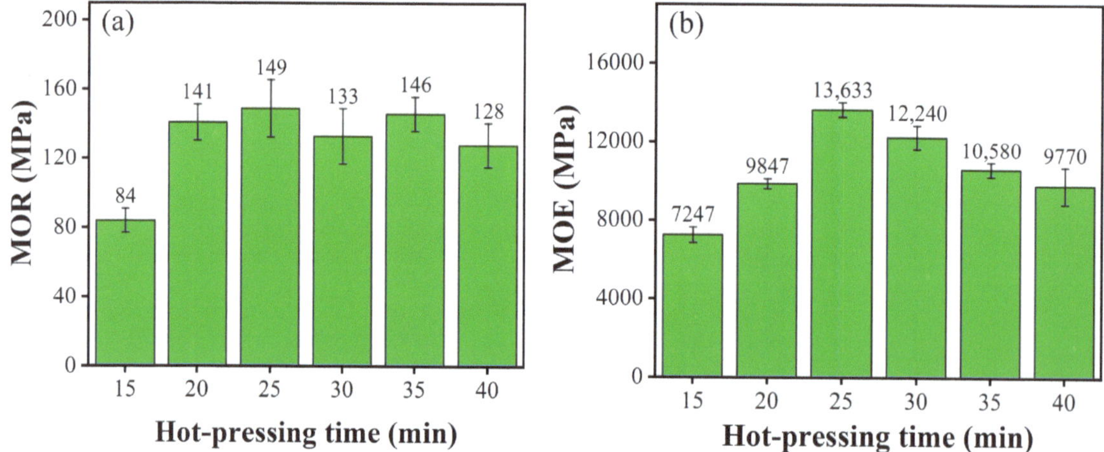

Figure 4. Effect of hot-pressing time on the (**a**) MOR and (**b**) MOE of bamboo scrimber.

3.1.3. Effect of Adhesive Solid Content

Figure 5 shows the influence of adhesive solid content on MOR and MOE, carried out under similar conditions with a hot-pressing temperature of 145 °C and a time of 20 min. From Figure 5a,b, it can be observed that the MOR and MOE of bamboo scrimber were relatively less affected by the solid content of the diluted phenolic resin. When the solid content increased from 15% to 40% (undiluted resin solid content), there was an overall trend of first increasing and then decreasing. When the solid content increased from 15% to 25%, the MOR increased by 19.1% and the MOE increased by 26.6%. This indicates that as the solid content increases, more resin macromolecules enter the bamboo scrimber. However, when the solid content increased from 25% to 40%, both indicators showed an overall decreasing trend, indicating that excessive solid content hindered the entry of resin molecules into the bamboo and affected the cross-linking of bamboo and resin [11]. Moreover, when the solid content reached 40%, the resin adhering to the bamboo surface would experience "resin exudation" during hot-pressing, significantly impacting product quality and production efficiency.

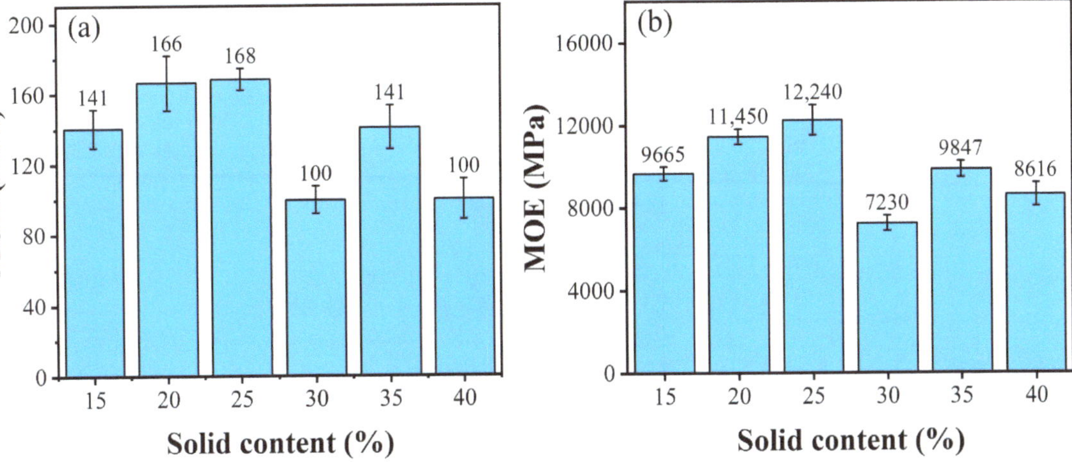

Figure 5. Effect of adhesive solid content on the (**a**) MOR and (**b**) MOE of bamboo scrimber.

3.2. Multi-Factor Experiment

3.2.1. Variance Analysis

The test protocol and results of the BBD are shown in Table 3. The larger the multivariate correlation coefficient R^2, the better the correlation [34]. The coefficient of variation (CV) < 10%, indicates high confidence and precision of the experiment [35]. As can be seen in Table 4, the fitted regression equation conformed to the above test principles and had good adaptability.

Table 3. Experiment scheme and results of Box-Behnken design.

Run	A	B	C	MOR (MPa)	MOE (MPa)
1	0	1	−1	117.35	7300
2	0	0	0	151.56	12,733
3	1	0	1	131.76	12,155
4	1	1	0	133.73	8960
5	1	0	−1	107.57	9112
6	−1	0	−1	104.40	7067
7	−1	−1	0	113.69	7613
8	−1	0	1	115.74	10,488
9	0	0	0	156.60	13,030
10	0	1	1	145.45	10,580
11	0	0	0	145.32	11,327
12	0	0	0	158.52	12,828
13	0	0	0	142.04	13,260
14	1	−1	0	129.69	11,380
15	0	−1	−1	140.52	9665
16	−1	1	0	137.13	8296
17	0	−1	1	140.92	9847

Table 4. Model Adequacy Indicators for Each Modeled Response of bamboo scrimber.

Response Variables	R^2-Value	CV (%)
MOR	0.9136	6.18
MOE	0.9492	6.84

Variance Analysis of MOR

The results of the ANOVA of MOR are shown in Table 5. The effects of hot-pressing temperature (A), hot-pressing time (B), and resin solid content (C) on MOR were significant ($p < 0.05$) and lack of fit was not significant ($p > 0.05$). A^2 and C^2 had an extremely significant effect ($p < 0.01$), while B^2 had no significant effect ($p > 0.05$). The results show that the hot-pressing temperature, hot-pressing time, and solid content had significant effects on MOR, and their influences decreased in the order of resin solid content, hot-pressing temperature, and hot-pressing time. The regression equation for MOR is as follows:

$$MOR = -716.02489 + 10.62728 \times A + 2.10551 \times B + 3.50796 \times C - 0.025864 \times AB + 0.012850 \times AC + 0.092333 \times BC - 0.034753 \times A^2 - 0.009387 \times B^2 - 0.142200 \times C^2 \tag{4}$$

Table 5. Analysis of Variance for MOR Regression Equation.

Source	Sum of Square	DF	Mean Square	F-Value	Prob > F	Significance
Model	3985.76	9	442.86	6.48	0.0111	significant
A	126.34	1	126.34	1.85	0.2160	
B	9.76	1	9.76	0.14	0.7166	
C	512.48	1	512.48	7.50	0.0289	
AB	94.07	1	94.07	1.38	0.2789	
AC	41.28	1	41.28	0.60	0.4623	
BC	191.82	1	191.82	2.81	0.1377	
A^2	1986.44	1	1986.44	29.09	0.0010	
B^2	1.17	1	1.17	0.017	0.8994	
C^2	851.40	1	851.40	12.47	0.0096	
Residual	478.05	7	68.29			
Lack of Fit	277.46	3	92.49	1.84	0.2795	not significant
Pure Error	200.58	4	50.15			
Cor Total	4463.80	16				

Variance Analysis of MOE

The results of the ANOVA of MOE are shown in Table 6. The effects of three factors on MOR were significant ($p < 0.05$), and lack of fit was not significant ($p > 0.05$). F_C, F_A, and F_B indicated that the effects decreased in the order of C, A, B, A^2, B^2, and C^2 and had an extremely significant effect on MOE ($p < 0.01$). The regression equation for MOE is as follows:

$$MOE = -94372.34631 + 910.01887 \times A + 2207.34556 \times B + 555.98917 \times C - 4.13733 \times AB - 0.37800 \times AC + 10.32667 \times BC - 2.57268 \times A^2 - 34.94089 \times B^2 - 13.22175 \times C^2 \tag{5}$$

Table 6. Analysis of Variance for MOE Regression Equation.

Source	Sum of Square	DF	Mean Square	F-Value	Prob > F	Significance
Model	6.529×10^7	9	7.254×10^6	14.52	0.0010	significant
A	8.289×10^6	1	8.289×10^6	16.59	0.0047	
B	1.419×10^6	1	1.419×10^6	2.84	0.1358	
C	1.232×10^7	1	1.232×10^7	24.65	0.0016	
AB	2.407×10^6	1	2.407×10^6	4.82	0.0642	
AC	3.572×10^4	1	3.572×10^4	0.072	0.7969	
BC	2.399×10^6	1	2.399×10^6	4.80	0.0645	
A^2	1.089×10^7	1	1.089×10^7	21.79	0.0023	
B^2	1.626×10^7	1	1.626×10^7	32.56	0.0007	
C^2	7.361×10^6	1	7.361×10^6	14.73	0.0064	
Residual	3.497×10^6	7	4.996×10^5			
Lack of Fit	1.192×10^6	3	3.975×10^5	0.69	0.6040	not significant
Pure Error	2.304×10^6	4	5.761×10^5			
Cor Total	6.879×10^7	16				

3.2.2. Effects of Interactions between Factors

In order to predict and optimize the response values and to analyze the interaction of two factors, the regression equations for the response values were used to obtain the corresponding response surfaces and contour plots. A contour consists of multiple response value lines that are connected by closed curves of points with the same response value. Response surface plots are three-dimensional forms of contour lines [36]. Response surfaces and contour plots provide a visual response to the extent to which the interaction affects the response values. The steeper the surface and the denser the contours, the more significant the effect, and the closer the contours are to an oval, the stronger the interaction between the two factors [37].

Effect of the Interactions on MOR

The three-dimensional response surfaces and two-dimensional contour plots of MOR by the interaction of different factors from the regression equation are shown in Figure 6. The contour lines in Figure 6d were close to the ellipse, indicating that the interaction between the hot-pressing temperature and the solid content is the most significant of the three factors; Figure 6b had a smaller number of contour lines and slower change compared with Figure 6f, indicating that the interaction between the hot-pressing temperature and the solid content was weaker, which was consistent with $F_{BC} > F_{AB}$ in Table 4.

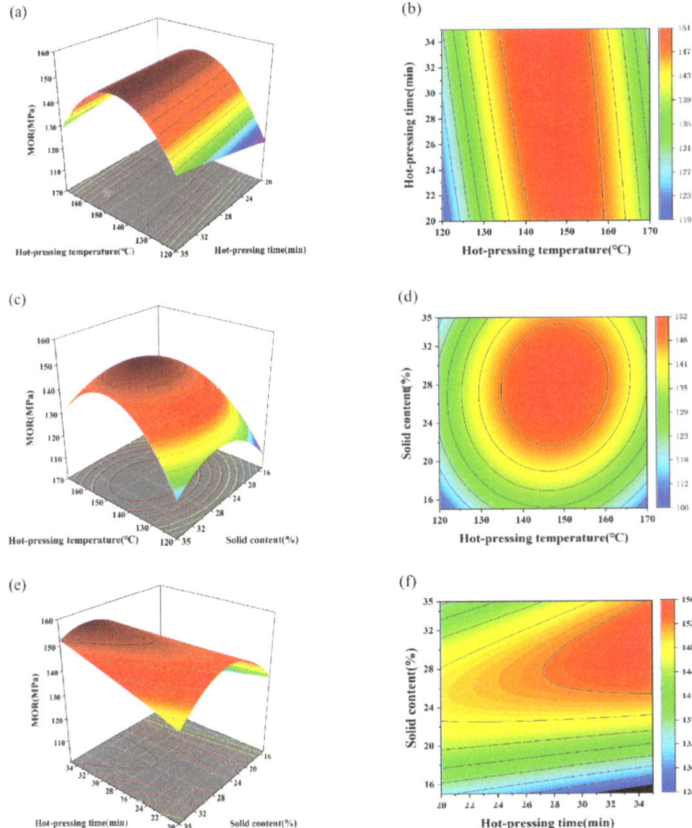

Figure 6. Contour plots and 3D surface for MOR of bamboo scrimber: (**a**,**b**) hot-pressing temperature and hot-pressing time; (**c**,**d**) hot-pressing temperature and solid content; (**e**,**f**) hot-pressing time and solid content.

In Figure 6a, there was a prominent "arch" surface where the hot-pressing time wa mostly a gentle straight line. The hot-pressing temperature was an "arch" line with a large surface inclination and curvature. This suggests that the effect of hot-pressing temperatur on MOR was more significant, which was consistent with the result that the F-value fo hot-pressing temperature was more significant than that for hot-pressing time (Table 4). A shown in Figure 6b, when both the hot-pressing temperature and the hot-pressing time wer increased simultaneously, the MOR first increased and then decreased. The optimum rang of hot-pressing temperatures was about 140 °C to 160 °C. With the increase of hot-pressin temperature, the MOR first increased and then slowly decreased, and the curve reached th highest response value at roughly 150 °C. When the hot-pressing temperature was lowe the flowability and bonding of the adhesive were weak and a strong bonding interfac could not be formed [33]. When the temperature increased, the cementing performanc was improved and the adhesive could penetrate into the bamboo bundle cells to form . good bonding interface. At this time, the macroscopic manifestation was the increase o MOR [11]. At the same time, the high temperature accelerated the evaporation of water i the adhesive, which reduced the fluidity of the adhesive and prevented the adhesive from entering the interface cells to form the interface. As the hot-pressing time continued t increase, the MOR gradually decreased because the longer the time, the longer the adhesiv would be aging, which would also lead to a decline in adhesive performance. In the actua production process, the longer the hot-pressing time, the greater the increase in energy consumption and decrease in productivity. As shown in Figure 6c, the change rule fo the hot-pressing temperature was consistent with Figure 6a, which still rose first and the decreased. As the solid content of the BPF resin adhesive increased, the MOR first rose an then slowly decreased, and the curve reached the maximum response value at around 29% When the solid content was low, the flowability of the adhesive was strong, and it coul easily enter the interstices of the bamboo bundle cells and the gaps between the bundle and the gluing effect is manifested in the increase of MOR [26]. However, too high a soli content will reduce the flowability of the adhesive and prevent the resin from entering th bamboo material. As shown in Figure 6e, the slope of the hot-pressing time-MOR curv was smaller than that of the solid content-MOR curve. Within the set parameter range, th hot-pressing time was less than the effect of solid content on the MOR. The top and botton contour lines of the contour plot shown in Figure 6f were asymmetric, with a slightly large color change in the lower part; the contour line at about 150 MPa was an elliptical shape There was a tendency for the red area on the right side to expand, indicating that there wa a significant interaction between hot pressing time and the solid content on the MOR o the bamboo scrimber. The optimum range of solid content was 26%-35%, and the MOF gradually increased whenever the hot-pressing time was prolonged and the BPF was full cured in the bamboo bundle. For the MOR of bamboo scrimber, the adhesive solid conten had the greatest influence, followed by the hot-pressing temperature, and the hot-pressing time had the least influence.

Effect of the Interaction on MOE

The three-dimensional response surfaces and two-dimensional contour plots of MOE by the interaction of different factors from the regression equation are shown in Figure 7 The contours of Figure 7b,f were both close to an elliptical shape, and the contour of Fig ure 7d was close to a circular shape, indicating that the interaction between the hot-pressing temperature and the solid content was weak, which was consistent with $F_{AB} > F_{BC} > F_{AC}$ in Table 4. As can be seen from Figure 7, the effects of adhesive solid content, hot-pressing temperature, and hot-pressing time on the MOE of bamboo scrimber were basically the same as those on the MOE. In Figure 7a,b, when the hot-pressing temperature and hot pressing time were increased simultaneously, the MOE of the bamboo scrimber showed an obvious pattern of increasing first and then decreasing. The curve of hot-pressing time with MOE showed a tendency to first increase and then decrease, and the maximum appeared at about 26 min. As the hot-pressing temperature increased, the MOE first increased and then slowly decreased, and the curve reached its maximum at about 155 °C. Figure 7c,d showed that the hot-pressing temperature and MOE change rule were still the first rise and then decline, with the curve reaching the maximum response value at about 155 °C. With the increase in the solid content of BPF resin adhesive, MOE first rose and then slowly decreased; the curve reached the maximum response value at about 29%. As shown in Figure 7e,f, the change rule curve of hot-pressing time and MOE reached the maximum response value at about 27 min. As the solid content of BPF resin adhesive increased, the MOE first increased and then slowly decreased, and the curve reached the maximum response value at about 29%. For the MOE of bamboo scrimber, the adhesive solid content had the greatest influence, followed by the hot-pressing temperature, and the hot-pressing time had the least influence, and its law was consistent with the MOE.

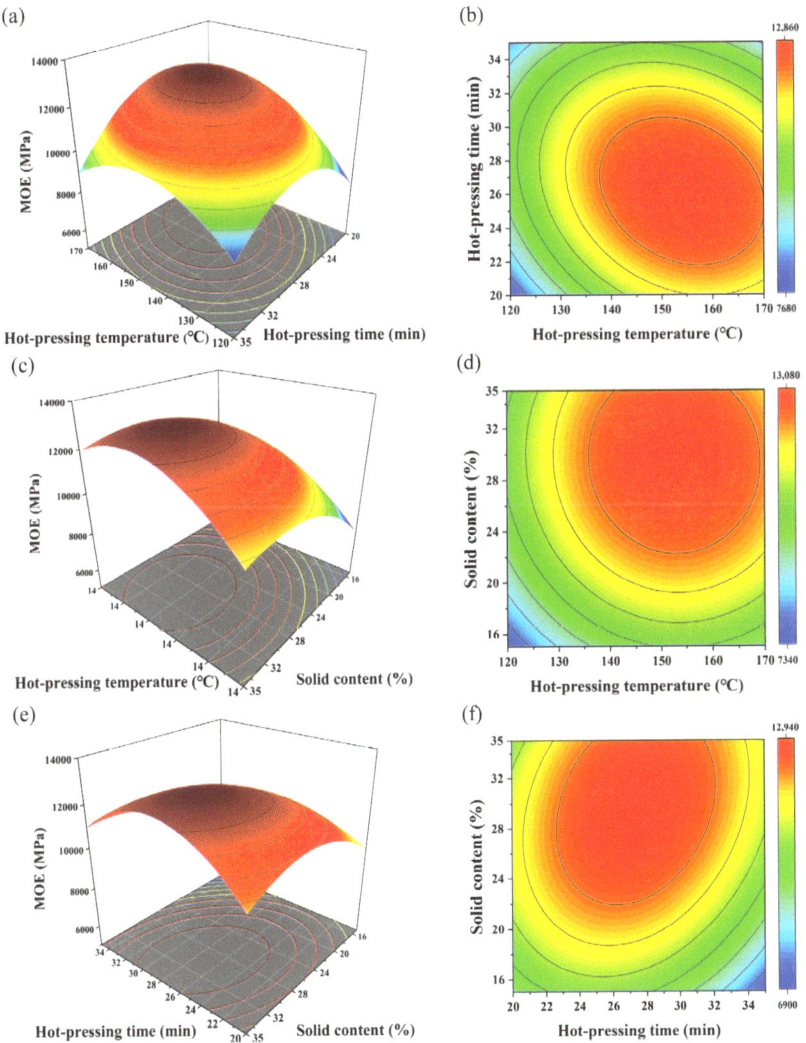

Figure 7. Contour plots and 3D surface for MOE of bamboo scrimber: (**a**,**b**) hot-pressing temperature and hot-pressing time; (**c**,**d**) hot-pressing temperature and solid content; (**e**,**f**) hot-pressing time and solid content.

3.3. Optimization and Validation

The optimal process was obtained by Design-Expert software: a hot-pressing temperature of 149.62 °C, a hot-pressing time of 27.65 min, and a resin solid content of 28.76%. Taking into consideration the laboratory conditions and practical feasibility, the process was adjusted to a hot-pressing temperature of 150 °C, a hot-pressing time of 27.5 min, and a resin solid content of 29%. Three confirmatory tests were carried out, and the MOR and MOE results are shown in Table 7. The average MOR was 150.05 MPa and the MOE was 12,802 MPa, which met the requirement of 120 E_b level according to the Chinese national standard (GB/T 40247-2021) [38]. Moreover, the difference between the theoretical value of MOR of 152.08 MPa and MOE of 13,036 MPa was not much, which was within the error range of 5%. It showed that the equation fits well with the actual situation, and the model was correct. Meanwhile, MOR and MOE were higher than the values of other studies [24,38], indicating that the optimized process has a beneficial effect on the enhancement of mechanical properties with bamboo scrimber. Figure 8 shows the morphological characteristics of the bamboo scrimber cross-section under optimal and non-optimal preparation process parameters. Figure 8a,b shows that the cross-sectional surface of bamboo scrimber was flat and evenly filled with BPF resin compared to Figure 8c,d. The TSR and WSR of the samples after a 24-h immersion period in room-temperature water are displayed in Table 7. According to the Chinese national standard (GB/T 40247-2021), the average WSR of bamboo scrimber was 1.96%, which met the requirement of a W2.0 level; the average TSR was 8.68%, which met the requirement of a T9.0 level. This result showed that the optimized bamboo scrimber has good water resistance and can be used as flooring or desktop decorative panels in the indoor environment.

Table 7. MOR, MOE, WSR, and TSR of optimized bamboo scrimber.

Sample	MOR (MPa)	MOE (MPa)	WSR (%)	TSR (%)
1	150.40 ± 11.95	11,547 ± 378.00	2.19 ± 0.26	11.72 ± 1.11
2	153.04 ± 15.70	13,560 ± 153.04	1.86 ± 0.12	6.18 ± 1.15
3	146.70 ± 7.11	13,300 ± 146.70	1.82 ± 0.09	8.14 ± 0.91

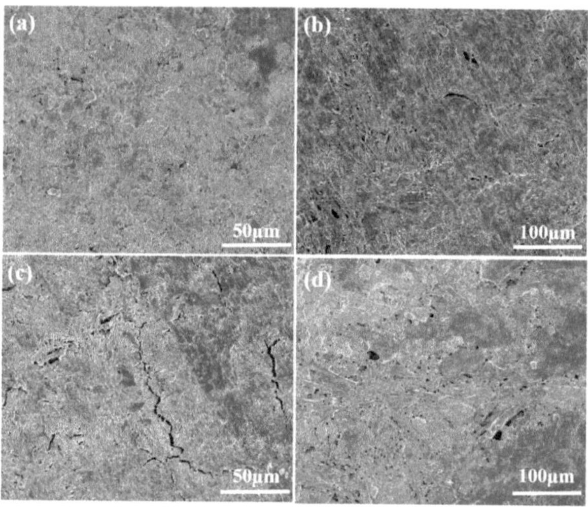

Figure 8. SEM images of bamboo scrimber cross-section, (**a,b**) bamboo scrimber prepared under optimized conditions, (**c,d**) bamboo scrimber prepared under non-optimized conditions. Among them, (**a,c**) represent images magnified by ×500 times, and (**b,d**) represent images magnified by ×1000 times.

4. Conclusions

Bamboo scrimber was prepared successfully using bio-oil phenolic resin. The RSM was used to optimize the hot-pressing process of bamboo scrimber, which could effectively predict the optimal process conditions. Combined with the optimization scheme given by the Design-Expert software and the actual production situation, the optimized process conditions are as follows: a hot-pressing temperature of 150 °C, a hot-pressing time of 27.5 min, and a solid content of impregnated bio-oil phenolic resin of 29%. Under these conditions, the MOR and MOE of bamboo scrimber were 150.05 MPa and 12,802 MPa, respectively. The order of influence of each variable on MOR and MOE was resin solid content, hot-pressing temperature, and hot-pressing time. Among them, the hot-pressing temperature and time, as well as the hot-pressing time and resin solid content, had significant interactive effects on the MOR and MOE. It demonstrated that RSM is applicable for regression analysis and parameter optimization of the process in the preparation of bamboo scrimber. The preparation of bamboo scrimber using bamboo waste bio-oil phenolic resin promotes the full use of bamboo components, and the optimized process has a certain guiding significance for the actual production in the bamboo industry.

Author Contributions: Conceptualization, Y.L. and X.R.; writing—original draft preparation, Y.L.; writing—review and editing, C.L. and X.R.; supervision, C.L. and F.C.; resources, L.C. All authors have read and agreed to the published version of the manuscript.

Funding: This research was supported by the fund of National Key Research and Development Program (2022YFD2200904), the Research and Demonstration of Key Technologies for "Bamboo as a Substitute for Plastic" in Pilot Member States of the International Bamboo and Rattan Organization (INBAR)—Development of New Bamboo-based Acoustical Materials as an Alternative to Plastic and Environmental Benefit Evaluation of Its Products, and the fund of Hebei Province Central Finance Forest and Grass Science and Technology Promotion Demonstration Project (JI-TG [2022] 004).

Data Availability Statement: Data will be made available on request.

Conflicts of Interest: The authors declare no conflicts of interest.

References

1. Zheng, Z.Y.; Yan, N.A.; Lou, Z.C.; Jiang, X.Z.; Zhang, X.M.; Chen, S.; Xu, R.; Liu, C.; Xu, L. Modification and Application of Bamboo-Based Materials: A Review-Part I: Modification Methods and Mechanisms. *Forests* **2023**, *14*, 2219. [CrossRef]
2. Adier, M.F.V.; Sevilla, M.E.P.; Valerio, D.N.R.; Ongpeng, J.M.C.; Sadowski, T. Bamboo as Sustainable Building Materials: A Systematic Review of Properties, Treatment Methods, and Standards. *Buildings* **2023**, *13*, 2449. [CrossRef]
3. Aizuddin, K.; Lai, K.S.; Baharum, N.A.; Yong, W.T.L.; Hoon, L.N.; Hamid, M.Z.A.; Cheng, W.H.; Abdullah, J.O. Bamboo for Biomass Energy Production. *Bioresources* **2023**, *18*, 2386–2407. [CrossRef]
4. Kaur, P.J.; Yadav, P.; Gupta, M.; Khandegar, V.; Jain, A. Bamboo as a Source for Value Added Products: Paving Way to Global Circular Economy. *Bioresources* **2022**, *17*, 5437–5463. [CrossRef]
5. Kunyong, Y.; Jian, L.; Terefe, R. Role of Bamboo Forest for Mitigation and Adaptation to Climate Change Challenges in China. *J. Sci. Res. Rep.* **2019**, *24*, 1–7.
6. Pan, C.; Zhou, G.; Shrestha, A.K.; Chen, J.; Kozak, R.; Li, N.; Li, J.; He, Y.; Sheng, C.; Wang, G. Bamboo as a Nature-Based Solution (NbS) for Climate Change Mitigation: Biomass, Products, and Carbon Credits. *Climate* **2023**, *11*, 175. [CrossRef]
7. Shu, B.; Xiao, Z.; Hong, L.; Zhang, S.; Li, C.; Fu, N.; Lu, X. Review on the Application of Bamboo-Based Materials in Construction Engineering. *J. Renew. Mater.* **2020**, *8*, 1215–1242. [CrossRef]
8. Sun, X.; He, M.; Li, Z. Novel engineered wood and bamboo composites for structural applications: State-of-art of manufacturing technology and mechanical performance evaluation. *Constr. Build. Mater.* **2020**, *249*, 118751. [CrossRef]
9. Bala, A.; Gupta, S. Engineered bamboo and bamboo-reinforced concrete elements as sustainable building materials: A review. *Constr. Build. Mater.* **2023**, *394*, 132116. [CrossRef]
10. Xu, J.Y.; Zhou, Z.Z.; Zhang, X.C.; Xu, Y.T. A Simple and Effective Method to Enhance the Mechanical Properties, Dimensional Stability, and Mildew Resistance of Bamboo Scrimber. *Polymers* **2023**, *15*, 4162. [CrossRef]
11. Wei, J.G.; Xu, Y.; Bao, M.Z.; Yu, Y.L.; Yu, W.J. Effect of Resin Content on the Surface Wettability of Engineering Bamboo Scrimbers. *Coatings* **2023**, *13*, 203. [CrossRef]
12. Tang, S.Y.; Zhou, A.P.; Li, J.N. Mechanical Properties and Strength Grading of Engineered Bamboo Composites in China. *Adv. Civ. Eng.* **2021**, *2021*, 666059. [CrossRef]

13. Li, X.Y.; Rao, F.; Li, N.; Lei, W.C.; Bao, M.Z.; Bao, Y.J.; Li, L.M.; Duan, Z.J.; Zu, Q.; Zhang, Y.H.; et al. High-performance bamboo scrimber composite prepared from heat-treated *Bambusa chungii* units with different resin contents for outdoor use. *Ind. Crop Prod.* **2023**, *205*, 117503. [CrossRef]

14. Xie, J.L.; Chen, L.; Yang, L.; Jiang, Y.Z.; Chen, Q.; Qi, J.Q. Physical-mechanical properties of bamboo scrimbers with response to surface layer modification: Thermal treatment and resin dosage. *Eur. J. Wood Wood Prod.* **2023**, *82*, 321–328. [CrossRef]

15. Gao, M.; Wu, W.H.; Wang, Y.H.; Wang, Y.X.; Wang, H. Phenolic foam modified with dicyandiamide as toughening agent. *J. Therm. Anal. Calorim.* **2016**, *124*, 189–195. [CrossRef]

16. Hendrik, J.; Hadi, Y.S.; Massijaya, M.Y.; Santoso, A.; Park, B.D. Penetration and Adhesion Strength of Phenol-Tannin Formaldehyde Resin Adhesives for Bonding Three Tropical Woods. *For. Prod. J.* **2018**, *68*, 256–263.

17. Karthäuser, J.; Biziks, V.; Mai, C.; Militz, H. Lignin and Lignin-Derived Compounds for Wood Applications—A Review. *Molecule* **2021**, *26*, 2533. [CrossRef] [PubMed]

18. Zhang, W.Z.; Jiang, N.; Zhang, T.T.; Li, T.S. Thermal stability and thermal degradation study of phenolic resin modified by cardanol. *Emerg. Mater. Res.* **2020**, *9*, 180–185. [CrossRef]

19. Yu, Y.X.; Li, C.; Jiang, C.X.; Chang, J.M.; Shen, D.N. Aging Behaviors of Phenol-Formaldehyde Resin Modified by Bio-Oil under Five Aging Conditions. *Polymers* **2022**, *14*, 1352. [CrossRef]

20. Yildiz, D.; Yorgun, S. Catalytic pyrolysis of Paulownia wood and bio-oil characterization. *Energy Sources Part A Recovery Util Environ. Eff.* **2024**, *46*, 1626–1643. [CrossRef]

21. Kato, Y.; Kohnosu, T.; Enomoto, R. Chemical properties of bio-oils produced by fast pyrolysis of bamboo. *Trans. Mater. Res. Soc. Jpn.* **2014**, *39*, 491–498. [CrossRef]

22. Li, Q.; Liu, X.S.; Su, H.D.; Mao, A.; Wan, H. Improving Performance of Phenol-Formaldehyde Resins Modified/Blended with Phenol-Rich Pyrolysis Bio-Oil. *For. Prod. J.* **2020**, *70*, 387–395. [CrossRef]

23. Khanal, A.; Manandhar, A.; Adhikari, S.; Shah, A. Techno-economic analysis of novolac resin production by partial substitution of petroleum-derived phenol with bio-oil phenol. *Biofuels Bioprod. Biorefining Biofpr* **2021**, *15*, 1611–1620. [CrossRef]

24. Li, C.M.; Ren, X.Y.; Han, S.Y.; Li, Y.X.; Chen, F.M. The Preparation and Performance of Bamboo Waste Bio-Oil Phenolic Resin Adhesives for Bamboo Scrimber. *Forests* **2024**, *15*, 79. [CrossRef]

25. Lu, T.; Ge, Y.L.; Zhou, C.F.; Ren, Y.M.; He, M.Y.; Xu, K.; Li, X.J. Effects of physical parameters on the temperature and vapor pressure behavior of bamboo scrimber during hot-pressing. *Wood Mater. Sci. Eng.* **2023**, *18*, 1641–1649. [CrossRef]

26. Ji, Y.H.; Lei, W.C.; Huang, Y.X.; Wu, J.Y.; Yu, W.J. Influence of Resin Content and Density on Water Resistance of Bamboo Scrimber Composite from a Bonding Interface Structure Perspective. *Polymers* **2022**, *14*, 1865. [CrossRef] [PubMed]

27. Kumari, P.; Kaur, P.; Kumar, V.; Pandey, B.; Nazir, R.; Katoch, K.; Dwivedi, P.; Dey, A.; Pandey, D.K. Response surface methodology and artificial neural network modeling for optimization of ultrasound-assisted extraction and rapid HPTLC analysis of asiaticoside from Centella asiatica. *Ind. Crops Prod.* **2022**, *176*, 114320. [CrossRef]

28. Hadiyat, M.A.; Sopha, B.M.; Wibowo, B.S. Response Surface Methodology Using Observational Data: A Systematic Literature Review. *Appl. Sci.* **2022**, *12*, 10663. [CrossRef]

29. Niu, X.Y.; Pang, J.Y.; Cai, H.Z.; Li, S.; Le, L.; Wu, J.H. Process Optimization of Large-Size Bamboo Bundle Laminated Veneer Lumber (BLVL) by Box-Behnken Design. *Bioresources* **2018**, *13*, 1401–1412. [CrossRef]

30. *GB/T 17657-2022*; Test Methods for Physical and Chemical Properties of Artificial Boards and Decorative Artificial Boards. GB Standards: Beijing, China, 2022.

31. Kiho, J.; Jung, K.Y.; Roh, J.; Jin, P.S. Effect of Hot-Pressing Time and Temperature on Properties of Bamboo Zephyr Boards. *J. Korean Wood Sci. Technol.* **2003**, *31*, 77–83.

32. *GB/T 40247-2021*; Bamboo Scrimber. GB Standards: Beijing, China, 2021.

33. Li, H.; Chen, M.L.; Hu, L.X.; Wang, X.; Gu, Z.C.; Li, J.Z.; Yang, Z.B. Optimization Design of Bamboo Filament Decorated Board Process Based on Response Surface. *Bioresources* **2023**, *18*, 73–86. [CrossRef]

34. Kazemian, A.; Basati, Y.; Khatibi, M.; Ma, T. Performance prediction and optimization of a photovoltaic thermal system integrated with phase change material using response surface method. *J. Clean. Prod.* **2021**, *290*, 125748. [CrossRef]

35. Watts, E.S.; Rose, S.P.; Mackenzie, A.M.; Pirgozliev, V.R. Investigations into the chemical composition and nutritional value of single-cultivar rapeseed meals for broiler chickens. *Arch. Anim. Nutr.* **2021**, *75*, 209–221. [CrossRef] [PubMed]

36. Omranian, S.R.; Hamzah, M.O.; Yee, T.S.; Hasan, M.R.M. Effects of short-term ageing scenarios on asphalt mixtures' fracture properties using imaging technique and response surface method. *Int. J. Pavement Eng.* **2020**, *21*, 1374–1392. [CrossRef]

37. Zhang, R.B.; Liu, T.; Zhang, Y.M.; Cai, Z.L.; Yuan, Y.Z. Preparation of spent fluid catalytic cracking catalyst-metakaolin based geopolymer and its process optimization through response surface method. *Constr. Build. Mater.* **2020**, *264*, 120727. [CrossRef]

38. Wang, H.Y.; Yuan, S.F.; Zhang, J.; Li, Q. Influence of Four Ageing Methods on the Mechanical Properties of Bamboo Scrimber. *Int. J. Polym. Sci.* **2021**, *2021*, 2478525. [CrossRef]

forests

MDPI

Article

The Novel Applications of Bionic Design Based on the Natural Structural Characteristics of Bamboo

Siyang Ji [1], Qunying Mou [2], Ting Li [1,3], Xiazhen Li [1,*], Zhiyong Cai [4] and Xianjun Li [1,*]

1   College of Material Science and Engineering, Central South University of Forestry and Technology, Changsha 410004, China; jisiyang1998@163.com (S.J.); liting0907thj@163.com (T.L.)
2   College of Electronic Information and Physics, Central South University of Forestry and Technology, Changsha 410004, China; t20030981@csuft.edu.cn
3   Hunan Taohuajiang Bamboo Science & Technology Company, Yiyang 413400, China
4   USDA Forest Products Laboratory, Madison, WI 3726-2398, USA; zcai@fs.fed.us
*   Correspondence: t20192438@csuft.edu.cn (X.L.); lxjmu@csuft.edu.cn (X.L.);
    Tel.: +86-181-7311-4006 (Xiazhen Li)

Abstract: The unique composite gradient structure of bamboo has made it widely recognized as an extremely efficient natural structure and material, endowing it with exceptional flexibility and resilience. This enabled bamboo to withstand the forces of wind and snow without fracturing. In this paper, the inherent structural characteristics of bamboo were examined in order to extract its biological advantages through experimental methods. Then, the structural characteristics of bamboo in its vertical and radial directions served as the respective inspiration for two bionic applications, which were further analyzed and optimized using finite element analysis to accurately evaluate their bearing capacities. It can be found that the density of vascular bundles increased proportionally with the height of the bamboo stem, while the circumference exhibited a linear decrease. The wall thickness of the bamboo decreased and stabilized after reaching a height of 10 m. The distribution of nodes exhibited a nearly symmetrical pattern from the base to the top of the bamboo stem. The tapering of the bamboo culm exhibited a non-linear pattern with height, characterized by an initial decrease followed by a slight increase ranging from 0.004 to 0.010. The vascular bundles in bamboo exhibited a functional gradient distribution, which had a 6:3:2 distribution ratio of vascular bundles in the wall's dense, transition, and sparse areas, respectively. The bionic cantilever beam incorporated characteristics of a hollow structure, a non-uniform distribution of nodes, and a certain amount of tapering, which effectively enhanced its flexural performance compared to the traditional ones. The thin-wall tube, featuring a "dendritic" partial pressure structure, demonstrated exceptional lateral compressive performance in transverse compression, particularly when the tube incorporated a gradient distribution of partition numbers and layer spacing.

Keywords: bamboo bionic design; variations of structural characteristics; gradient distribution; bionic applications

Citation: Ji, S.; Mou, Q.; Li, T.; Li, X.; Cai, Z.; Li, X. The Novel Applications of Bionic Design Based on the Natural Structural Characteristics of Bamboo. *Forests* **2024**, *15*, 1205. https://doi.org/10.3390/f15071205

Academic Editor: Chunping Dai

Received: 25 June 2024
Revised: 8 July 2024
Accepted: 11 July 2024
Published: 12 July 2024

1. Introduction

The principle of natural selection and survival of the fittest was widely acknowledged to be universally adhered to by all living organisms in nature. Biological structures gradually evolved over time to attain a state of optimal equilibrium [1], resulting in the development of an organism's own optimal structure and the acquisition of distinctive functionalities to adapt to the intricate and dynamic natural environment [2]. The applications of bionic thinking have been extensive throughout history, spanning from ancient times to the present day, and have significantly contributed to the development of numerous innovative products (Figure 1a) [3–6]. Therefore, a comprehensive exploration of the biological structures and functions in nature not only facilitated a profound comprehension of the natural world but also held the potential to address significant challenges and complexitie in science and technology.

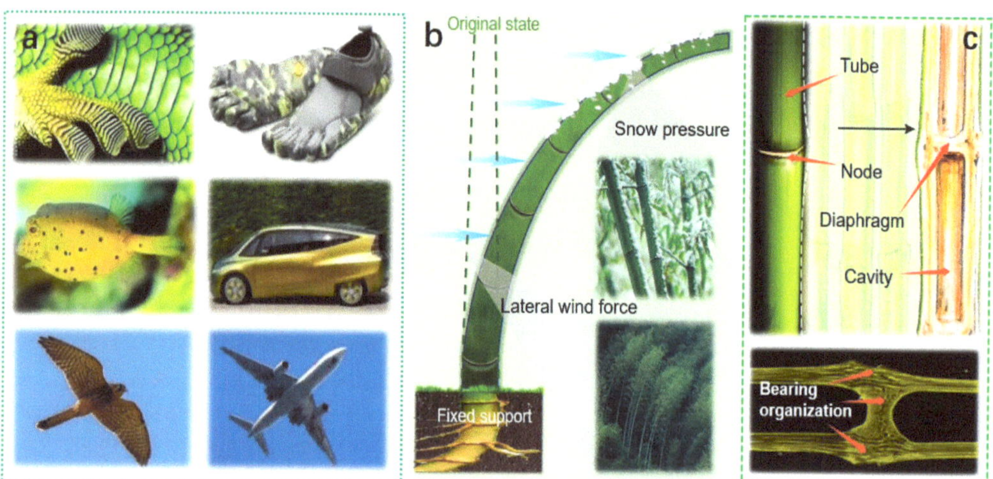

Figure 1. Bionic design examples and structural characteristics of the bamboo culm. (**a**) Bion examples; (**b**) structural advantages of bamboo; (**c**) macro-structural characteristics of bamboo.

Bamboo was characterized by a conical hollow shape that gradually tapered from th base to the top (Figure 1b), consisting of alternating internodes and nodes in the stem in macroscopic structure (Figure 1c) [7,8]. The distribution pattern showed a higher concer tration of fibers in the outer region and a sparser arrangement in the inner region [9–11]. microscopic analysis revealed that bamboo was characterized as a biocomposite materia wherein the thick-walled fibers were arranged axially to function as the reinforcement phas while the thin-walled tissue acted as the matrix phase [12–15]. This unique composite gr dient structure of bamboo contributed to its exceptional flexibility and resilience, enablin it to withstand the forces of wind and snow without fracturing (Figure 1b) [16,17]. Cons quently, bamboo has been widely recognized as one of nature's most efficient structure and materials [8,15,18–21]. Moreover, bamboo embodied profound principles of materia design and structural mechanics [22,23], making it highly applicable in engineering and a excellent candidate for the bionic design of composite materials [10,17,24].

The scarcity of resources and the implementation of the two-carbon strategy hav made lightweight design an inevitable trend in engineering material development [25 This trend played a pivotal role in modern society by providing essential load-bearin and functional capabilities across various sectors, including construction, the automotiv industry, and other areas [26,27]. One of the primary approaches to achieve this wa through the reduction in wall thickness, incorporating hollow design, component miniatu ization, and the utilization of composite materials [28]. These advancements also pose challenges for conventional structural materials, such as natural wood and bamboo. Fort nately, the emerging interdisciplinary field of bionics offered robust technical, ideologica and methodological support for the research and production of innovative materials an products [29,30].

This study was inspired by the gradient structural characteristics of bamboo that served as bionic thinking. Firstly, variations of the vertical and radial structural characteristics of bamboo were investigated experimentally. Next, the bamboo structure was inspired to develop two novel bionic applications based on the macro- and meso-scale structure of the bamboo culm, respectively, and the bionic cantilever beam was developed by utilizing the macro-structure of the bamboo culm, which featured hollow tubes, an array of nodes, and specific longitudinal tapering. This design optimization aimed to enhance stiffness and stability when subjected to lateral load bearing. The other bionic structure was a thin-walled tube, which was designed to mimic the characteristics of vascular bundle distribution density in order to enhance its lateral load-bearing capacity. Finally, the feasibilities of both the novel bionic applications were theoretically analyzed through numerical analysis by varying the structural parameters in order to obtain the optimal designing schemes. The findings are advantageous for the development of lightweight and high-strength advanced materials for engineering applications.

2. Materials and Methods

2.1. Material

This study involved the collection of 5 Moso bamboo (*Phyllostachys pubescens*) from Taojiang County in Hunan Province, China, each with a growth cycle of 4 years. Each culm was precisely cut to a length of 2 m, spanning from the base to the top of the stems, and was subsequently labeled with sequential numerical identifiers.

2.2. Method

2.2.1. Culm Structural Characteristics Determination

The length and circumference of each internode were measured using a vernier caliper and tapeline, with an accuracy of 0.01 mm and 0.1 mm, respectively, from the base to the top of each culm. To ensure precise measurements, the diameters of each culm were measured in two perpendicular directions at both ends, and the average value of these measurements was taken as the final diameter for each end (Equations (1) and (2)). Next, the wall thicknesses at both ends of each culm were measured at four designated locations and then averaged using Equations (3) and (4) to obtain the final thickness. Finally, the tapering of each culm can be determined by employing Equation (5), which is associated with the diameters at both ends of the culm. The purpose was to examine the variations and correlations of culm structural characteristics along the height of the bamboo. The measurement process is illustrated in Figure 2a–f.

$$D = (D_1 + D_2)/2 \tag{1}$$

$$d = (d_1 + d_2)/2 \tag{2}$$

$$T = (T_1 + T_2 + T_3 + T_4)/4 \tag{3}$$

$$t = (t_1 + t_2 + t_3 + t_4)/4 \tag{4}$$

$$\alpha_e = (D - d)/L \tag{5}$$

where D and d are the mean diameters of the large and small end of the culm (mm), respectively; D_1 and D_2, d_1 and d_2 are the measured diameters at the large and small ends (mm), respectively. T_1, T_2, T_3, and T_4 are the measured wall thicknesses in the four locations at the large end (mm), while t_1, t_2, t_3, and t_4 are for the small ends (mm); α_e is the tapering of the culm and L is the culm's length.

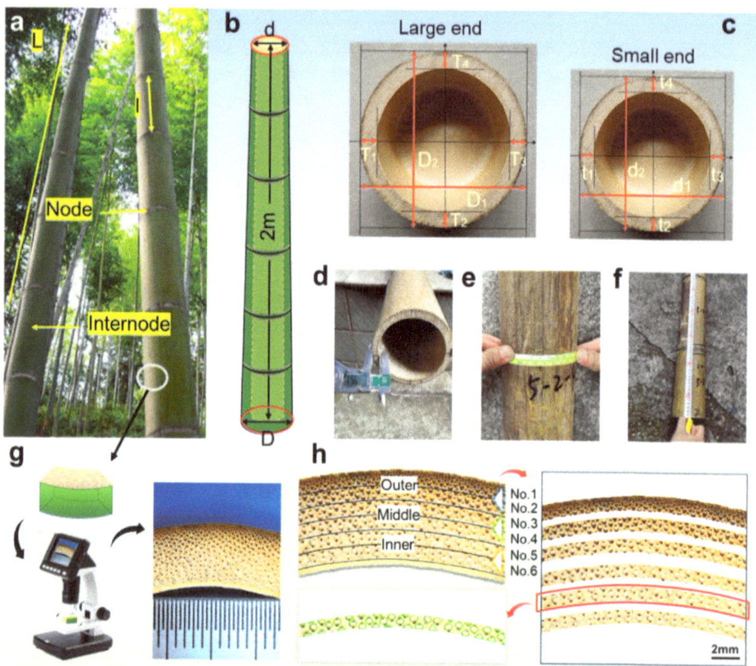

Figure 2. Measurement processes of the structural characteristics of the bamboo. (**a**) Configuration of the bamboo (L = height, I = internode length); (**b**) 2 m truncated culm; (**c**) specific sizes of the large and small ends of the bamboo culm; (**d**–**f**) determination of wall thickness, circumference, and internode length, respectively; (**g**) microstructure sampling observation; (**h**) image processing. The image was partitioned into 6 equal sections from the outside to the inside (No.1–No.6), and then the vascular bundle area of each section was manually annotated.

2.2.2. Vascular Bundle Distribution Measurement

The samples were cut into cubic shapes measuring approximately 4 mm and subsequently underwent a drying process to prepare them for the scanning electron microscope (SEM) experiment. The samples were coated with a layer of gold using sputtering techniques to enhance their electrical conductivity. The binding areas between the vascular bundle tissue and parenchyma cells were observed using an SEM (Tescan mira LMS, Brno, Czech Republic) operating at 3.0 kV.

The cross-sectional surface area of the samples was polished to obtain images using a digital microscope (Figure 2g). They were subsequently divided into six equal segments along the wall thickness direction using AutoCAD 2017 and Photoshop 25.9.1 software. The division was performed to facilitate the utilization of Image-Pro Plus 6.0 software (IPI 6.0, Media Cybernetics, Inc., Rockville, MD, USA) for accurately calculating the proportion of fiber tissue for each segment. As shown in Figure 2h, the layers of No. 1 and No. 2 were identified as outer layers, the middle layers were No. 3 and No. 4, and the remaining two considered the inner layers. The distribution of vascular bundle density (n/mm^2) was determined by quantifying the number of vascular bundles (n) in each layer, revealing the radial gradient distribution law of the vascular bundles.

2.2.3. Bionic Application of the Bamboo Structure
Bionic Bamboo Cantilever Beam

(1) Structural configuration. The cantilever beam models were designed based on the biological structural features of a bamboo stem, which included the distribution of nodes and a tapered shape that gradually narrowed from the bottom to the top along the

culm. Our study involved the design of four types of constitutive models for cantilever beam loading and investigated the impact of the culm's structural characteristics on their bending performance through finite element analysis. Moreover, the impact of nodes on the bearing performance of the beams was further examined by incorporating an equal number of diaphragms into each beam to simulate the nodes, with varying distribution patterns along the stem (Figure 3). The control model consisted of diaphragm plates that were evenly spaced, while different spacing configurations were utilized for the distribution of a diaphragm plate in the experimental group. As shown in Figure 3, Model I demonstrated a higher density of partitioning at both ends and a lower density in the middle. In contrast, Model II initially exhibited dense partitioning and gradually transitioned to sparse partitioning. The behavior of Model III mirrored that of Model II but had an increased longitudinal tapering value of 0.01, according to the result determined in Section 3.1. The bionic inspiration will be detailed in the Inspiration and Similarity Analysis Section.

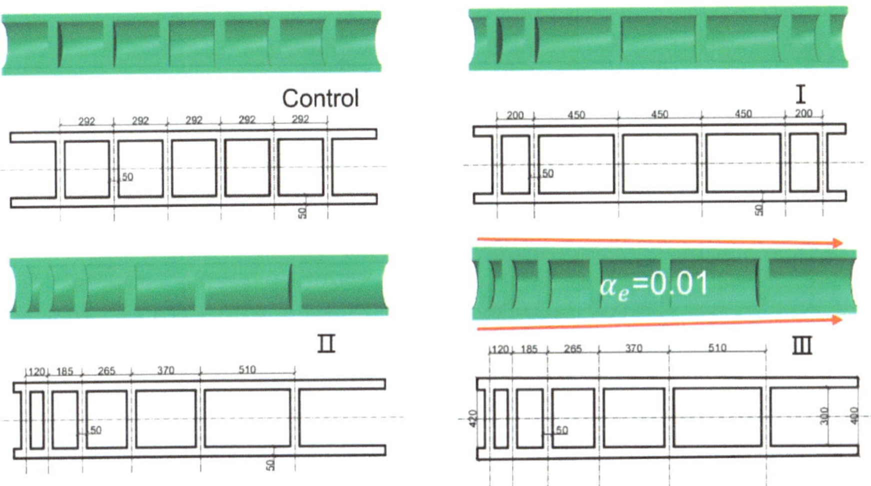

Figure 3. The details of the bionic bamboo cantilever beams.

(2) Numerical simulation. ABAQUS 6.14 was used for structural simulation analysis in this study. Firstly, the adoption of a solid element with stretchable and deformable characteristics was considered to ensure model stability while optimizing computing resources. Steel, being a conventional structural material, exhibited isotropic characteristics with a Young's modulus (E) of 200 GPa and a Poisson ratio (ν) of 0.3, which were utilized to replicate real-world loading conditions. In the estimation process, a concentrated force of 1500 N was exerted on one end of the cantilever beam structure and was aligned parallel to its cross-section while ensuring complete fixation at the opposite end. The seed mesh density was used with an approximate global size of 57 while maintaining the default value of 0.1 for the curvature control and minimum size factor. The proper segmentation was achieved by utilizing C3D10 tetrahedral elements for constructing solid components and supplementing them with mapped triangular meshes on boundary surfaces when necessary. The comprehensive analysis step was incorporated with a time period of 1, non-linear geometry was disabled, a maximum number of increments was set to 100, and the minimum increment step size was established as 10^{-5}. The analysis step parameters and mesh sizes were uniformly configured across all groups.

Bionic Bamboo Thin-Walled Tube

(1) Structural configuration. The multi-layer inner tubes were designed as a load bearing support for traditional thin wall steel tubes based on the distribution pattern of vascular bundles and the "dendritic" partial pressure structure formed in the thin wall. The design of all bionic bamboo tubes aimed to simulate the natural culm shape of bamboo, featuring a length of 2000 mm, an outer ring diameter of 120 mm, and an inner ring diameter of 60 mm. The bionic thin-walled tubes consisted of three layers, each with distinct geometric dimensions and wall thicknesses as shown in Figure 4. The ratios of biomimetic units in vascular bundles were set to 12:12:12 and 16:12:8 between layers, respectively, ensuring uniform distribution while maintaining equal thickness for all bionic units. Furthermore, in order to optimize the cross-sectional effect even further, the spacing of the inner tubes was determined based on Sec. 1 and 3 using a ladder distribution to simulate the density distribution characteristics of the vascular bundle in bamboo. The bionic inspiration of the tubes will be detailed in the Inspiration and Similarity Analysis Section.

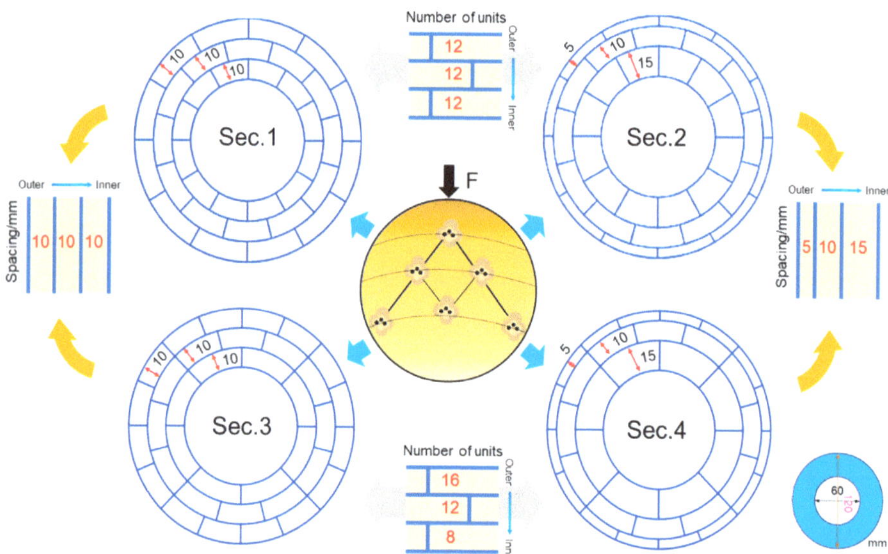

Figure 4. Four bionic sections of dendritic partial pressure for the thin-walled tube.

(2) Numerical simulation. After considering the trade-off between model stability and computational resources, the bionic bamboo tubes those simulated using ABAQUS 6.14 software and employed S4R shell elements for accurate representation. Additionally, the primary shape of the quadrilateral elements was utilized through appropriate partitioning. The seed mesh density was also set to approximately 57 for global size while maintaining the default value of 0.1 for the curvature control and minimum size factor. The rigid plate adopted an R3D4 discrete rigid element and utilized an advanced algorithm. The lower surface of the tube was constrained against a fixed rigid wall and then compressed by the upper rigid wall to achieve a deformation of 5 mm. The lower rigid plate was constrained in all six degrees of freedom, and the displacement boundary condition of the upper rigid plate in the diameter direction of the thin-walled pipe was set to 5. The "surface-to-surface" contact algorithm was utilized to simulate the contact behavior during the deformation progress. The contact surface employed penalty contact with a friction coefficient of 0.2 for the tangential direction while adopting rigid contact in the normal direction to allow for post-contact separation and enable finite sliding. The parameter configuration for the analysis step was identical to the setups utilized in the stimulation

process of the bionic bamboo cantilever beam referred to above. The steel Q235 alloy material utilized in this study exhibited a yield strength of 158 MPa, an ultimate stress of 208 MPa, a Young's modulus of 200 GPa, and a Poisson ratio of 0.3. It was important to note that the consideration of material failure for the steel tube was excluded in this study.

3. Results and Discussions

3.1. Structural Characteristics of the Bamboo Culm

The results depicted in Figure 5a,b demonstrate a decreasing trend in both the wall thickness and circumference of the internode and an increase in height. Specifically, the wall thickness gradually reduced as the height increased [31,32], eventually reaching a plateau at 10 m. This variation can be accurately described by a quadratic equation, exhibiting a coefficient of determination (R^2) greater than 0.922. In contrast, the circumferences demonstrated a linear decrease with height and displayed high R^2 values exceeding 0.986. The variation in the wall thickness and circumference of the culm can also be visually observed by examining the cross-sections at the bottom, middle, and top locations (Figure 5d), where there was an increased density of vascular bundles distributed with the culm height. The findings indicated a potential for a height-based prediction of wall thickness and diameter at a specific location in the stem. Moreover, the length of internodes demonstrated an initial increase, followed by a subsequent decrease upon reaching a height of approximately 6~8 m along the stem (Figure 5c). It was noteworthy that the internode length along the culm displayed a nearly symmetrical pattern throughout the height of bamboo stems and was characterized by a parabolic shape [33], which reflected an intriguing biological trait. Furthermore, a notable correlation was observed between the internode circumference and length for each individual stem, which can also be accurately represented by a quadratic equation, with R^2 values exceeding 0.922 (Figure 5e). The correlations facilitated the establishment of a quantitative model linking the length and diameter of the bamboo internodes at any height.

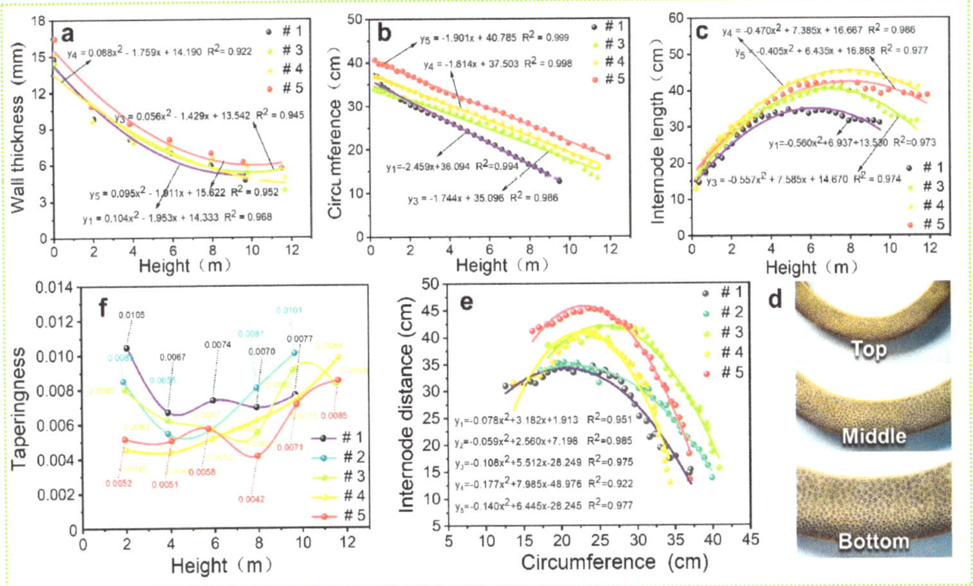

Figure 5. Structural size parameters of bamboo. (**a–c**) Correlations of wall thickness, circumference, and internode distance with height, respectively; (**d**) cross-section surfaces of bamboo at different locations; (**e**) correlation between internode length and circumference; (**f**) variation of tapering with height.

As shown in Figure 5f, the tapering values of bamboo exhibited a non-linear trend characterized by an initial decrease followed by a slight increase with increased height, ranging from 0.004 to 0.010. The variation in tapering can be attributed to the decrease in the wall thickness and circumferences of the internodes from the bottom to the top of the culm, as illustrated in Figure 5a,b. It can be found that the tapering at a height of 4–6 m in #1 and #5 exhibited slight irregularities for a slightly higher degree compared to the others. The observed deviation in the end diameter within the culms at a height of 4~6 m, which coincided with the cross-section at the node, can be attributed to an increased tapering effect. Additionally, the taperingness at the top and bottom of the culms exceeded the specified requirements in ISO 22156 and Indian building standards for the bamboo culm (less than 0.0058) [34] but fell within the acceptable limit set by Colombian NSR 10 (less than 0.01) [35], which were higher compared to the middle location. Although the parallel misalignment of bamboo stems may be adequate for general use, it is advisable to establish regulations governing the length of the culms in order to mitigate the impact of tapering when they are utilized as structural components.

3.2. Microstructure and Vascular Bundle Distribution

3.2.1. Microstructure

As depicted in Figure 6a, a two-phase composite structure composed of vascular bundle fibers and parenchyma cells was formed through a series-parallel connection. The vascular bundles consisted of numerous lignified cells with densely packed tissues, while the parenchyma tissue was characterized by interconnected yet relatively dispersed hollow cell structures. The fibrous femoral region was defined by the top and bottom locations of the vascular bundles, where densely packed lignified cell structures provided exceptional strength and toughness to the bundles. Furthermore, the presence of vascular bundles contributed to the overall higher density in comparison to the parenchyma tissue that consists of thin-walled cells. Additionally, significant variations were confirmed in the chemical compositions at these locations, thereby exerting a profound influence on the mechanical properties on the macro level. These vascular bundles were essential for maintaining stability and resistance to deformation, especially under additional loading owing to their abundant lignified cells, while the parenchyma tissues, characterized by their thin wall thickness and low density, facilitated the connection and transfer of loads [36].

3.2.2. Vascular Bundle Distribution

The vascular bundles were uniformly distributed within the flexible parenchyma cells in a gradient, forming a natural parallel arrangement resembling "island chains" [37]. They displayed a functional gradient distribution in the radial direction, characterized by compact and diminutive outer layers while progressively increasing in size and becoming more porous for the inner layers (Figure 6a). Moreover, the ratio of fiber tissue gradually increased from the inner to the outer wall in a linear pattern, with a coefficient of determination R^2 equal to 0.978 (Figure 6b). From the perspective of chemical composition, the distribution of cellulose, hemicellulose, and lignin in the radial direction of the bamboo wall exhibited unevenness [38] and also showed a gradual pattern, which greatly enhanced the mechanical properties. The bamboo wall can be classified into three regions based on the distribution pattern of the vascular bundles: dense, transition, and sparse. The ratios of the vascular bundles in these regions were 6.48 n/mm^2, 3.01 n/mm^2, and 1.96 n/mm^2 respectively (Figure 6c). It was worth noting that the significant proportionate distribution of the vascular bundles among the three regions had a ratio of 6:3:2, which held substantial academic significance. The gradual decrease in the volume fraction and density of the vascular bundles along the radial direction from the outer to inner layers of the bamboo wall also significantly contributed to variations in physical and mechanical properties, as well as chemical composition. The variation of the vascular bundles in the radial direction would serve as a fundamental for the bionic design in Section 3.3.

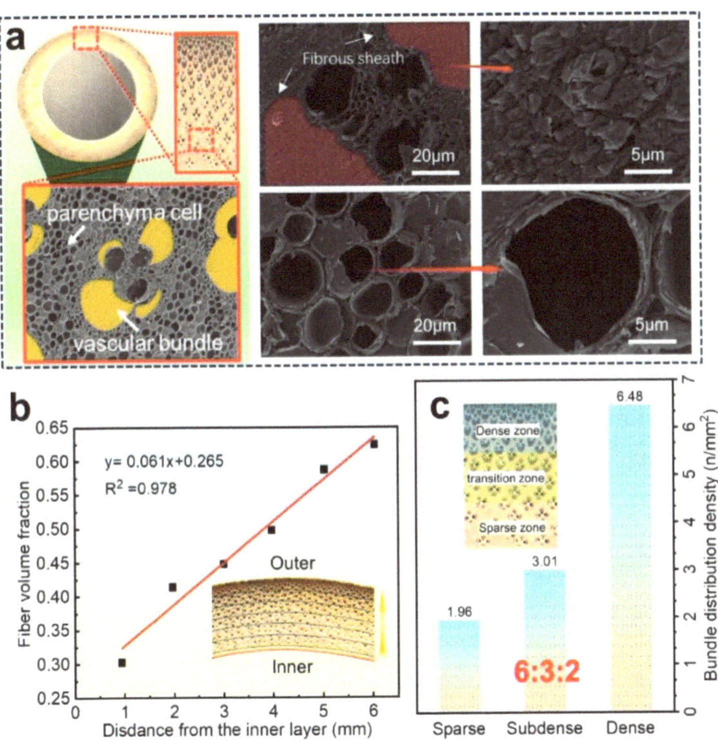

Figure 6. The structural characteristics of bamboo in a cross-section. (**a**) Microstructure characteristics; (**b**) fiber volume fraction; (**c**) variation of vascular bundle density.

3.3. Bionic Application of the Bamboo Structure

3.3.1. Bionic Cantilever Beam

The Analysis of Inspiration and Similarity in Cantilever Beam

The bearing behavior of bamboo was analogous to that of the cantilever beam loading configuration when subjected to lateral pressures during its growth cycle [17], with the root system serving as an immovable support fixed within the terrestrial medium (Figure 1b). The prototype of the bionic design can be considered as an optimal solution for enhancing the lateral bearing capacity improvement in a cantilever beam, which was attributed to the thin wall and hollowness of the bamboo stem. These not only contributed to its excellent energy efficiency during the bending bearing processes but also shared similarities in the shape and structural bearing behavior of a bamboo stem. Furthermore, the variations in both the wall thickness and length of each internode along the stem were also crucial factors in achieving a nearly uniform distribution of compressive stresses when subjected to its own weight, thereby ensuring that each cross-section displayed comparable resistance to bending deformation caused by lateral loads [39]. Additionally, the presence of nodes and diaphragms further enhanced the bending capacity and stability of bamboo, effectively serving as a dense lateral reinforcement mechanism. This contributed to the exceptional natural composite structural beam composed of internodes and nodes (Figure 1c). Also, the uneven distribution of nodes along the stem not only enhanced its stiffness but also contributed to its overall structural integrity [40].

Bending Performances

The midpoint of the free end section of the model was selected as the control poir (RP-1) (Figure 7a), allowing for the plotting of a load–displacement curve. As depicte in Figure 7b, the four beams demonstrated elastic deformation during the numerica simulation process. It was worth noting that Model III exhibited significantly smalle displacement compared to the others when subjected to an identical force. The findin further confirmed that incorporating longitudinal tapering and partition distribution, alon, with gradient distribution in line with the bamboo stem, can effectively enhance the flexura performance of traditional cantilever beams.

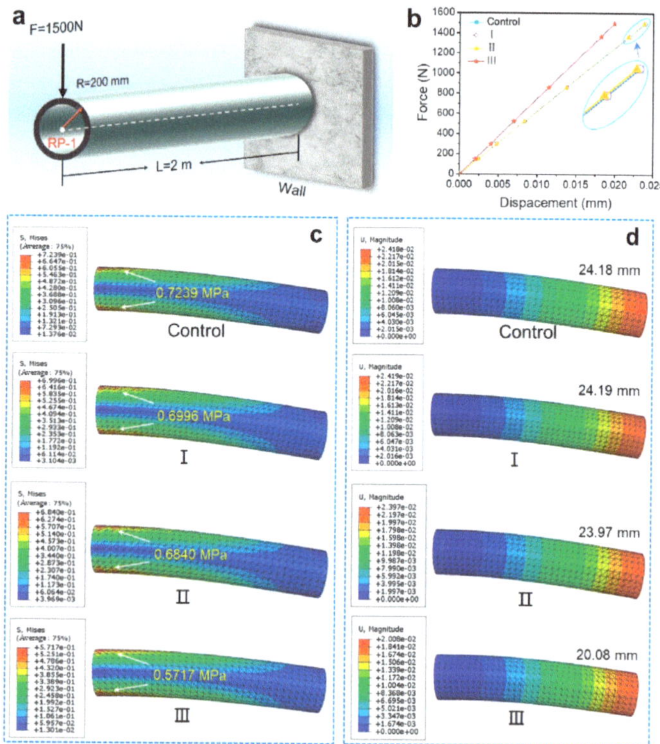

Figure 7. The bearing behavior of the Bionic cantilever beam. (**a**) Working diagram; (**b**) load–displacemen curve; (**c**) stress distribution contour plot; (**d**) displacement contour plot.

The stress distribution cloud maps of the bionic beams were presented in Figure 7c illustrating that the stress concentration areas of all the cantilever beams were situatec above and below the junction with the wall. Additionally, it can be observed that the stres: gradually decreased as the distance of the concentrated load end reduced. The maximum stresses of Models I (0.6996 MPa), II (0.6840 MPa), and III (0.5717 MPa) were all lowe than that of the control group (0.7239 MPa). This indicated that non-uniformly distributec diaphragm plates effectively enhanced the bending performance of the beams compared tc the uniformly spaced ones, where the distribution of diaphragm plates was simulated tc that of the nodes. The performance of Model II, characterized by a sparse-to-dense patterr of diaphragm plates, outperformed that of Model I with narrow ends and a wide middle Furthermore, the introduction of a certain degree of longitudinal tapering can furthe enhance the bending properties of cantilever bamboo beams. All the displacement clouc maps of the bamboo beams are shown in Figure 7d, illustrating the maximum displacemen at each load end. The displacement change sequence under the same lateral force was as follows: Model I > Control > Model II > Model III. It can be concluded that the incorporation of transverse diaphragms and appropriately increasing taper can effectively enhance the flexural performance of cantilever beams, leading to an optimal bending performance when arranging the diaphragm plates from dense to sparse.

3.3.2. Bionic Thin-Walled Tube

The Analysis of Inspiration and Similarity in Thin-Walled Tube

The exceptional mechanical properties of bamboo, including compressive resistance, toughness, and impact resistance, can be ascribed to the unique thin-walled hollow structure and microstructural distribution characteristics inherent in bamboo [41]. The vascular bundles in the cross-section of bamboo exhibited a predominantly rhomboid shape, with an increasing size gradient and a sparser distribution towards the inner layer. This arrangement of the structure divided the parenchyma within the vascular bundles system into a dendritic structure, facilitating stress transmission along the parenchyma cell tissue while deflecting and offsetting it within the rigid region of vascular bundle tissue. This resulted in a "dendritic" partial pressure structure formed by interconnecting parenchyma cells within the bamboo wall (Figure 8a). The external force that was laterally applied to the bamboo resulted in its penetration into the thin-walled structure, with a higher concentration observed in the outer regions and a lower degree within the inner regions where the vascular bundles were presented (Figure 8b). The presence of these vascular bundles weakened the external load by dividing it into two forces, ultimately reaching the inner layer of the bamboo. The circular structure of the bamboo generated converging forces that counterbalanced each other, effectively decomposing external impact loads layer by layer and achieving exceptional impact resistance and energy absorption (Figure 8c). This further indicated that the gradient structure of the wall can enhance the strength, toughness, and deformation resistance of the bamboo.

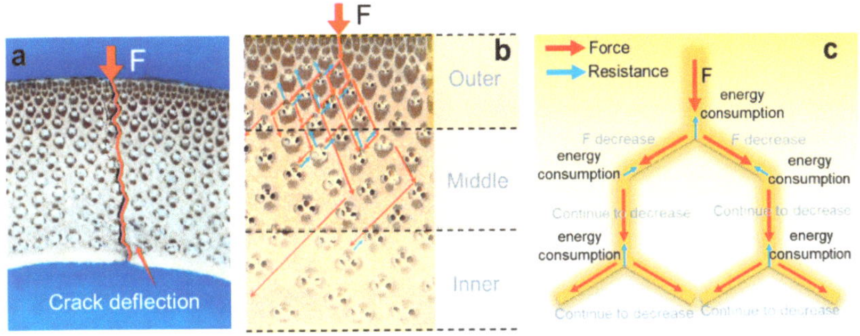

Figure 8. Exceptional distribution characteristic of bamboo bundles. (**a**) Macroscopic crack growth of bam-boo; (**b**) radial stress partial pressure path of bamboo wall section; (**c**) partial pressure mechanism.

Radial Compressive Performance

By designating the center point of the contact between the bionic tubes and upper-end plate as the control point (RP-2, Figure 9a), we ascertained the Y-axis reaction force of the beam from the field output and graphed load–displacement curves. As depicted in Figure 9b, the compressive reaction force of Sec. 1 was significantly higher compared to others when subjected to an equal displacement. Conversely, Sec. 4 exhibited a reaction force of less than 25 N, indicating minimal resistance. The aforementioned observation emphasized the exceptional transverse compressive performance and a certain level of transverse energy absorption effect exhibited by the bionic thin-wall tubes. Furthermore, it suggested that Sec. 4, which incorporated a gradient distribution of partition elements and layer spacing within the tubes, demonstrated superior lateral resistance performance.

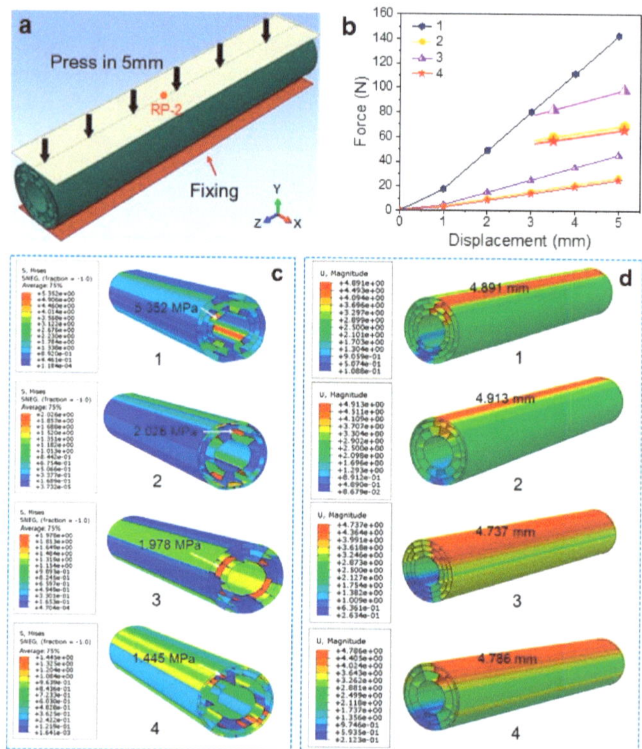

Figure 9. Lateral bearing behavior of the bionic thin-walled tube. (**a**) Working diagram; (**b**) load displacement curve; (**c**) stress distribution contour plot; (**d**) displacement contour plot.

The variations of the deformations and stresses of the tubes are shown in Figure 9c,d. The results revealed that the maximum stress of all the tubes remained below the yield stress of steel, indicating that only elastic deformation occurred. This observation suggested that the bionic tubes exhibited excellent radial bearing capacity. It was worth noting that the maximum stress on the innermost surface in Sec. 1 reached a value of 5.352 MPa. In Sec. 2, both the inner and second layers exhibited a maximum stress of 2.026 MPa. The wall surface of the second layer reached a maximum stress of 1.978 MPa in Sec. 3. Both the second- and third-layer wall surfaces obtained a maximum value of 1.445 MPa in Sec. 4. Consequently, it can be concluded that Sec. 4 demonstrated a superior lateral compressive capacity among the designed tubes. The design of the internal pressure cylinder and radial partition enabled the bionic tubes to withstand a higher lateral pressure while maintaining a low internal stress state. The radial load performance of the tubes can be significantly enhanced by incorporating a certain proportion of divider units while keeping the total number unchanged. Also, the compression performance can be further improved by adjusting the interlayer spacing, which was first narrow and then wide.

The maximum displacements of the tubes were all located beneath the rigid plane where the displacement was applied, as depicted in Figure 9d. The deformation of the internal structure of the tubes was found to be significantly influenced by the spacing between the layers, as evidenced by comparing the displacement cloud maps of all the tubes. The reduction in layer spacing, as opposed to variations in the number of partition elements, resulted in a relative increase in partition density. This increase contributed to further decomposition and weakening of the external pressure. The distribution of partition elements in the cross-sections was similar to that of vascular bundles, where the latter functioned as nodes with high hardness. When subjected to lateral pressure, the outer layer of the tubes would primarily resist it. As a result, it can be concluded that Sec. 2 and 4 exhibited a more stable internal structure compared to Sec. 1 and 3.

It is widely acknowledged that with the advancement of green and sustainable construction practices, increasing attention is being directed towards utilizing original bamboo for engineering purposes due to its excellent structural characteristics [16]. However, the enhancement of bamboo's durability posed a significant challenge when considering its application as a building material, primarily due to its susceptibility to degradation under special levels of humidity and temperature. Based on the aforementioned investigation, it has been determined that the mechanical advantages of bamboo structures have been analyzed through theoretical approaches, further validating their value for bionic design in engineering applications. Therefore, developing new sustainable materials with high durability and strong mechanical performance by leveraging the typical structure characteristics of bamboo becomes crucial for future endeavors.

4. Conclusions

The robustness and consistency of bamboo were well-known due to its unique multi-scale structure, which contributed to exceptional mechanical performance and bearing stability. The present study aimed to investigate the variations of structural parameters in natural bamboo, with the objective of extracting bionic structural characteristics. The biological structural characteristics of bamboo were utilized to develop two bionic applications, which underwent finite element analysis to simulate bearing conditions. This approach aimed to optimize the practice applications by leveraging the exceptional structure of bamboo. The main conclusions derived from this research were as follows:

(1) The distribution density of vascular bundles increased proportionally with the height of the bamboo stem, while the circumference exhibited a linear decrease. The wall thickness decreased and stabilized after reaching 10 m. The distribution of nodes on the stem exhibited a nearly symmetrical pattern in the vertical direction of the stem, characterized by a parabolic shape. There was an initial increase followed by a subsequent decrease in internode distance upon reaching a height of approximately 6~8 m along the stem. The culm tapering of bamboo exhibited a non-linear pattern with height, characterized by an initial decrease followed by a slight increase ranging from 0.004 to 0.010. The wall thickness, circumference, internode distance, and height were closely interrelated.

(2) Bamboo represented a two-phase composite structure comprising reinforcement provided by vascular bundles and a basic body composed of parenchyma cells. The vascular bundles in bamboo exhibited a functional gradient distribution, characterized by smaller and denser outer bundles and gradually increasing size and looseness in the inner ones and were arranged radially. The distribution ratio of vascular bundles in the wall's dense, transition, and sparse area of the wall was 6:3:2, respectively.

(3) The structure of the culm that conferred excellent flexural behavior without fracture was identified and utilized to optimize the design of the conventional cantilever beam. The bionic cantilever beam incorporated the characteristics of a hollow structure, a non-uniform distribution of nodes, and a certain amount of longitudinal tapering, which effectively enhanced its flexural performance compared to the traditional ones.

(4) The thin-wall steel tubes were designed with a "dendritic" partial pressure structure and were optimized based on the gradient distribution of vascular bundles and the arrangement in a multi-layer manner. These tubes exhibited exceptional lateral compressive performance in transverse compression, particularly when the tubes with a gradient distribution of partition numbers and layer spacing experienced reduced stress levels. They possessed a more stable structure and demonstrated superior bearing performance when they were under an identical lateral compression deformation.

Highlight

1. The structural characteristics of bamboo was systematically summarized.
2. The quantitative relationships between wall thickness, circumference, internode distance and height were established.

3. An optimized bionic cantilever beam featured a hollow structure, non-uniform distri bution of nodes, and a gradual longitudinal tapering. Vertical structural characteriza tions were also developed.

4. An optimized thin-walled tube exhibited a "dendritic" configuration of partial pres sure characterized by radial distribution of vascular bundles.

Author Contributions: S.J.: investigation and writing—original draft. Q.M.: investigation and writing—original draft. T.L.: investigation. X.L. (Xiazhen Li): methodology and writing—review and editing. Z.C.: writing—review and editing. X.L. (Xianjun Li): conceptualization and supervision. All authors have read and agreed to the published version of the manuscript.

Funding: This work was supported by the Natural Science Foundation of Hunan Province (2023JJ6016, 2024JJ8250), the Scientific Research Fund of Hunan Provincial Education Department (23A0234), and the Technological Innovation Fund of Hunan Forestry Department (XLKY202452).

Data Availability Statement: The data presented in this study are available on request from the corresponding author due to privacy.

Conflicts of Interest: Author Ting L. was employed by the company Hunan Taohuajiang Bamboo Science & Technology. The remaining authors declare that the research was conducted in the absence of any commercial or financial relationships that could be construed as a potential conflict of interest.

References

1. Zou, M.; Song, J.; Xu, S.; Liu, S.; Chang, Z. Bionic Design of the Bumper Beam Inspired by the Bending and Energy Absorption Characteristics of Bamboo. *Appl. Bionics. Biomech.* **2018**, *2018*, 8062321. [CrossRef] [PubMed]
2. Xing, D.; Chen, W.; Ma, J.; Zhao, L. Structural bionic design for thin-walled cylindrical shell against buckling under axial compression. *Proc. Inst. Mech. Eng. Part C J. Mech. Eng. Sci.* **2011**, *225*, 2619–2627. [CrossRef]
3. Xu, Z.; Gao, W.; Bai, H. Silk-based bioinspired structural and functional materials. *Science* **2022**, *25*, 103940. [CrossRef] [PubMed]
4. Mao, A.; Zhao, N.; Liang, Y.; Bai, H. Mechanically Efficient Cellular Materials Inspired by Cuttlebone. *Adv. Mater.* **2021**, *3*, 202007348. [CrossRef]
5. Chen, S.M.; Gao, H.L.; Zhu, Y.B.; Yao, H.B.; Mao, L.B.; Song, Q.Y.; Xia, J.; Pan, Z.; He, Z.; Wu, H.A.; et al. Biomimetic twisted plywood structural materials. *Natl. Sci. Rev.* **2018**, *5*, 703–714. [CrossRef]
6. Xu, P.; Guo, W.; Yang, L.; Yang, C.; Ruan, D.; Xu, J.; Yao, S. Crashworthiness analysis of the biomimetic lotus root lattice structure. *Int. J. Mesh. Sci.* **2024**, *263*, 108774. [CrossRef]
7. Ray, A.K.; Mondal, S.; Das, S.K.; Ramachandrarao, P. Bamboo-A functionally graded composite-correlation between microstructure and mechanical strength. *J. Mater. Sci.* **2005**, *40*, 5249–5253. [CrossRef]
8. Li, J.; Ma, R.; Lu, Y.; Wu, Z.; Liu, R.; Su, M.; Jin, X.; Zhang, R.; Bao, Y.; Chen, Y.; et al. Bamboo-inspired design of a stable and high-efficiency catalytic capillary microreactor for nitroaromatics reduction. *Appl. Catal. B Environ.* **2022**, *310*, 121297. [CrossRef]
9. Kanzawa, E.; Aoyagi, S.; Nakano, T. Vascular bundle shape in cross-section and relaxation properties of Moso bamboo (*Phyllostachys pubescens*). *Mate. Sci. Eng.* **2011**, *31*, 1050–1054. [CrossRef]
10. Wang, Y.-Y.; Wang, X.-Q.; Li, Y.-Q.; Huang, P.; Yang, B.; Hu, N.; Fu, S.-Y. High-Performance Bamboo Steel Derived from Natural Bamboo. *ACS Appl. Mater. Interfaces* **2020**, *13*, 1431–1440. [CrossRef]
11. Chen, Q.; Wang, G.; Ma, X.X.; Chen, M.L.; Fang, C.H.; Fei, B.H. The effect of graded fibrous structure of bamboo (*Phyllostachys edulis*) on its water vapor sorption isotherms. *Ind. Crops Prod.* **2020**, *151*, 112467. [CrossRef]
12. Sanchez, C.; Arribart, H.; Giraud, M.M. Biomimetism and bioinspiration as tools for the design of innovative materials and systems. *Nat. Mater.* **2005**, *4*, 277–288. [CrossRef] [PubMed]
13. Ghavami, K. Bamboo as reinforcement in structural concrete elements. *Cem. Concr. Comp.* **2005**, *27*, 637–649. [CrossRef]
14. Palombini, F.L.; Nogueira, F.M.; Kindlein, W.; Paciornik, S.; Mariath, J.; Oliveira, B.F. Biomimetic systems and design in the 3D characterization of the complex vascular system of bamboo node based on X-ray microtomography and finite element analysis. *Mater. Res.* **2019**, *35*, 842–854. [CrossRef]
15. Dixon, P.G.; Gibson, L.J. The structure and mechanics of Moso bamboo material. *J. R. Soc. Interface* **2014**, *11*, 0321. [CrossRef] [PubMed]
16. Han, S.; He, Y.; Ye, H.; Ren, X.; Chen, F.; Liu, K.; Shi, S.Q.; Wang, G. Mechanical Behavior of Bamboo, and Its Biomimetic Composites and Structural Members: A Systematic Review. *J. Bionic. Eng.* **2023**, *21*, 56–73. [CrossRef]
17. Xu, H.; Li, J.; Wang, H.; Xu, X. The unique flexibility feature of bamboo: Cantilever-beam loading form the coupling bending-shear effects. *Ind. Crops Prod.* **2023**, *205*, 117494. [CrossRef]
18. Palombini, F.L.; Mariath, J.; Oliveira, B. Bionic design of thin-walled structure based on the geometry of the vascular bundles of bamboo. *Thin Wall Struct.* **2020**, *155*, 106936. [CrossRef]
19. Chen, C.; Li, Z.; Mi, R.; Dai, J.; Xie, H.; Pei, Y.; Li, J.; Qiao, H.; Tang, H.; Yang, B.; et al. Rapid Processing of Whole Bamboo with Exposed, Aligned Nanofibrils toward a High-Performance Structural Material. *ACS Nano* **2020**, *14*, 5194–5202. [CrossRef]

20. Han, S.; Chen, F.; Li, H.; Wang, G. Effect of off-axis angle on tension failures of laminated moso bamboo-poplar veneer composites: An in situ characterization. *Mater. Des.* **2021**, *212*, 110254. [CrossRef]
21. Abdul, H.; Bhat, I.; Jawaid, M.; Zaidon, A.; Hermawan, D.; Hadi, Y.S. Bamboo fibre reinforced biocomposites: A review. *Mater. Des.* **2012**, *42*, 353–368. [CrossRef]
22. Chen, P.Y.; McKittrick, J.; Meyers, M.A. Biological materials: Functional adaptations and bioinspired designs. *Prog. Mater. Sci.* **2012**, *57*, 1492–1704. [CrossRef]
23. Ortiz, C.; Boyce, M.C. Bioinspired Structural Materials. *Science* **2008**, *319*, 1053–1054. [CrossRef] [PubMed]
24. Chen, Y.; Han, L.; Zhang, H.; Dong, L. Improvement of the strength and toughness of biodegradable polylactide/silica nanocomposites by uniaxial pre-stretching. *Int. J. Biol. Macromol.* **2021**, *190*, 198–205. [CrossRef] [PubMed]
25. Peng, X.; Zhang, B.; Wang, Z.; Su, W.; Niu, S.; Han, Z.; Ren, L. Bioinspired Strategies for Excellent Mechanical Properties of Composites. *J. Bionic. Eng.* **2022**, *19*, 1203–1228. [CrossRef]
26. Xia, M.; Sanjayan, J. Method of formulating geopolymer for 3D printing for construction applications. *Mater. Des.* **2016**, *110*, 382–390. [CrossRef]
27. Li, X.Z.; Ji, S.Y.; Li, T.; Liu, Z.X.; Hao, X.F.; Chen, Z.J.; Zhong, Y.; Li, X.J. The physical, mechanical and fire performance of bamboo scrimber processed with thermal-treated bamboo bundles. *Ind. Crop. Prod.* **2023**, *205*, 117549. [CrossRef]
28. Wei, C.G. *Crashworthiness Bionic Design and Analysis of Thin-Walled Tube Inspired by Bamboo Structure: Jilin University*; Jilin University: Jilin, China, 2014. (In Chinese)
29. Yong, L. Significance and Progress of Bionics. *J. Bionic Eng.* **2018**, *1*, 1–3. [CrossRef]
30. Yuan, Y.; Yu, X.; Yang, X.; Xiao, Y.; Xiang, B.; Wang, Y. Bionic building energy efficiency and bionic green architecture: A review. *Renew. Sustain. Energy Rev.* **2017**, *74*, 771–787. [CrossRef]
31. Chaowana, K.; Wisadsatorn, S.; Chaowana, P. Bamboo as a Sustainable Building Material-Culm Characteristics and Properties. *Sustainability* **2021**, *13*, 7376. [CrossRef]
32. Yang, X.M.; Chai, Y.; Liu, H.R.; Sun, Z.J.; Jiang, Z.H. Study on size classification of bamboo stem and round bamboo. *J. For. Eng.* **2019**, *4*, 6. (In Chinese)
33. Song, J.F.; Xu, S.C.; Wang, H.X.; Wu, X.Q.; Zou, M. Bionic design and multi-objective optimization for variable wall thickness tube inspired bamboo structures. *Thin Walled Struct.* **2018**, *125*, 76–88. [CrossRef]
34. de León, A.E.D. National Building Code of India and the International Building Code: An Introduction. In Proceedings of the Indo-US Forensic Engineering Workshop, Tiruchirappalli, India, 15 December 2010.
35. Wang, Q.P.; Liu, X.E.; Zhang, G.L.; Yang, S.M.; Ma, J.F.; Tian, G.L. Research progress in measurement methods and variation patterns of bamboo density. *World For. Res.* **2016**, *29*, 5. (In Chinese)
36. Xu, H.; Zhang, Y.; Wang, J.; Li, J.; Zhong, T.; Ma, X.; Wang, H. A universal transfer-learning-based detection model for characterizing vascular bundles in Phyllostachys. *Ind. Crops Prod.* **2022**, *180*, 114705. [CrossRef]
37. Wu, P. Biomimetic Design of Column Structure Based on Macro and Micro Characteristics of Bamboo. Master's Dissertation, Yanshan University, Qinhuangdao, China, 2015. (In Chinese).
38. Li, P.; Ji, S.; Mou, Q.; Zheng, X.; Li, X.; Li, X. Variations of the fundamental chemical components in Moso bamboo (*Phyllostachys pubescens*). *Wood Mater. Sci. Eng.* **2024**, *4*, 1–9. [CrossRef]
39. Zhang, T.; Wang, A.; Wang, Q.; Guan, F. Bending characteristics analysis and lightweight design of a bionic beam inspired by bamboo structures. *Thin Walled Struct.* **2019**, *142*, 476–498. [CrossRef]
40. Amada, S.; Ichikawa, Y.; Munekata, T.; Nagase, Y.; Shimizu, H. Fiber texture and mechanical graded structure of bamboo. *Compos. Part. B Eng.* **1997**, *28*, 13–20. [CrossRef]
41. Liang, H.; Hao, W.; Sun, H.; Pu, Y.; Zhao, Y.; Ma, F. On design of novel bionic bamboo tubes for multiple compression load cases. *Int. J. Mech. Sci.* **2022**, *218*, 107067. [CrossRef]

Article

Effects of Heat Treatment on the Chemical Composition and Microstructure of *Cupressus funebris* Endl. Wood

Jianhua Lyu [1,†], Jialei Wang [2,†] and Ming Chen [2,*]

1   Department of Environmental Design, College of Landscape Architecture, Sichuan Agricultural University, Chengdu 611130, China; ljh@sicau.edu.cn
2   Department of Product Design, School of Arts and Media, Sichuan Agricultural University, Ya'an 625014, China
*   Correspondence: chenming@sicau.edu.cn
†   These authors contributed equally to this work.

Abstract: The effects of heat treatment on *Cupressus funebris* Endl. wood were examined under different combinations of temperature, time, and pressure. The chemical composition, crystallinity, and microstructure of heat-treated wood flour and specimens were characterized using scanning electron microscopy (SEM), Fourier transform infrared spectroscopy (FTIR), and X-ray diffraction (XRD). Vacuum heat treatment led to changes in the functional groups and microstructure of *C. funebris* wood, and the relative lignin content decreased with increasing treatment temperature, which was significant at lower negative pressures. Cellulose crystallinity showed a change rule of first increasing and then decreasing throughout the heat treatment range, and the relative crystallinity ranged from 102.46% to 116.39%. The cellulose treated at 120 °C for 5 h at 0.02 MPa had the highest crystallinity of 44.65%. These results indicate that although heat treatment can improve cellulose crystallinity, very high temperatures can lead to decreased crystallinity. The morphology and structure of the cell wall remained stable throughout the heat treatment range; however, at elevated temperatures, slight deformation occurred, along with rupture of the intercellular layer.

Keywords: *Cupressus funebris* Endl. wood; heat treatment; chemical composition; microstructure

Citation: Lyu, J.; Wang, J.; Chen, M. Effects of Heat Treatment on the Chemical Composition and Microstructure of *Cupressus funebris* Endl. Wood. *Forests* **2024**, *15*, 1370. https://doi.org/10.3390/f15081370

Academic Editor: Anuj Kumar

Received: 25 June 2024
Revised: 26 July 2024
Accepted: 5 August 2024
Published: 6 August 2024

1. Introduction

Wood is one of the four main building materials, and it has renewable, recyclable and biodegradable properties [1]. China is rich in *Cupressus funebris* Endl. resources widely distributed in the Yangtze River Basin and the southern part of China. It is also an economically and environmentally important species in the hilly areas of central Sichuan where it is the largest tree species in plantation forests [2], and it is an irreplaceable main tree species in the calcareous purple soil and limestone soil areas of Sichuan Province [3]. There are more than one million hectares of *C. funebris* in Sichuan Province alone; therefore, it is important to study effective and feasible ways to improve the comprehensive utilization of this tree species [4]. *C. funebris* wood from Sichuan is also known as cedar wood, and it is commonly used as timber for construction, furniture, shipbuilding, and handicrafts, owing to its hardness, density, fine and uniform structure, and strong texture. The *C. funebris* wood exhibits a straight or sloping grain, with a medium and uniform structure. The weight and hardness properties are of a medium to large value. The degree of wood shrinkage is slight to moderate, with a medium strength and impact toughness [5], although it contains many knots [6]. In addition, the entire tree can also be utilized to extract rich chemical products. From different parts of *C. funebris*, the oils, called cedar wood oil, which is an important raw material of synthetic perfume, could be extracted. A total of 33 volatile components were identified from the wood of *C. funebris*, and the content of ferruginol and cedrol was highest in essential oil [7,8].

Thermal modification or heat treatment is effective in improving the dimensional stability and biological durability of wood, and it has received considerable attention as an environmentally friendly method compared to other methods [9]. Heat-treated modified wood is usually obtained by heating at temperatures ranging from 150 °C to 280 °C under oxygen-deficient conditions using media such as steam, nitrogen, or thermal oil and maintaining such conditions for a certain period. Heat treatment causes the chemical composition and structure of the wood to undergo thermal degradation and crosslinking reactions, thus significantly enhancing the ability of the material to maintain dimensional homeostasis and resistance to microbial attacks [10–12]. The magnitude of the color change in heat-treated wood is closely related to its chemical composition, which is the underlying cause of changes in its mechanical properties. In the production process, the technology of heat-treated lumber is determined by elucidating the mechanism of the effect of heat treatment on each strength index and selecting appropriate mechanical indices and treatment media according to the field of application [13–15]. A study has found that optimal process parameters for the mid-temperature vacuum heat modification of *C. funebris* wood were determined based on the mass loss rate and modulus of rupture (MOR), resulting in a modification temperature of 120 °C, holding time of 5 h, and a pressure intensity of 0.1 [16]. However, there is a lack of studies on the changes in functional groups, cellulose crystallinity, and microstructure of *C. funebris* wood caused by vacuum heat treatment. Heat-treated lumber is widely used in residential decorations, furniture manufacturing, outdoor fencing, and the cladding of building facades [17].

Crystallinity is an important property of wood, having an effect on its physical, mechanical, and chemical properties. In general, a high degree of crystallinity of cellulose results in the high tensile strength, bending strength, and dimensional stability of wood. During heat treatment, the chemical components of the wood cell wall undergo pyrolysis; the pyrolysis of hemicelluloses and cellulose occurs rapidly, with the weight loss of hemicelluloses occurring mainly at 220–315 °C and that of cellulose at 315–400 °C. However, lignin was more difficult to decompose because its weight loss occurred over a wide temperature range, from 160 to 900 °C [18,19].

The increase in temperature and time during heat treatment aggravates the degradation and crosslinking of wood, causing significant changes in the cell wall structure, affecting the physical and mechanical properties of wood and color changes [11,20–23]. Heat treatment does not damage ray parenchyma pit membranes, bordered pits, large window pit membranes, or Margo fibrils [24]. It is necessary to analyze the effects of heat treatment on the chemical components and microstructure of wood to understand how vacuum heat treatment alters its physical and mechanical properties. The objective of this study was to investigate the effects of the chemical composition and structural characteristics of the cell walls of *C. funebris* wood that was subjected to vacuum heat treatment. Changes in the functional groups, cellulose crystallinity, and microstructure were investigated under different treatment conditions, e.g., different temperatures, time, and vacuum pressure.

2. Materials and Methods

2.1. Test Material

Thirty-three-year-old *Cupressus funebris* Endl. trees from pure forest plantations in Yongxin Town, Jingyang District, Deyang City, Sichuan Province, China (31°1′–31°19′ North and 104°15′–104°35′ East), were used in this study. Five *C. funebris* trees with normal growth, complete trunks, straightness, and no obvious defects were randomly selected from the sample area marked with north–south directions. One log was cut from 1.3 m (diameter at breast height, DBH) to 3.3 m from each sample tree, and processed into 20 × 20 mm × 20 mm (R × T × L) specimens. There was a total of 11 groups of heat-treated specimens, with 10 replicates in each group. To eliminate the influence of wood moisture in the heat treatment process on the test results, the specimens were pre-treated in an electric

blast drying oven (101A-3), and the specimens were dried until they reached a moistur
content of approximately 12.69%.

2.2. Vacuum Heat Treatment Process

In this experiment, the vacuum heat treatment of *C. funebris* wood was conducte
in an intelligent electric vacuum drying oven (Shanghai Kuntian Laboratory Instrument
Co., Ltd., Shanghai, China, DZF-6050AB). The range of heat treatment temperatures con
monly used in the industry and the standard technical conditions for modified wood wer
employed [16]. The heat treatment temperature, time, and vacuum pressure ranges wer
set to 120–180 °C, 1–5 h, and 0.02 MPa to 0.1 MPa, respectively. A full factorial exper
mental design was used to generate a balanced orthogonal table, with a total of 11 group
of heat-treated specimens and ten replicates in each group. The control group did no
undergo treatment. Minitab software (Version 19, Minitab Inc., State College, PA, USA), a
efficient statistical and analysis tool, was used in the experimental design to analyze facto
effects and formulate parameter design, and the high and low levels of each factor wer
systematically planned. The specific experimental factors, levels, and codes are presente
in Table 1, and the experimental design is presented in Table 2. The vacuum heat treatmen
process was divided into the following three steps (Figure 1):

Step 1: The specimen was placed in an electric blast drying oven for pretreatment a
room temperature. The temperature was set to 103 °C, and the sample was oven-dried.

Step 2: The pre-treated wood was placed into an intelligent vacuum drying oven anc
pumped to vacuum. The temperature was increased to the target temperature at a rate o
1.3 °C/min and maintained for the corresponding time with the vacuum negative pressur
degree needed accordingly.

Step 3: For the vacuum heat treatment time to meet the requirements of holding tim
wait until the intelligent vacuum drying oven box temperature drops to 50 °C, release th
pressure to open the oven to remove the specimen, and save it in the desiccator cooled t
room temperature for subsequent tests.

Table 1. Encoding and levels of test factors.

Factor	Encoding		Level	
			Low	**High**
Heating Temperature (°C)	H1	A	120	180
Holding Time (h)	T2	B	1	5
Vacuum Pressure (MPa)	V3	C	0.10	0.02

Table 2. Experimental design for heat treatment experiments.

Standard Sequence	Operational Sequence	Heating Temperature (°C)	Holding Time (h)	Vacuum Pressure (MPa)
1	1	120	1	0.02
7	2	120	5	0.10
9	3	150	3	0.06
5	4	120	1	0.10
8	5	180	5	0.10
3	6	120	5	0.02
10	7	150	3	0.06
6	8	180	1	0.10
4	9	180	5	0.02
11	10	150	3	0.06
2	11	180	1	0.02

Figure 1. Process flowchart showing negative pressure heat treatment of *C. funebris* wood samples.

2.3. Fourier Transform Infrared Spectroscopy Test

Treated wood specimens with a vacuum negative pressure of 0.1 MPa at 180 °C and 120 °C for 5 h and at 180 °C for 1 h and treated wood specimens with a vacuum negative pressure of 0.02 MPa at 180 °C and 120 °C for 5 h, and at 180 °C for 1 h and untreated wood specimens were selected. They were then cut into small pieces and ground into wood powder to a 200-mesh particle size using a mill. The samples were oven-dried at a temperature of (103 ± 2) °C. All analyses were performed using a Scientific Nicolet iS20 Fourier Transform Infrared spectrometer (Thermo Fisher Scientific, Waltham, MA, USA), the scan range was 4000–400 cm^{-1} with a resolution of 4.00 cm^{-1}, and each spectrum was obtained from 32 scans.

The FTIR spectra were analyzed in terms of spectral band positions to identify the signatures of the main functional groups. An assignment of the main bands was made by analyzing the acquired spectra and comparing them with those available in the scientific literature [25].

2.4. X-ray Diffraction Test

Treated wood specimens with a vacuum negative pressure of 0.1 MPa at 180 °C and 120 °C for 5 h and at 180 °C for 1 h and treated wood specimens with a vacuum negative pressure of 0.02 MPa at 180 °C and 120 °C for 5 h, and at 180 °C for 1 h and untreated wood specimens were selected. They were then cut into small pieces and ground into wood powder to a particle mesh size between 80 and 100 using a mill. The samples were oven-dried at a temperature of (103 ± 2) °C. To analyze the effect of superheated steam treatment on the crystalline structure of cellulose, the powder was examined with an X-ray diffractometer (X′ Pert PRO-30X; Philips, Amsterdam, The Netherlands)) with Cu Kα radiation (λ = 0.154 nm). The XRD pattern was determined using the following parameters: a scanning range of 5°–40°; a voltage of 40 kV; and a scan rate of 2°/min. The crystallinity index (C$_r$I) of each specimen was calculated according to the Segal method [26], as follows:

$$CrI = \frac{I_{002} - I_{am}}{I_{am}} \times 100\% \tag{1}$$

where C$_r$I is relative crystallinity (%); I_{002} is the maximum intensity of the lattice diffraction angle of 002; and I_{am} is the minimum intensity corresponding to amorphous cellulose fraction.

2.5. Scanning Electron Microscopy Test

Treated wood specimens with a vacuum negative pressure of 0.1 MPa at 180 °C, 150 °C, and 120 °C for 5 h and untreated wood specimens were placed in liquid nitrogen for 5–10 min. Cross and tangential section slices (20 μm thick) were then cut using a sliding microtome LM2010R (Leica, Wetzlar, Germany) with a classical microtome blade (Leica). Scanning electron microscopical (SEM) analysis was conducted using a ZEISS Sigma 300 SEM (ZEISS Sigma 300, Zeiss, Germany). After spraying gold on the surface, a 20 kV acceleration voltage was selected for SEM analysis.

3. Results and Discussion

3.1. Fourier Transform Infrared Spectroscopy Analysis

The identified changes in the functional groups before and after heat treatment reflected the chemical changes in *C. funebris* wood during the heat treatment process. Figure 2 shows that the localized spectra of vacuum thermally modified *C. funebris* wood changed significantly compared to those of untreated wood, indicating differences in the chemical components of the thermally modified wood. Although various functional group changes were observed in the spectra of the heat-treated wood, only the functional groups that directly affected the mechanical and dimensional stability were analyzed in this study. During negative vacuum heat treatment, various chemical reactions occurred within the wood, including the degradation of cellulose, pyrolysis of lignin, and extensive decomposition of hemicelluloses. These chemical changes led to variations in the contents of different functional groups in the wood, such as the hydroxyl, methyl, and carboxyl groups, which affected the physical and chemical properties.

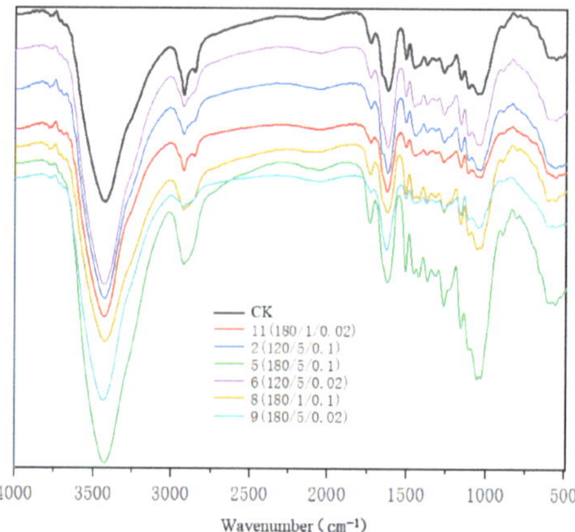

Figure 2. Infrared spectra of *C. funebris* wood samples under different heat treatment conditions (CK indicates control check group).

As illustrated in Figures 2–4, 3431 cm^{-1} is the O-H group stretching vibration peak, and the absorption peak here was enhanced with negative pressure and temperature increases in the range of 120–180 °C during intermediate temperature heat treatment. This mainly occurred via changes in the internal structure of the material in relation to the heat treatment and the increase in the crystallinity of the wood fibers [27]. Essentially all cellulose hydroxyl groups function as hydrogen bonds. The increase in crystallinity was attributed to the formation of strong hydrogen bonds, resulting in a more ordered fiber structure and the absence of free hydroxyl groups in cellulose, either crystalline or amorphous cellulose. In the spectrogram, 2925 cm^{-1} characterizes the strength of the stretching vibrations of C-H in methyl and methylene groups. This indicates that when the polysaccharides in hemicelluloses were thermally decomposed, the number of groups decreased; therefore, the absorption peaks diminished with increasing temperature and time, which agrees with the findings of Liang et al. based on poplar hemicelluloses [28]. The intermolecular interaction forces in hemicelluloses were enhanced at higher pressures, which may have slightly enhanced the lightness of the methyl group peaks; this result is consistent with that of Chang et al. in relation to interactions between cellulose at high pressures [29,30]. In the spectrum, 1732 cm^{-1} is mainly the C=O bond stretching vibration peak, and it represents the xylan in hemicelluloses, as well as the acetyl, carboxyl, and alcohols in lignin, and the peak value of the absorption peaks at this time does not change much, which may be due to the oxidative decomposition of some unstable compounds caused by the elevated temperature of the heat treatment and the reduction of the carbonyl group and the content of the other groups [21].

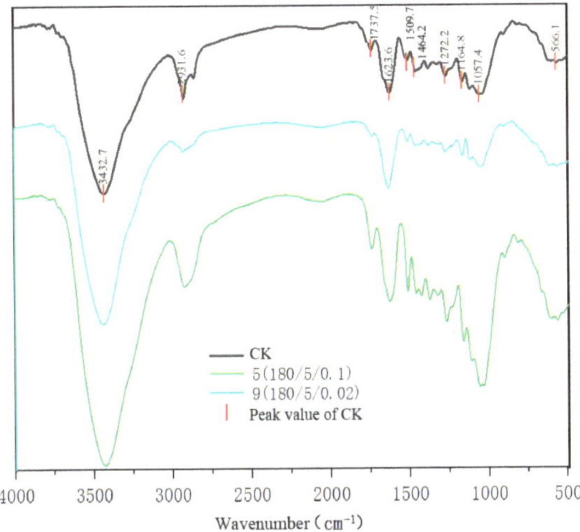

Figure 3. Effect of heat treatment and pressure (CK indicates control check group).

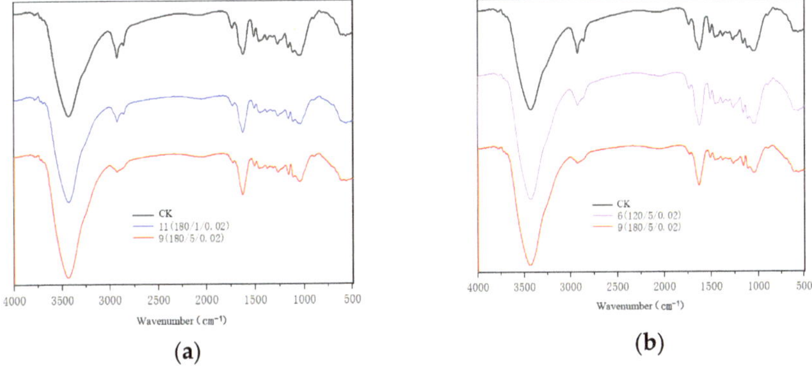

(**a**)

(**b**)

Figure 4. Effects of (**a**) heat treatment time and (**b**) heat treatment temperature (CK indicates control check group).

However, the higher heat treatment temperature conditions allowed the hydrolysis of the acetyl group of hemicelluloses to generate acetic acid, which created an acidic environment and promoted the esterification reaction of lignin. As this subsequently increased the number of carbonyl groups to a certain degree [31], the amplitude of the wave peak change was weaker. The wave peak at 1509 cm^{-1} was related to the vibration of the aromatic skeleton in lignin, and a significant change in the peak only occurred when the heat treatment was elevated to 180 °C [32]. This was because the structure of lignin began to change at 180 °C, and the peak was weakened by the decreased lignin content under the low-pressure environment. In contrast, the thermal decomposition of hemicelluloses was promoted by the higher pressure so that the relative content of lignin increased, and the peak was enhanced [33]. Similar to the findings of other studies, lignin polycondensation reactions with other cell wall components, resulting in further crosslinking, contribute to an apparent increase in lignin content [9]. Despite the chemical changes in the cell walls of the wood induced by heat treatment under vacuum conditions, the FTIR spectra showed few overall changes.

3.2. Crystallinity Analysis of Heat-Treated Cellulose

Figure 5 demonstrates the changes in the cellulose crystallinity of *C. funebris* under different heat treatment conditions. The results show that the heat treatment had a limited effect on the crystalline region, and the diffraction position was basically stable. Although there was no significant change in the crystalline structure, the heat treatment adjusted the cellulose crystallinity of *C. funebris*.

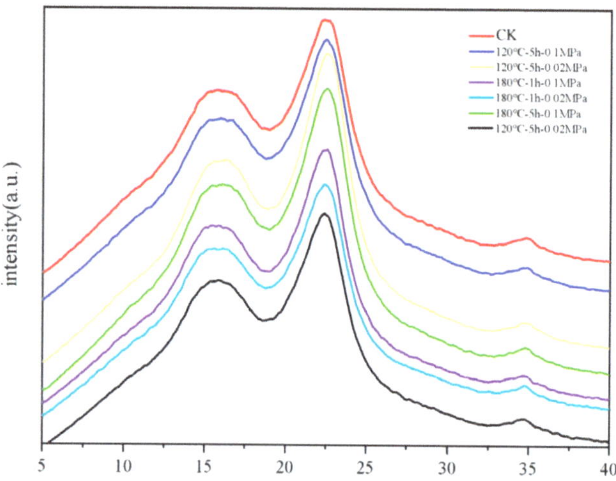

Figure 5. Effect of heat treatment on crystallinity of cellulose.

Table 3 illustrates relatively limited changes in the relative crystallinity of *C. funebris* wood under different vacuum heat treatment conditions. Under 0.02 MPa, the cellulose crystallinity at 180 °C for 1 h and 5 h was 39.58% and 39.30%, respectively, and the cellulose crystallinity at 120 °C for 5 h was 44.65%. Under 0.1 MPa, the cellulose crystallinity at 180 °C for 1 h and 5 h was 42.52%, respectively, 41.95% at 0.1 MPa, and 41.24% at 120 °C for 5 h. Compared with the untreated specimens, with a crystallinity of 38.36, the crystallinity increased by 3.19%, 2.46%, and 16.39% with a vacuum negative pressure of 0.02 MPa at 180 °C for 5 h, respectively. Furthermore, compared with the untreated specimens, the crystallinity increased by 10.84%, 9.36%, and 7.50% with a vacuum negative pressure of 0.1 MPa at 180 °C for 1 h and 5 h and at 120 °C for 5 h, respectively. The crystallinity of cellulose increased by 10.84%, 9.36%, and 7.50% in the 1 h and 5 h treatments at 180 °C and in the 5 h treatment at 120 °C under 0.1 MPa negative pressure. It can be seen that negative pressure had the most significant effect on the crystallinity of cellulose, while at lower temperatures, the crystallinity decreased with an increase in negative pressure. At higher temperatures, the crystallinity increased with an increase in negative pressure. Some studies have found that because the high crystallinity of cellulose is usually positively correlated with the strength of wood, this finding indicated that the degree of negative vacuum pressure was the most significant factor affecting the flexural strength of wood in mechanical property tests and that treating wood at 120 °C and 0.1 MPa maximized its flexural strength [16,34]. The microfilament angle can influence the density, dimensional stability, and modulus of elasticity, strength, and creep properties of wood. Additionally, the microcrystalline morphology affects the wettability and fiber strength, as well as other properties, of wood. It can be observed that the cell wall microfilament angle and cellulose crystallinity serve as crucial parameters for assessing the quality and performance of wood. This may be because the increased pressure weakened the intramolecular interatomic repulsive forces, and the cellulose molecules were closer together. Through a dynamic simulation, Jiang et al. also determined that pressurized heat treatment increased the stiffness of wood [29].

Table 3. Relative crystallinity of *C. funebris* wood under different heat treatment conditions.

Process Conditions	I_{002}	I_{am}	Crystallization Index	Relative Crystallization Index
Untreated group	19,793.06	12,200.68	38.36	100
120 °C/5 h/0.1 MPa	20,227.44	11,886.3	41.24	107.50
180 °C/5 h/0.1 M Pa	22,941.49	13,317.8	41.95	109.36
120 °C/5 h/0.02 MPa	23,425.59	12,966.84	44.65	116.39
180 °C/1 h/0.1 MPa	20,144.46	11,579.86	42.52	110.84
180 °C/5 h/0.02 MPa	18,621.11	11,302.88	39.30	102.46
180 °C/1 h/0.02 MPa	18,318.12	11,067.17	39.58	103.19

The relative crystallinity of cellulose in *C. funebris* wood tended to increase first and then decrease with increasing heat treatment temperature. At the beginning of the heat treatment, the increase in cellulose crystallinity may have originated from hemicellulose degradation, leading to an increase in the proportion of crystallization. Meanwhile, the bridging reaction in the amorphous region of cellulose contributed to a more orderly arrangement of microfibrils, shortening the molecular spacing and tending toward the crystalline region, which increased the possibility of hydrogen bond formation, thus increasing the degree of crystallinity. Increased crystallinity reduces the hygroscopicity and swelling of the cell walls, thus enhancing the dimensional stability of *C. funebris* wood [35]. When the heat treatment was increased to 180 °C, the crystallinity decreased, mainly due to hemicelluloses degrading to produce acetic acid. This led to an acidic environment and promoted the degradation of trace microfilaments in the amorphous regions of cellulose, resulting in glycosidic bond breaking and depolymerization decreasing the degree of crystallinity. However, the crystallinity decreased during the heat treatment. This was mainly due to the acetyl groups on hemicelluloses falling off and acetic acids forming, which resulted in the partial acidolysis of the cellulose molecular chain at high temperatures and further destroyed the cellulose aggregation to reduce the polymerization degree of cellulose and, accordingly, led to the crystallinity reduction and thus to a decrease in the physical and mechanical properties of the wood [36]. Changes in cellulose crystallinity directly affect the physicomechanical properties of wood, such as its dimensional stability, hardness, and stiffness, and Birinci et al. obtained similar conclusions in their study on camphor pine and beech [37]. Understanding these changes is important for optimizing the wood heat treatment process and its performance in various applications.

3.3. Microstructural Analysis of Heat-Treated Wood

The morphology and microstructure of wood at different heat treatment temperatures were investigated with scanning electron microscopy (ZEISS Sigma 300, Zeiss, Germany). Figure 6 shows the cross-section of wood after heat treatment at 150 °C, in which the surface is seen to be relatively flat and smooth, and it can be seen that all cell walls were cut clear at 4000× magnification. However, slight shrinkage and the deformation of the cell wall were observed, indicating that the temperature began to affect the wood fiber. Despite the slight compression of the cell wall, the wood microstructure remained relatively stable, with a high degree of surface smoothness.

Changes in the cross-section were evident after heat treatment at 180 °C. The surface of the cell wall exhibited a rougher texture, cracks in the intercellular layer were pronounced, and the wall became thin. These cracks may have resulted from lignin and cellulose degradation. Some studies have found that it is known that hemicelluloses were the wood cell components most degraded by the heat treatment due to their amorphous nature, low-molecular weight, and branched structure. Zauer et al. also found that cell wall compression produced cracks after the thermal modification of spruce [38]. Radial cracks occurred mainly in impermeable wood, such as Norway spruce, caused by large stresses in the wood structure during treatment [24]. Such structural changes suggest that chemical physical changes occurred within the wood at 180 °C, which may have led to a decrease in strength and stability. In summary, when heat treatment was conducted at negative pressure above 180 °C, significant changes in the cell wall structure began to occur, which affected the properties of heat-treated timber.

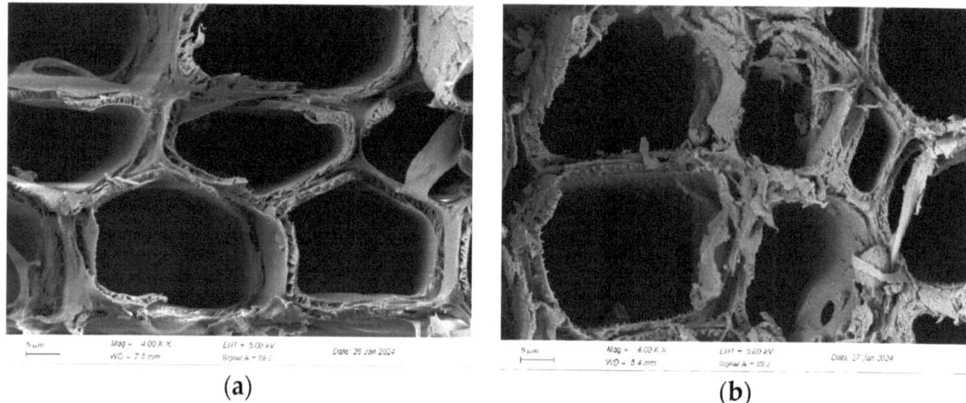

(a)   (b)

Figure 6. SEM images of cross-section of *C. funebris* wood after heat treatment at (**a**) 150 °C and (**b**) 180 °C.

Figure 7 shows the microstructure of *C. funebris* wood in a tangential section after heat treatment obtained using SEM at 2000× magnification (Figure 7a,b) and 10,000× magnification (Figure 7c,d). The morphology of the wood rays is shown in detail. In the samples treated at 150 °C, the wood rays have a smooth and flat appearance, and the integrity of the cellular structure is basically maintained, indicating that the microstructure of the wood was not obviously damaged (Figure 7a,c). However, after heat treatment at 180 °C, the structural integrity of the cell walls of the wood rays is maintained, but the cell lumens are narrowed owing to thermal stress (Figure 7b,d).

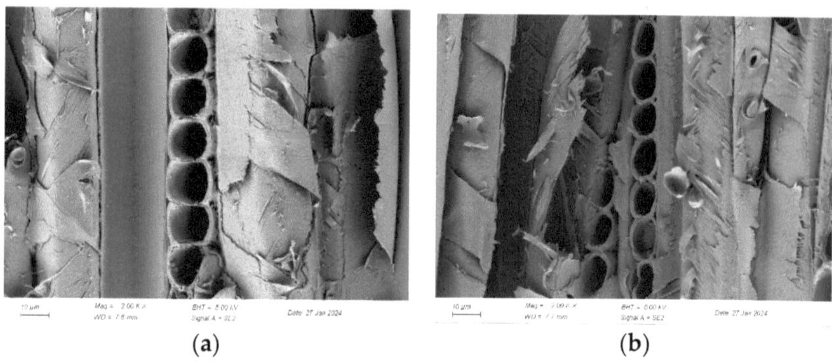

(a)   (b)

Figure 7. *Cont.*

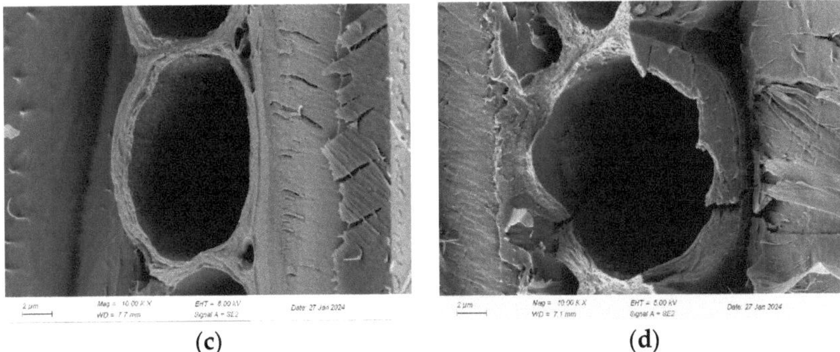

(**c**)   (**d**)

Figure 7. SEM images of tangential section of *C. funebris* wood after heat treatment (**a,c**) 150 °C and (**b,d**) 180 °C.

This change reflects a significant increase in the vapor pressure within the wood, owing to the conversion of internal moisture to vapor during heat treatment. When the vapor pressure exceeded the structural stability of the cell walls composed of cellulose and lignin in wood, the cell walls ruptured, and the connections between the original fibers broke, resulting in cracks. These cracks may also have been caused by the high vapor pressure generated by the resistance to moisture movement when the wood was heated, as the stomata of such wood were clogged with gum and methylcellulose. This further restricted the escape of moisture, thus exacerbating the internal pressure [39].

In addition, as the heat treatment temperature increased, the components of the wood cell wall, such as lignin, hemicelluloses, and cellulose, degraded to produce volatile organic compounds, further increasing the pressure inside the wood. When the pressure inside the cell lumen exceeded the tolerance limit of the cell wall, the wood ray membrane deformed, leading to crack formation, cell wall destruction, and tissue detachment. The effect of temperature led to changes in the chemical composition of heat-treated *C. funebris* wood, which was mainly attributed to the degradation and volatilization of carbohydrates in hemicelluloses (pentoses and hexoses). Some studies have found that the extractives content increased significantly as a function of the temperature of treatment [40]. At temperatures above 150 °C, heat treatment led to an increase in the total extractive content because of the presence of low-molecular weight substances, whereas heat treatment promoted the degradation of cellulose when the temperature was increased to 180 °C. However, most of the extractives disappear or degrade during the heat treatment, especially the most volatile ones [41].

4. Conclusions

This study explored the changes in functional groups of heat-treated *C. funebris* wood using Fourier transform infrared spectroscopy analysis. At the same time, the crystallinity of cellulose and the microstructure of the wood cell wall were characterized using X-ray diffraction and scanning electron microscopy. These analyses revealed the major changes in the microstructure of *C. funebris* wood induced by vacuum heat treatment. The principal findings can be summarized as follows:

The cellulose crystallinity of the wood increased, owing to the decomposition of hemicelluloses. It was not easy to form hydrogen bonds under negative pressure, and the number of free hydroxyl groups increased. Cellulose contains crystalline and amorphous regions, and crystallinity is a measure of the weight fraction of the crystalline regions. The decrease in the amorphous region in wood after heat treatment results in increased crystallinity [42]. The decomposition of polysaccharides in the cell wall caused a decrease in the absorption peak of the hydrocarbon groups. At the same time, the degradation products of hemicelluloses promoted the reaction of other chemical components. The relative content of lignin decreased with an increase in treatment temperature, and the performance was significant under an environment of lower negative pressure. The higher the heat treatment temperature, the greater the degradation of chemical components and the greater the impact of vacuum heat-treated *C. funebris* wood physics and mechanic. Heat treatment changed the chemical composition of the wood cell wall, but only minimal changes were observed via FTIR spectroscopy. After the vacuum heat treatment of *C. funebris* wood between 120 and 180 °C, the crystallinity showed the change rule of increasing first and then decreasing and was higher than the untreated wood specimen; the relative crystallinity range is 102.46–116.39. The negative pressure is a significant factor affecting the changes in cellulose crystallinity, thus affecting the mechanical strength and dimensional stability of wood. At lower temperatures, crystallinity decreased with increasing negative pressure, whereas at higher temperatures, crystallinity increased with increasing negative pressure. The SEM results showed that the overall morphology and structure of the cell wall remained stable under the medium-temperature vacuum heat treatment conditions; when the temperature rose to 150 °C, the structure of the cell wall was slightly deformed by the extrusion of the cell wall, and the intercellular layer produced cracks. When the temperature reached 180 °C, the cell lumen appeared to be a significantly narrowing phenomenon.

Future research could further deepen the influence of cell wall chemical components and microstructural changes on the physical and mechanical properties of *C. funebris* wood to reveal the specific mechanism of *C. funebris* properties changes at the cell wall level and provide more scientific guidance for *C. funebris* wood modification and application.

Author Contributions: Conceptualization, J.L. and M.C.; methodology, J.L. and M.C.; writing—original draft preparation, J.W. and J.L.; writing—review and editing, J.L. and M.C. All authors have read and agreed to the published version of the manuscript.

Funding: This research was supported by the Sichuan Science and Technology Program (grant no. 2023YFS0462), the Double Support Plan of Sichuan Agricultural University (grant no. 2022SYZD06).

Data Availability Statement: The original contributions presented in the study are included in the article, further inquiries can be directed to the corresponding author.

Conflicts of Interest: The authors declare no conflicts of interest.

References

1. Wang, L.; Toppinen, A.; Juslin, H. Use of wood in green building: A study of expert perspectives from the UK. *J. Clean. Prod.* **2014**, *65*, 350–361. [CrossRef]
2. Pan, Y. Influence of Different Gap Reconstruction Mode on the Understory Plant and Soil Animal Biodiversity of Low-Efficiency Cypress Forest in Hilly Area of Central Sichuan, China. Master's Thesis, Sichuan Agricultural University, Ya'an, China, 2014. (In Chinese)
3. Lyu, Q.; Luo, Y.; Liu, S.Z.; Zhang, Y.; Li, X.J.; Hou, G.R.; Chen, G.; Zhao, K.J.; Fan, C.; Li, X.W. Forest gaps alter the soil bacterial community of weeping cypress plantations by modulating the understory plant diversity. *Front. Plant Sci.* **2022**, *13*, 92090. [CrossRef] [PubMed]
4. Li, D.; Li, X.; Su, Y.; Li, X.; Yin, H.; Li, X.; Guo, M.; He, Y. Forest gaps influence fungal community assembly in a weeping cypress forest. *Appl. Microbiol. Biotechnol.* **2019**, *103*, 3215–3224. [CrossRef]
5. Wagenführ, R.; Scheiber, C. *Holzatlas*; Fachbuchverlag Leipzig: Munchen, Germany, 1974.
6. Lyu, J.; Qu, H.; Chen, M. Influence of Wood Knots of Chinese Weeping Cypress on Selected Physical Properties. *Forests* **2023**, *14*, 1148. [CrossRef]
7. Yu, M.; Peng, F.; Xu, G.; Liu, Q.; Fan, X.; Yu, S. Empirical Research on Application of *Cupressus funebris* Multi-objective Uses in Thousand-isle Lake Scenic Spot. *J. Green Sci. Technol.* **2013**, *5*, 138–139. (In Chinese) [CrossRef]
8. Lyu, J.; Zhao, J.; Xie, J.; Li, X.; Chen, M. Distribution and composition Analysis of Essential Oils Extracted from Different Parts of *Cupressus funebris* and *Juniperus chinensis*. *BioResources* **2018**, *13*, 5778–5792. [CrossRef]
9. Esteves, B.; Pereira, H. Wood modification by heat treatment: A review. *BioResources* **2009**, *4*, 370–404. [CrossRef]
10. Xing, D.; Hu, J.; Yao, L. Research review of material prediction and quality control of heat-treated wood. *J. Zhejiang A&F Univ.* **2020**, *37*, 793–800. (In Chinese) [CrossRef]
11. Xu, J.; Zhang, Y.; Shen, Y.; Li, C.; Wang, Y.; Ma, Z.; Sun, W. New Perspective on Wood Thermal Modification: Relevance between the Evolution of Chemical Structure and Physical-Mechanical Properties, and Online Analysis of Release of VOCs. *Polymers* **2019**, *11*, 1145. [CrossRef]

22. Kozakiewicz, P.; Drożdżek, M.; Laskowska, A.; Grześkiewicz, M.; Bytner, O.; Radomski, A.; Mróz, A.; Betlej, I.; Zawadzki, J. Chemical Composition as a Factor Affecting the Mechanical Properties of Thermally Modified Black Poplar (*Populus nigra* L.). *BioResources* **2020**, *15*, 3915–3929. [CrossRef]

23. Tjeerdsma, B.; Boonstra, M.; Pizzi, A.; Tekely, P.; Militz, H. Characterisation of thermally modified wood: Molecular reasons for wood performance improvement. *Holz. Roh. Werkst.* **1998**, *56*, 149. [CrossRef]

24. Wu, J.Y.; Zhong, T.H.; Zhang, W.F.; Shi, J.J.; Fei, B.H.; Chen, H. Comparison of colors, microstructure, chemical composition and thermal properties of bamboo fibers and parenchyma cells with heat treatment. *J. Wood Sci.* **2021**, *67*, 56. [CrossRef]

25. Wu, Z.; Chen, Y.; Huang, C.; Li, J.; Bao, Y.; Sun, F. A Review of Effects of Heat Treatment on Wood Mechanical Properties. *World For. Res.* **2019**, *32*, 59–64. (In Chinese) [CrossRef]

26. Wang, J.; Lyu, J.; Li, X.; Jiang, Y.; Chen, M. Optimization of parameters for vacuum heat modification of *Cupressus funebris* wood. *BioResources* **2023**, *18*, 5531. [CrossRef]

27. Zelinka, S.L.; Altgen, M.; Emmerich, L.; Guigo, N.; Keplinger, T.; Kymäläinen, M.; Thybring, E.E.; Thygesen, L.G. Review of Wood Modification and Wood Functionalization Technologies. *Forests* **2022**, *13*, 1004. [CrossRef]

28. Kawamoto, H. Lignin pyrolysis reactions. *J. Wood Sci.* **2017**, *63*, 117–132. [CrossRef]

29. Yang, H.; Yan, R.; Chen, H.; Lee, D.H.; Zheng, C. Characteristics of hemicellulose, cellulose and lignin pyrolysis. *Fuel* **2007**, *86*, 1781–1788. [CrossRef]

30. Barlović, N.; Čavlović, A.O.; Pervan, S.; Klarić, M.; Prekrat, S.; Španić, N. Chemical changes and environmental issues of heat treatment of wood. *Drvna Ind.* **2022**, *73*, 245–251. [CrossRef]

31. Esteves, B.; Ayata, U.; Cruz-Lopes, L.; Brás, I.; Ferreira, J.; Domingos, I. Changes in the content and composition of the extractives in thermally modified tropical hardwoods. *Maderas Cienc. Tecnol.* **2022**, *24*, 22. [CrossRef]

32. Kubovský, I.; Kačíková, D.; Kačík, F. Structural changes of oak wood main components caused by thermal modification. *Polymers* **2022**, *12*, 485. [CrossRef]

33. Wang, D.; Fu, F.; Lin, L. Molecular-level characterization of changes in the mechanical properties of wood in response to thermal treatment. *Cellulose* **2022**, *29*, 3131–3142. [CrossRef]

34. Boonstra, M.J.; Rijsdijk, J.F.; Sander, C.; Kegel, E.; Tjeerdsma, B.; Militz, H.; Acker, J.V.; Stevens, M. Microstructural and physical aspects of heat treated wood. Part 1. Softwoods. *Maderas Cienc. Tecnol.* **2006**, *8*, 193–208. [CrossRef]

35. Di Lena, G.; Sanchez del Pulgar, J.; Lucarini, M.; Durazzo, A.; Ondrejíčková, P.; Oancea, F.; Frincu, R.-M.; Aguzzi, A.; Ferrari Nicoli, S.; Casini, I.; et al. Valorization Potentials of Rapeseed Meal in a Biorefinery Perspective: Focus on Nutritional and Bioactive Components. *Molecules* **2021**, *26*, 6787. [CrossRef] [PubMed]

36. Segal, L.; Creely, J.J.; Martin, A.E.; Conrad, C.M. An Empirical Method for Estimating the Degree of Crystallinity of Native Cellulose Using the X-Ray Diffractometer. *Text. Res. J.* **1959**, *29*, 786–794. [CrossRef]

37. Bhuiyan, R.; Hirai, T.; Sobue, N.N. Effect of intermittent heat treatment on crystallinity in wood cellulose. *J. Wood Sci.* **2001**, *47*, 336–341. [CrossRef]

38. Liang, T.; Wang, L. Thermal treatment of poplar hemicelluloses at 180 to 220 C under nitrogen atmosphere. *BioResources* **2017**, *12*, 1128–1135. [CrossRef]

39. Jiang, X.; Wang, W.; Guo, Y.; Dai, M. Effect of Pressurized Hydrothermal Treatment on the Properties of Cellulose Amorphous Region Based on Molecular Dynamics Simulation. *Forests* **2023**, *14*, 1208. [CrossRef]

40. Chang, H.C.; Zhang, R.L.; Hsu, D.T. The effect of pressure on cation–cellulose interactions in cellulose/ionic liquid mixtures. *Phys. Chem. Chem. Phys.* **2015**, *17*, 27573–27578. [CrossRef] [PubMed]

41. Shen, Y.; Gao, Z.; Hou, X.; Chen, Z.; Jiang, J.; Sun, J. Spectral and thermal analysis of Eucalyptus wood drying at different temperature and methods. *Drying Technol.* **2019**, *38*, 313–320. [CrossRef]

42. Cheng, S.; Huang, A.; Wang, S.; Zhang, Q. Effect of Different Heat Treatment Temperatures on the Chemical Composition and Structure of Chinese Fir Wood. *BioResources* **2016**, *11*, 4006–4016. [CrossRef]

43. Zhang, J. Physical and Mechanical Properties and Thermal Degradation Characteristics of Heat-Treated Chinese Fir. Master's Thesis, Beijing Forestry University, Beijing, China, 2021. (In Chinese) [CrossRef]

44. He, Z.; Qi, Y.; Zhang, G.; Zhao, Y.; Dai, Y.; Liu, B.; Lian, C.; Dong, X.; Li, Y. Mechanical Properties and Dimensional Stability of Poplar Wood Modified by Pre-Compression and Post-Vacuum-Thermo Treatments. *Polymers* **2022**, *14*, 1571. [CrossRef] [PubMed]

45. Gao, Y.; Zhou, Y.; Fu, Z.; Van Den Bulcke, J.; Van Acker, J. Swelling behavior of thermally modified timber from a cellular and chemical perspective. *Holzforschung* **2023**, *77*, 713–723. [CrossRef]

46. Brosse, N.; Hage, R.E.I.; Chaouch, M.; Pétrissans, M.; Dumarçay, S.; Gérardin, P. Investigation of the chemical modifications of beech wood lignin during heat treatment. *Polym. Degrad. Stab.* **2010**, *95*, 1721–1726. [CrossRef]

47. Birinci, E.; Karamanoğlu, M.; Kesik, H.İ.; Kaymakci, A. Effect of heat treatment parameters on the physical, mechanical, and crystallinity index properties of Scots pine and beech wood. *BioResources* **2022**, *17*, 4713–4729. [CrossRef]

48. Zauer, M.; Meissner, F.; Plagge, R.; Wagenführ, A. Capillary pore-size distribution and equilibrium moisture content of wood determined by means of pressure plate technique. *Holzforschung* **2016**, *70*, 137–143. [CrossRef]

49. Lengowski, E.C.; Bonfatti Júnior, E.A.; Nisgoski, S.; Bolzon de Muñiz, G.I.; Klock, U. Properties of thermally modified teakwood. *Maderas Cienc. Tecnol.* **2021**, *23*, 10. [CrossRef]

50. Cademartori, P.H.G.; dos Santos, P.S.B.; Serrano, L.; Labidi, J.; Gatto, D.A. Effect of thermal treatment on physicochemical properties of gympie messmate wood. *Ind. Crop Prod.* **2013**, *45*, 360–366. [CrossRef]

41. Mecca, M.; D'Auria, M.; Todaro, L. Effect of heat treatment on wood chemical composition, extraction yield and quality of th extractives of some wood species by the use of molybdenum catalysts. *Wood Sci. Technol.* **2019**, *53*, 119–133. [CrossRef]
42. Phuong, L.X.; Shida, S.; Saito, Y. Effects of heat treatment on brittleness of *Styrax tonkinensis* wood. *J. Wood Sci.* **2007**, *53*, 181–18 [CrossRef]

Article

A Study on the Effects of Vacuum, Nitrogen, and Air Heat Treatments on Single-Chain Cellulose Based on a Molecular Dynamics Simulation

Yuna Hua, Wei Wang *, Jingying Gao, Ning Li and Zening Qu

College of Mechanical and Electrical Engineering, Northeast Forestry University, Harbin 150040, China; yolohua@nefu.edu.cn (Y.H.); lkxgjy@163.com (J.G.); lero@nefu.edu.cn (N.L.); zeningqu@nefu.edu.cn (Z.Q.)
* Correspondence: vickywong@nefu.edu.cn; Tel.: +86-133-1361-3588

Abstract: Employing molecular dynamics software, three models—vacuum–cellulose, nitrogen–cellulose, and air–cellulose—were built to clarify, via a microscopic perspective, the macroscopic changes in single-chain cellulose undergoing vacuum, nitrogen, and air heat treatments. Kinetic simulations were run following model equilibrium within the NPT system of 423, 443, 463, 483, and 503 K. The energy variations, cell parameters, densities, mean square displacements, hydrogen bonding numbers, and mechanical parameters were analyzed for the three models. The findings demonstrate that as the temperature climbed, the cellular characteristics among two models—the nitrogen and vacuum models—decreased and subsequently increased. The nitrogen model reached its lowest value at 443 K. In contrast, the vacuum model reached its minimum value at 463 K. The vacuum heat treatment may enhance the structural stability of the single-chain cellulose more effectively than the nitrogen and air treatments because it increases the number of hydrogen bonds within the cellulose chain and stabilizes the mean square displacement. Furthermore, the temperature has an impact on the mechanical characteristics of the cellulose amorphous zone; the maximum values of E and G for the vacuum and nitrogen models are found at 463 and 443 K, respectively. The Young's modulus and shear modulus were consistently more significant for the vacuum model at either temperature, and the Poisson's ratio was the opposite. Therefore, the vacuum heat treatment can better maintain wood stiffness and deformation resistance, thus improving wood utilization. These findings provide an essential theoretical basis for wood processing and modification, which can help optimize the heat treatment and enhance wood's utilization and added value.

Keywords: heat treatment; cellulose; molecular dynamics; mechanical properties

Citation: Hua, Y.; Wang, W.; Gao, J.; Li, N.; Qu, Z. A Study on the Effects of Vacuum, Nitrogen, and Air Heat Treatments on Single-Chain Cellulose Based on a Molecular Dynamics Simulation. *Forests* **2024**, *15*, 1613. https://doi.org/10.3390/f15091613

Academic Editors: Bruno Esteves and Antonios Papadopoulos

Received: 14 July 2024
Revised: 22 August 2024
Accepted: 11 September 2024
Published: 13 September 2024

1. Introduction

Cellulose is one of the most abundant organic polymer materials in nature, and its unique physical and chemical properties make it play a crucial role in the fields of wood, paper, textiles, and biodegradable materials [1]. The structure of cellulose is complex, as many glucose groups are closely connected to form a long and thin molecular chain. These molecular chains are divided into highly ordered crystal regions and relatively loose amorphous regions, and this composite structure gives cellulose materials diversified performance characteristics [2]. Among them, the amorphous region is not as regular as the crystal region due to its irregular structure and high degree of freedom, and it is particularly sensitive to changes in external environmental factors such as temperature, pressure, and gas atmosphere. Therefore, studying the properties of the amorphous region of cellulose is of great significance to gain an in-depth understanding of the overall properties of cellulose materials.

There is a close and profound relationship between the cellulose chain and wood. The properties of wood, as the main natural carrier of cellulose, are largely determined by the arrangement of cellulose chains, how they bind, and how they interact with other

components in the wood. As the growth years of trees increase, the lignin content gradua[
decreases, while the cellulose content shows a steady rising trend [3]. Cellulose chains a[
interwoven into the cell walls of wood, providing the necessary mechanical strength ar[
determining key properties such as hygroscopicity, thermal stability, and processabili[
Therefore, the study of the behavior of cellulose chains in wood and the influence [
external conditions such as heat treatment is the key to understanding the changes in wo[
properties and optimizing wood processing and utilization strategies.

The heat treatment is an advanced process that uses water vapor, inert gasses, air, h[
oil, or water as a heat transfer medium to treat materials at high temperatures rangir[
from 423 to 513 K. The heat treatment process leads to significant changes in the ce[
wall structure, which can realize the structural control and performance optimization [
cellulosic materials. This method is not only an effective way to optimize the physical ar[
chemical properties of wood, but also significantly improves the durability of wood [[
The internal structure and chemical makeup of heat-treated wood change for the bett[
This is demonstrated by the wood expanding and contracting less, becoming significant[
less hygroscopic, and having more water in it at equilibrium, all of which contribute to tl[
wood's improved stability in terms of dimension. Furthermore, applying a heat treatme[
strengthens the wood's resistance to corrosion and insects, giving a more reliable assuran[
for a variety of wood uses [5].

The selection of heat treatment media profoundly influences the shaping of woc[
properties, and each medium shows its unique advantages and potential limitations in tl[
treatment process. In their investigation into using different gas and oil media in the he[
processing of bamboo, Yang et al. [6] methodically examined the impacts of various med[
on the mechanical, appearance, dimensional stability, and thermodynamic characteristi[
of the material. A thorough analysis was conducted to determine how multiple med[
affected the mechanical, dimensional stability, quality of the exterior, and thermal an[
physical features of bamboo. They discovered that the density of bamboo treated wit[
linseed oil as a heat treatment medium increased significantly. Additionally, they foun[
that the bamboo's moisture removal and shrinkage resistance significantly improved via[
gradual rise in the treatment temperature, giving the bamboo more excellent dimensiona[
stability. The thermal alteration from black pine wood was investigated by Bal et al. [7] i[
three distinct atmospheres: nitrogen, air, and vacuum. The experiment encompassed thre[
temperature points: 453 K, 473 K, and 493 K. According to the research results, the vacuur[
environment has the least harmful effect on the mechanical properties of wood, whic[
indicates that vacuum has a promising application in wood heat treatment. In additio[
the study by Candelier et al. [8] focused on beech wood and observed the changes in it[
physicochemical properties in vacuum and nitrogen heat treatment environments. Accorc[
ing to the experimental data, beech wood treated in a vacuum environment demonstrate[
a more moderate decrease in important mechanical property indexes, such as the modulu[
of rupture (MOR) and modulus of elasticity (MOE) along with the Brinell hardness (HB[
when compared to the heat treatment with nitrogen. This further supports the vacuur[
treatment's superiority in maintaining timber's mechanical characteristics.

Molecular simulation technology provides a new idea for the study of wood hea[
treatments, and this method provides a good theoretical basis for existing macroscopi[
experiments. Molecular dynamics, as an advanced computer simulation technique, aims a[
accurately calculating and predicting the structure and properties of a system by modelin[
the trajectories of molecular motions. Alder and Wainwright [9] first proposed this concep[
in the late 1950s, and it was successfully applied in the exploration of hard-sphere interac[
tion mechanisms. Not only does it close the gap between theory and experiment, allowin[
us to test the precision of the model through contrasting empirical findings, but it als[
more intuitively reproduces the subtle change processes that are challenging to capture i[
macroscopic experiments in a computer environment [10]. Molecular dynamics simulatio[
has demonstrated excellent ability in material property prediction [11]. Tsai et al. [12] use[
this technique to deeply analyze the mechanical properties of graphite. They revealed th[

significant advantages of monolayer graphene in mechanical properties by comparing the key parameters of monolayer graphene, such as the Young's modulus, shear modulus, and Poisson's ratio, with those of multilayer graphene. With an eye toward their possible uses, Mao et al. [13] investigated the changes in the characteristics of carbon nanotubes using molecular dynamics simulations. They examined the dispersion process in tiny molecules like methane inside them. Cao et al. [14] created a carbon dioxide–cellulose model using the Materials Studio 2020 software. They carefully examined the thermal and mechanical features of the cellulose composite structure at different pressures. This microscale numerical simulation technique not only reveals the microscopic-level information that is difficult to capture by traditional experimental means, but also dramatically simplifies the complexity of the experimental operation and effectively circumvents the various limitations and unfavorable conditions that may be encountered in conventional experiments.

Therefore, based on molecular dynamics simulation technology, an in-depth study of the effects of vacuum, nitrogen, and air heat treatments on the properties of a cellulose amorphous region not only helps to reveal the microscopic mechanism of cellulose materials during heat treatment, but also provides an important theoretical basis and technical guidance for the processing and utilization of cellulose materials. The choice of heat treatment media is the key to optimizing the process parameters, improving the materials' properties, reducing costs, and achieving environmental friendliness. Using the Materials Studio software, this study simulated the effects of vacuum, nitrogen, and air thermal treatments on single-chain cellulose at five different temperatures through a microscopic perspective. The models were analyzed based on their variations in energy, mechanical parameters, mean square displacements, number of hydrogen bonds, cell parameters, and densities. The changes in the macroscopic properties of the heat treatment process under different temperatures and media are explained from the microscopic level, which provides more theoretical support for wood heat treatment and the optimization of the wood heat treatment process.

2. Materials and Methods

2.1. Model Establishment

Cellulose is a linear polymer connected by β-1,4-glycosidic linkages [15]. Scientists have identified four crystalline forms of cellulose: cellulose I, II, III, and IV. In nature, natural sources of cellulose such as bacterial cellulose, algae, and higher plants fall under the category of cellulose type I. The most prevalent kind of cellulose is called type I cellulose. When further subdivided, natural cellulose type I exhibits two crystalline forms: cellulose Iα and cellulose Iβ [16]. A molecular chain is a long polymer chain that consists of many repeating units connected by covalent bonds. The structure of cellulose molecular chains consists of a series of crystalline regions alternating with amorphous regions. These two types of areas exhibit significant differences in the characteristic expression of the molecular chains, the arrangement patterns, the binding forces between molecules, and the materials' mechanical characteristics. Specifically, the molecular chains in the crystalline region are more compact and ordered, and the intermolecular bonding is tight, giving it excellent mechanical stability and resistance to the external environment. On the contrary, the amorphous region presents a looser molecular arrangement and weaker binding ability, and outside factors influence its mechanical characteristics. The amorphous area of cellulose Iβ was chosen as the paper's focal point in light of these variations.

In this study, simulations using the Materials Studio program (2020, San Diego, CA, USA) were conducted to characterize the effects of different temperatures and media on the properties of single-chain cellulose from a microscopic point of view. The degree of polymerization refers to the number of repeating units in the polymer molecule. It is a measure of the length of the polymer chain. It has been found that the simulation results are consistent with the actual situation when the polymerization degree of cellulose is greater than 10, and the different polymerization degrees of cellulose have little effect on the simulation results of the material properties. Although the actual chain length of cellulose is quite large, the increase in chain length during the simulation will increase the simulation time and complexity. Therefore, cellulose chains with a polymerization degree of 20 were selected in this study [17]. As a result, this study established a cellulose chain with a degree of polymerization of 20 [18]. Theodorou's [19] technique and Wang et al.'s [20] simulation model scale parameters were utilized in the construction of the amorphous zone polymers. Three simulation models were created, all with density settings of 1.5 g/cm^3. The three simulation models are vacuum–cellulose, nitrogen–cellulose, and air–cellulose models, as shown in Figure 1. In the vacuum–cellulose model, a chain of cellulose molecules with a degree of polymerization (DP) of 20 was added, and the pressure parameter was set to 20 kPa [8]. In the nitrogen–cellulose model, 20 nitrogen molecules and a chain of cellulose molecules with a degree of polymerization 20 were added. Sixteen nitrogen molecules, four oxygen molecules, and a cellulose molecular chain with a degree of polymerization of 20 were added to the air–cellulose model. The cellulose Iβ molecular chain is shown as a stick model, the nitrogen molecules as a blue ball-and-stick model, and the oxygen molecules as a red ball-and-stick model.

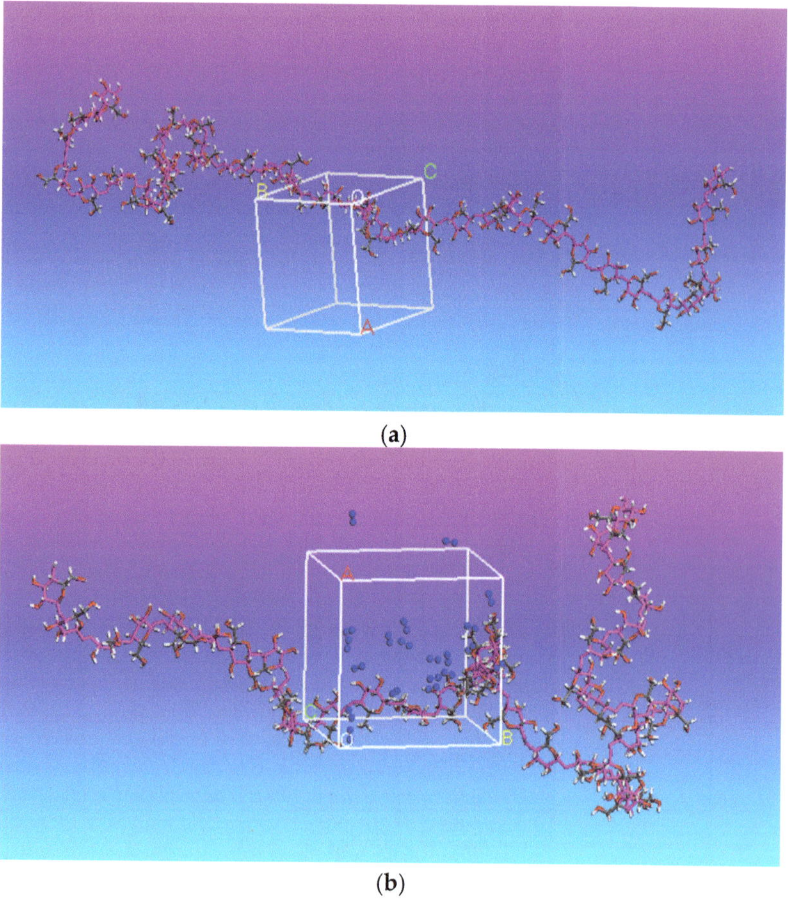

(a)

(b)

Figure 1. *Cont.*

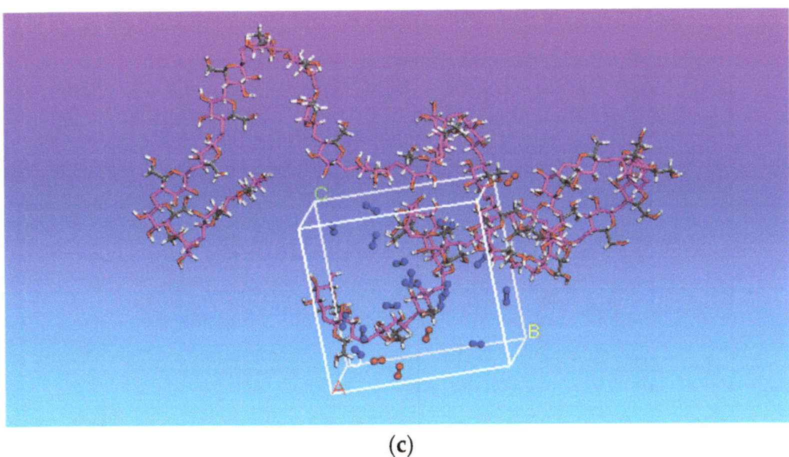

(c)

Figure 1. Diagrams of the three models: (**a**) the vacuum–cellulose model; (**b**) the nitrogen–cellulose model; and (**c**) the air–cellulose model.

2.2. Dynamic Simulation

After the models were constructed, they were subjected to an energy balance treatment. In this paper, geometric optimization and relaxation were performed on the three models, and the subsequent operations were carried out through the Forcite module. Firstly, geometry optimization was carried out using the Forcite Geometry Optimization module, and the intelligent algorithm was chosen to run for 5000 steps to minimize the model energy and achieve initial structural relaxation. Secondly, the Forcite Dynamics module was used to complete the dynamics relaxation, with the regular system molecular dynamics (NVT) simulations being selected for the system. The NVT ensemble can keep the number of particles (N), volume (V), and temperature (T) of the system constant. Through the NVT ensemble, the thermal motion of the system at a normal temperature can be simulated, and the system can reach the thermal equilibrium state. The temperature was selected to be 300 K at room temperature, the initial speed was set to random, the step size was set to 1 fs, the total step size was set to 1 ns, and a frame was output every 5000 steps [21]. Finally, the system energy of the three models was minimized, the local unreasonable structure was eliminated, and the equilibrium state was reached.

Molecular dynamics simulations were conducted following the energy balance treatment for all three models. The isothermal isobaric system molecular dynamics (NPT) simulations system was set, and the kinetic simulations were carried out at 423, 443, 463, 483, and 503 K, respectively. The NPT ensemble not only keeps the particle population and temperature of the system constant, but also allows the volume of the system to change. This is essential for studying properties such as the phase transition, thermal expansion, and compressibility of materials at different temperatures and pressures. The impacts of the vacuum and nitrogen heat treatments on the macroscopic characteristics of cellulose were confirmed by experimental investigations by Bal [7] and Candelier et al. [8], whose studies offer a solid theoretical foundation for the modeling parameters used in this work. Temperature control in the kinetic simulation was carried out using Andersen's method [22], pressure control was carried out using Berendsen's method [23], electrostatic interactions were carried out using Ewald's method [24], van der Waals interactions were carried out using an atom-based method [25], and the force field was selected from PCFF, which is appropriate for the computation of organic matter [26].

3. Results and Discussion

3.1. Energy Balance

The temperature and energy trends accumulated during the molecular dynamic simulations could be utilized to assess whether the system attained equilibrium [27 Figure 2 displays the system's time-varying energy and temperature changes following the vacuum–cellulose model's kinetic simulation, which lasted 1 ns. The system is considered to have reached equilibrium when the changes in temperature and energy are within the range of 5% to 10%. In the meantime, whether or not the system's energy is in equilibriur is ascertained using the convergence parameter [28]. Its formula is as follows:

$$\delta E = \frac{1}{N}\sum_{i=1}^{N}\left|\frac{E_i - E_0}{E_0}\right|$$

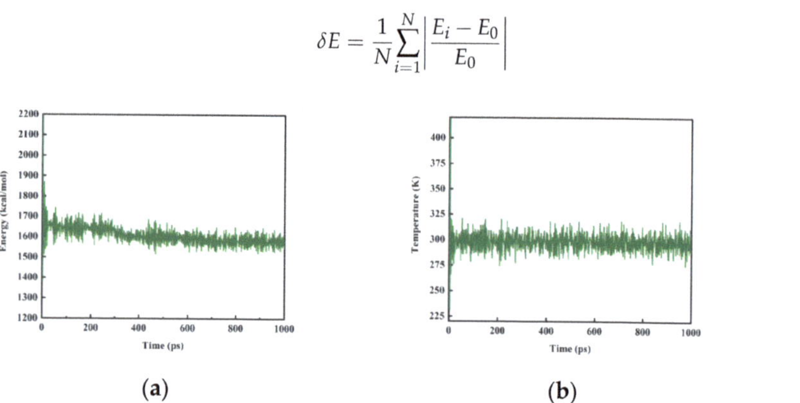

(a)   (b)

Figure 2. Energy versus temperature fluctuation plots for vacuum–cellulose model: (**a**) energy–tim variation; and (**b**) temperature–time variation.

Equation (1) uses the notation N to represent the total number of simulation steps, E to represent the starting energy value, and E_i to represent the energy value following the stage I of the simulation. The system tends to equilibrate, and the simulation findings are trustworthy while $0.001 < \delta E < 0.003$. It is found that the energy convergence parameter of the vacuum, nitrogen, and air models are 0.0023, 0.0028, and 0.0026 at the last 200 ps respectively, which indicates that the simulation results are reliable.

Figure 2b displays the trend plot of the vacuum–cellulose model's temperature ove time. At the beginning of the simulation, the system usually starts from a specific initia state, which can result in uneven energy and temperature distribution. Therefore, in the early stages of the simulation, significant fluctuations in energy and temperature can be observed, which are manifestations of the transition of the system from a non-equilibriun state to an equilibrium state. During the simulation, if the system gradually reache an equilibrium state, then the fluctuations in energy and temperature will stabilize and fluctuate around some average value. The model's temperature changes during the las 200 ps are within ±25 K, indicating that kinetic relaxation brought the model to equilibrium Both the nitrogen and air systems attained equilibrium, and their temperature–time chart resemble those of the systems.

By analyzing the energy and temperature variations in the three models, it can be concluded that the systems of all three models were stabilized after the initial geometrica optimization and kinetic relaxation, proving the present study's reliability and allowing subsequent kinetic simulations to be carried out.

3.2. Lattice Parameters and Density

Cell characteristics and model densities may also be utilized to describe the degree of densification of cellulose chains and to analyze the mechanical characteristics of cellu lose [29]. A cell is the smallest unit of a parallelepiped that can fully reflect the chemical and structural characteristics of atoms or ions in a crystal in three dimensions. It is importan to note that the cell structures of the vacuum and nitrogen or air models both exhibit a cubic appearance. In light of this, we decided to quantitatively characterize the cell size by documenting these model cells' lengths, breadths, and heights in three dimensions. Table 1 lists each model's precise temperature-dependent variations in cell characteristics and density.

Table 1. Cell parameters and densities of three models at different temperatures.

Temperature (K)	Medium	Cell Parameters (Å)			Density (g/cm^3)
		The Length	The Width	The Height	
423	Vacuum	20.24	20.24	20.24	1.363
	Nitrogen	20.99	20.99	20.99	1.322
	Air	21.42	21.42	21.42	1.313
443	Vacuum	20.15	20.15	20.15	1.382
	Nitrogen	20.89	20.89	20.89	1.341
	Air	21.45	21.45	21.45	1.306
463	Vacuum	20.13	20.13	20.13	1.386
	Nitrogen	20.93	20.93	20.93	1.334
	Air	21.53	21.53	21.53	1.289
483	Vacuum	20.20	20.20	20.20	1.372
	Nitrogen	21.05	21.05	21.05	1.311
	Air	21.57	21.57	21.57	1.276
503	Vacuum	20.21	20.21	20.21	1.369
	Nitrogen	21.11	21.11	21.11	1.300
	Air	21.69	21.69	21.69	1.271

According to the data presented in Table 1, we can observe that the cell sizes of both models, vacuum–cellulose and nitrogen–cellulose, show a decreasing trend and then an increasing trend with the increasing temperature, and the density changes in the opposite trend. In particular, when the temperature hits 463 K, the vacuum–cellulose model's cell size decreases to the lowest value of 20.13, and its density rises to the highest value of 1.386 g/cm^3. When the temperature hits 443 K, the density increases to the highest point of 1.341 g/cm^3, while the nitrogen–cellulose model simultaneously achieves a minimum density of 20.89. The cellulose chain's molecular mobility intensifies with the temperature, resulting in changes in the intermolecular contact forces (such as hydrogen bonding and van der Waals forces). Initially, the temperature rise could have encouraged a tighter intermolecular arrangement, which would have led to a reduction in cell size and a rise in density. However, when the temperature rises to a certain threshold, the thermal motion between molecules becomes too intense, leading to an increase in intermolecular distances and a subsequent increase in cell size, leading to a decrease in density. Despite the temperature change, the overall arrangement of the cellulose chains did not fundamentally change, resulting in a relatively small range of density changes. At the same time, different media affect cellulose chains differently, leading to various temperatures at which cell size minima are reached for the vacuum and nitrogen models. In contrast, the air environment contains more types of gas molecules, which interact with the cellulose chains more complexly, leading to different cell sizes and density trends with temperature than the vacuum and nitrogen environments. As the temperature rises, the air–cellulose model's cell size grows steadily while the density continuously falls, as Table 1 illustrates.

A further analysis shows that the cell volume of the vacuum model is the smallest compared to the other models at identical temperatures, and this property enables the vacuum model to exhibit a higher compression tightness at the same temperature. To be precise, in the vacuum environment, the arrangement of cellulose chains is mainly controlled by the internal solid intermolecular interaction forces, as the pressure interference from external gas molecules is wholly excluded. This undisturbed internal force field encourages a tighter and more organized arrangement of cellulose chains, which immediately results in a considerable decrease in cell volume and a rise in density. In addition, the vacuum conditions also limit the activity of heat transfer and molecular collisions, and the range and amplitude of thermal motion of the cellulose chain molecules in the vacuum condition are more limited at the same temperature than in the environment with a gaseous medium which further consolidates their tightly packed state.

3.3. Mean Square Displacement

Thermal stability is a parameter that indicates the retention of a material's properties in a thermal environment. The intensity of cellulose chain movement is closely related to the structural stability of the cellulose; the more intensely it moves, the less stable it is [30]. Thus, the thermal stability of materials at the macroscopic level can be characterized by the intensity of movement of the cellulose chains at the microscopic level. The mean square displacement (*MSD*), which may be used to accurately describe the migration routes and behaviors of molecules in a system, is the mean square summation of the increments of a molecule's position vectors after a simulated period. The *MSD* value in the following equation represents the deviation in the particle's location with respect to a reference point over time.

$$MSD = \sum_{i=1}^{n} \langle |\vec{r_i}(t) - \vec{r_i}(0)|^2 \rangle \tag{2}$$

In Equation (2), n represents the number of small molecules, $<>$ indicates the mean value, and $\left|\vec{r_i}(t) - \vec{r_i}(0)\right|^2$ symbolizes the center-of-mass displacement of a small molecule i between an instant, t, and the beginning moment.

Figure 3 displays the mean square displacement curves for the three models at various temperatures.

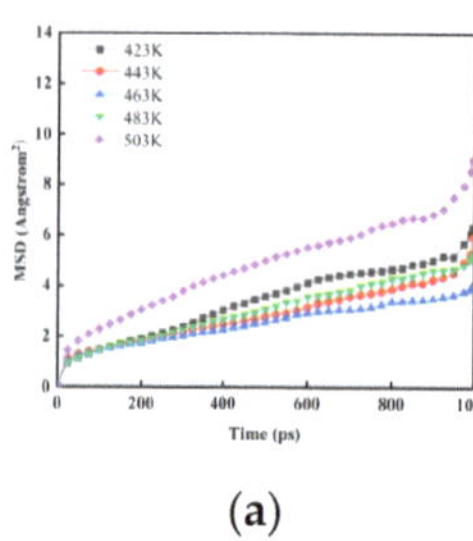

(a)

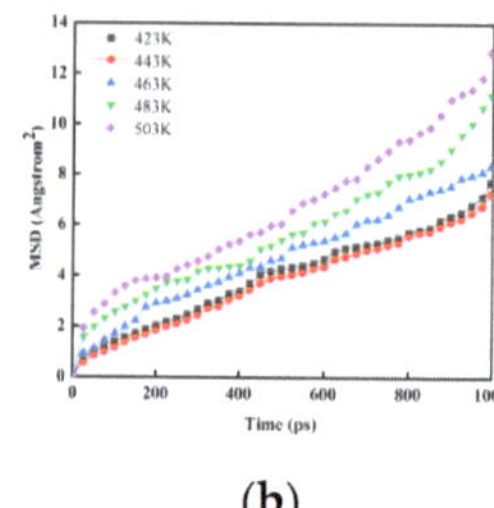

(b)

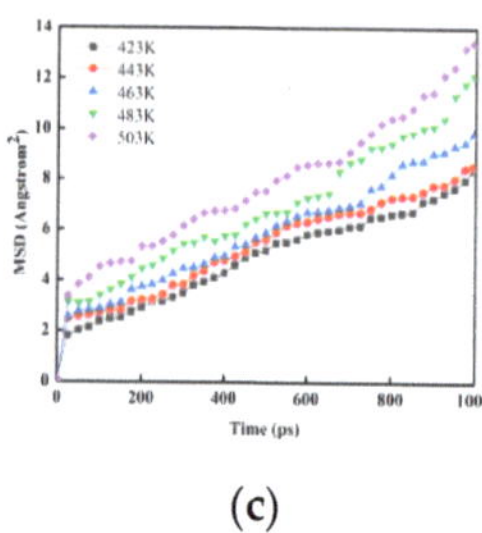

(c)

Figure 3. The mean square displacement curves for the three models: (**a**) the vacuum–cellulose model; (**b**) the nitrogen–cellulose model; and (**c**) the air–cellulose model.

The graphs of the mean square displacements for the air, nitrogen, and vacuum models at various temperatures are displayed in Figure 3. Taking 503 K as an example, the mean square displacement of the vacuum–cellulose model increases from 0 to 8.9, the mean square displacement of the nitrogen–cellulose model increases from 0 to 12.8, and the mean square displacement of the air–cellulose model increases from 0 to 13.4. It can be seen that the mean square displacement of the cellulose chain in the vacuum model fluctuates gently while that in the air model fluctuates sharply. The fluctuation in *MSD* values of cellulose chains in the vacuum model is smoother mainly because its environment is simple and stable, and cellulose chains are primarily affected by intermolecular interaction forces. On the other hand, the cellulose chains in the air model show more drastic fluctuations in their *MSD* curves due to the interference of multiple gas molecules. Meanwhile, the mean square displacement curves of the vacuum model show a smoother state than those of the nitrogen and air models at any temperature. At 463 K, the mean square displacement of the vacuum model increases from 0 to 3.9, and the mean square displacement curves of the nitrogen model fluctuates most gently at 443 K. One central element influencing molecular motion is temperature. As the temperature increases, the molecules' thermal motion intensifies and the vibration and displacement of the cellulose chains increase. However, at specific temperatures (e.g., 463 K and 443 K), the cellulose chains in the vacuum and nitrogen models may have reached some dynamic equilibrium, making the fluctuations in their *MSD* curves the flattest. This may be because, at these temperatures, the internal forces of the cellulose chains are optimally matched to the interaction with the external environment, reducing unwanted fluctuations.

3.4. Hydrogen Bonding

As a remarkable non-covalent interaction force, the essence of hydrogen bonding lies in forming a stable link between a hydrogen atom covalently bonded to an electronegative atom and another more electronegative atom [31]. This force is vital in polymer systems, where it constructs a network of hydrogen bonds that significantly enhances the robustness of the molecular structure and has a profound effect on the overall conformation of the molecule. The glycosidic bonds in cellulose chains and multiple hydroxyl groups in its smallest constituent unit, glucose, produce strong hydrogen bonding in cellulose [32].

Hydrogen bonding in cellulose composite models is divided into intra-chain and inter-chain bonding. Intra-chain hydrogen bonding refers to the bonding formed within the cellulose chain, enhancing thermal stability. In contrast, inter-chain hydrogen bonding spans different molecules and promotes interactions between cellulose chains, contributing significantly to the material's mechanical properties and overall strength [33].

The primary goal of this research is to compare how well the three models perform in different settings. The vacuum model is a single-chain model, and its only role is intra-chain hydrogen bonding. Therefore, the study on hydrogen bonding presented in this paper focuses on analyzing the effect of intra-chain hydrogen bonding on the thermal stability of cellulose chains, and the variation in the number of intra-chain hydrogen bonds at different temperatures for the models is shown in Figure 4.

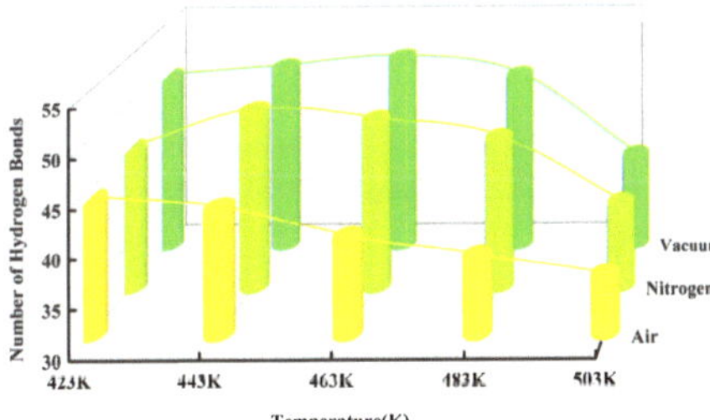

Figure 4. The number of hydrogen bonds at different temperatures for the three models.

The stability of cellulose's structure and characteristics are primarily dependent on intra-chain hydrogen bonding. The number of intra-chain hydrogen bonds in the vacuum model is much larger than that of the nitrogen and air models at any temperature, as Figure 4 illustrates. With the increase in temperature, the number of hydrogen bonds in both the vacuum and nitrogen models tended to increase and decrease. This could be because, in the interval of a lower heat treatment temperature, the cellulose chain has thermal movement with an increasing temperature, which produces more hydrogen bonds while reducing molecular spacing, thus enhancing molecular bonding and making the molecular structure more stable. With the increase in temperature, the speed and frequency of molecular movement will increase, resulting in more frequent intermolecular collisions and changes in the distance and relative positions between molecules, and the destruction or even fracture of hydrogen bonds within the chain occurs, resulting in a decrease in the overall number of hydrogen bonds. The above analysis corresponds to the mean orientation shift curve of the cellulose chain. Consequently, the cellulose chain's structure may become unstable at overly high temperatures, which may impact the structure's overall thermal stability.

According to a further data analysis, the total amount of intra-chain hydrogen bond in cellulose in the nitrogen and vacuum models peaked at 443 K and 463 K. This indicates that the formation and stabilization of intra-chain hydrogen bonding are optimized in this temperature interval, which significantly enhances the cellulose's intermolecular bonding. The intermolecular contacts increased as the amount of intrachain hydrogen bonds increased, limiting the amplitude of the thermal movement of cellulose chains and slowing down the changes in molecular structure brought about by thermal movement. This improved molecular bonding strengthened the stability of the cellulose chain itself and raised the structure's overall heat resistance. In particular, under vacuum conditions, the lack of interference and collisions with air molecules allows cellulose chains to form and maintain more intra-chain hydrogen bonds in a purer and more stable environment. As a result, cellulose chains under vacuum conditions exhibit stronger molecular bonding and superior thermal stability compared to other environments such as nitrogen or air. This property makes the vacuum heat treatment an effective means to better maintain and enhance heat-treated wood's structural strength and durability.

3.5. Mechanical Properties

The mechanical properties reveal the material's susceptibility to damage and distortion due to outside forces. The mechanical properties can be characterized in molecular dynamics simulations by many mechanical property parameters. Since amorphous cellulose is isotropic [34], it can be calculated using Formula (3).

$$[C_{ij}] = \begin{bmatrix} \lambda + 2\mu & \lambda & \lambda & 0 & 0 & 0 \\ \lambda & \lambda + 2\mu & \lambda & 0 & 0 & 0 \\ \lambda & \lambda & \lambda + 2\mu & 0 & 0 & 0 \\ 0 & 0 & 0 & \mu & 0 & 0 \\ 0 & 0 & 0 & 0 & \mu & 0 \\ 0 & 0 & 0 & 0 & 0 & \mu \end{bmatrix} \tag{3}$$

where λ and μ are known as Lamey's constants and are employed in the calculation of the Poisson's ratio (η), Young's modulus (E), shear modulus (G), and other values, and the calculation formula is as follows:

$$E = \frac{\mu(3\lambda + 2\mu)}{\lambda + \mu} \tag{4}$$

$$G = \mu \tag{5}$$

$$K = \lambda + \frac{2}{3\mu} \tag{6}$$

$$\gamma = \frac{\lambda}{2(\lambda + \mu)} \tag{7}$$

Each mechanical parameter of the vacuum, nitrogen, and air models is calculated with the help of the above equation, as shown in Table 2.

The modulus of elasticity may be thought of as a gauge for how difficult it is to cause elastic deformation in a material; the higher its value, the higher the tension required to cause elastic deformation depending on the stress, the Young's modulus (E), shear modulus (G), and bulk modulus (K). It was calculated that the K of the cellulose chain produces a small change with an increasing temperature, while E and G can first show a clear increasing trend and then a decreasing one. Therefore, to highlight the comparative effect of the three models, we choose E and G as the primary mechanical parameters for a comparative analysis.

Table 2. Each mechanical parameter of the three models at different temperatures.

Temperature (K)	Medium	λ	μ	E	G	γ
	Vacuum	6.14	4.16	10.8	4.16	0.30
423	Nitrogen	6.94	2.12	5.86	2.12	0.38
	Air	6.82	2.03	5.62	2.03	0.39
	Vacuum	7.15	4.75	12.35	4.75	0.30
443	Nitrogen	5.66	2.26	6.14	2.26	0.36
	Air	6.33	1.98	5.47	1.98	0.38
	Vacuum	6.97	6.05	15.34	6.05	0.27
463	Nitrogen	3.18	2.14	5.56	2.14	0.30
	Air	6.17	1.89	5.23	1.89	0.38
	Vacuum	4.1	4.82	11.86	4.82	0.23
483	Nitrogen	5.77	1.48	4.14	1.48	0.40
	Air	6.08	1.33	3.75	1.33	0.41
	Vacuum	3.63	3.83	9.52	3.83	0.24
503	Nitrogen	6.39	1.02	2.92	1.02	0.43
	Air	6.01	0.81	2.33	0.81	0.44

As shown in Figure 5, the E and G values of the vacuum model are significantly larger than those of the nitrogen and air models at any temperature. The maximum values of E and G for the vacuum and nitrogen models are reached at 463 K and 443 K, respectively. The conclusions of the above data on mean square displacement, hydrogen bonding, and densification degree are generally consistent with this. Cellulose chains are used in many heat treatment experiments for wood, and the analysis of the mechanical properties of cellulose chains is helpful for the relevant heat treatment experiments. The elastic modulus and density of heat-treated wood were found to correlate positively, as shown in the study conducted by Bao et al. [35]. A greater degree of hydrogen bonding among cellulose chains reinforces the van der Waals forces in the interchain contacts, strengthening the cellulose system as a whole and increasing the Young's modulus and shear modulus as well as timber's ability to withstand deformation [36]. The vacuum environment contributes to the rapid discharge of water inside the wood during the heat treatment process, thus promoting the densification of the wood. With less moisture, the fiber structure of the wood becomes tighter, reducing pores and defects, which helps to improve the overall mechanical properties of the wood. As a result, heat-treating wood in a vacuum preserves its strength and stiffness better than heat-treating it in a nitrogen environment, and the results are superior at 463 K.

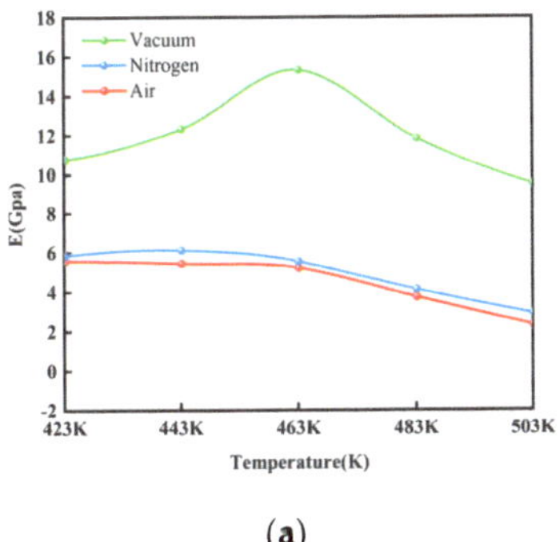

(a)

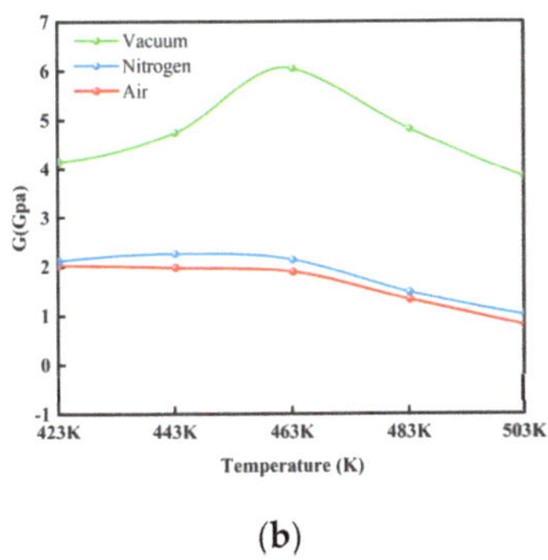

(b)

Figure 5. (a) Young's modulus and (b) shear modulus of three models at different temperatures.

The ratio of the axial positive strain to transverse positive strain in a material under unidirectional tension or compression is known as Poisson's ratio (γ). It is an elastic constant that represents the material's transverse deformation and is positively correlated with the material's plasticity; the greater the value, the more plastic the material is. The data in Table 2 show that the γ value of the air model is more significant than that of the vacuum and nitrogen models at any temperature. The oxygen in the air can promote the oxidation reaction on the surface of the wood and form an oxide layer, which can maintain the moisture and humidity inside the wood to a certain extent, and it is conducive to maintaining the plasticity of the wood. Therefore, air conditions can maintain wood plasticity better than vacuum and nitrogen conditions for heat treatments.

Wood is a widely used cellulose-based material; its properties can be changed by heat treatments in different environments so as to meet diversified application needs. In decoration and construction scenarios, the strength and stability of wood are of paramount importance. Compared with heat treatments in nitrogen or air, heat treatments in a vacuum environment can significantly improve the strength and deformation resistance of wood. When wood is used in art design, sculpture, and other fields that need to show its unique texture and shape change, heat treatments in an air environment may be more appropriate. Although the improvement in the strength of wood through a heat treatment in air may not be as significant as that in a vacuum environment, this treatment can better maintain the toughness and plasticity of wood. Therefore, when selecting a heat treatment environment, it is necessary to take into account the specific application scenarios and required characteristics of wood in order to maximize the potential of wood and meet the diverse needs of cellulose-based materials in different fields.

4. Conclusions

In this paper, three models, namely vacuum–cellulose, nitrogen–cellulose, and air–cellulose, were constructed using the Materials Studio software, and five temperature gradients were set for a molecular dynamics simulation. Following a detailed analysis of the models' energy balance, mechanical characteristics, intra-chain hydrogen bonding, mean square displacements, cell parameters, and densities, the following results were drawn:

1. The cell parameters and densities of the vacuum, nitrogen, and air models were compared at five different temperatures. Taking 463 K as an example, the cell size of the vacuum–cellulose model was 20.13, that of the nitrogen–cellulose model was 20.93, and that of the air–cellulose model was 21.53. In the vacuum model, the cell volume was smaller, and the average density was larger, which indicates that the cellulose cells treated by a vacuum at the same temperature were more compact. Therefore, a vacuum heat treatment can better improve the density of the material.

2. In the vacuum and nitrogen models, the *MSD* of cellulose chains first fell and subsequently increased as the temperature rose. In contrast, due to the interference of various gas molecules, the *MSD* of the cellulose chain in the air model increased from 0 to 13.4 at 503 K, and its *MSD* curve fluctuated more violently. Overly high temperatures weaken the structural integrity of cellulose chains by upsetting their internal structure. This also concerns how many hydrogen bonds are in the cellulose chain; more intra-chain hydrogen bonds bolster intermolecular links and solidify the cellulose chain's structural integrity.

3. At all temperatures, the vacuum model consistently exhibited bigger values for both the Young's modulus and shear modulus when compared to the other two models. This indicates that the cellulose chains treated with the vacuum heat treatment were stiffer and more deformation resistant. On the contrary, comparing the Poisson's ratios of the three models, the Poisson's ratio of the air model was always more considerable, which indicates that the plasticity of the air heat-treated cellulose chains was better. Thus, to maximize the use of wood, appropriate heat treatment temperatures and media may be chosen based on real-world requirements.

Author Contributions: Conceptualization, Y.H. and J.G.; methodology, Y.H. and Z.Q.; software, Y.H. and N.L.; validation, Y.H., W.W., N.L. and Z.Q.; formal analysis, Y.H.; investigation, Y.H.; resources, Y.H.; data curation, Y.H.; writing—original draft preparation, Y.H.; writing—review and editing, Y.H. and W.W.; visualization, Y.H.; supervision, W.W., Z.Q. and J.G.; project administration, W.W.; funding acquisition, W.W. All authors have read and agreed to the published version of the manuscript.

Funding: This research was supported by the Natural Scientific Foundation of Heilongjiang Province, grant number LC201407.

Data Availability Statement: The data are available upon request from the corresponding author.

Conflicts of Interest: The authors declare no conflicts of interest.

References

1. Credou, J.; Berthelot, T. Cellulose: From biocompatible to bioactive material. *J. Mater. Chem. B* **2014**, *2*, 4767–4788. [CrossRef] [PubMed]
2. Khazraji, A.C.; Robert, S. Interaction effects between cellulose and water in nanocrystalline and amorphous regions: A novel approach using molecular modeling. *J. Nanomater.* **2013**, *44*, 409676. [CrossRef]
3. Jiang, X.; Wang, W.; Guo, Y.; Dai, M. Molecular Dynamics Simulation of the Effect of Low Temperature on the Properties of Lignocellulosic Amorphous Region. *Forests* **2023**, *14*, 1208. [CrossRef]
4. Kamperidou, V. Chemical and Structural Characterization of Poplar and Black Pine Wood Exposed to Short Thermal Modification. *Drv. Ind.* **2021**, *72*, 155–167. [CrossRef]
5. Cao, Y.; Lu, J.; Huang, R.; Jiang, J. Increased Dimensional Stability of Chinese Fir through Steam-Heat Treatment. *Eur. J. Wood Prod.* **2012**, *70*, 441–444. [CrossRef]
6. Yang, T.-H.; Lee, C.-H.; Lee, C.-J.; Cheng, Y.-W. Effects of Different Thermal Modification Media on Physical and Mechanical Properties of Moso Bamboo. *Constr. Build. Mater.* **2016**, *119*, 251–259. [CrossRef]
7. Bal, B.C. A Comparative Study of Some of the Mechanical Properties of Pine Wood Heat Treated in Vacuum, Nitrogen, and Air Atmospheres. *BioRes* **2018**, *13*, 5504–5511. [CrossRef]
8. Candelier, K.; Dumarçay, S.; Pétrissans, A.; Gérardin, P.; Pétrissans, M. Comparison of Mechanical Properties of Heat Treated Beech Wood Cured under Nitrogen or Vacuum. *Polym. Degrad. Stab.* **2013**, *98*, 1762–1765. [CrossRef]
9. Alder, B.J.; Wainwright, T.E. Phase Transition for a Hard Sphere System. *J. Chem. Phys.* **1957**, *27*, 1208–1209. [CrossRef]
10. *Computational Soft Matter: From Synthetic Polymers to Proteins*; Lect: Lecture Notes; NIC series; NIC: Jülich, Germany, 2004.
11. Fan, H.B.; Yuen, M.M.F. Material Properties of the Cross-Linked Epoxy Resin Compound Predicted by Molecular Dynamics Simulation. *Polymer* **2007**, *48*, 2174–2178. [CrossRef]
12. Tsai, J.-L.; Tu, J.-F. Characterizing Mechanical Properties of Graphite Using Molecular Dynamics Simulation. *Mater. Des.* **2010**, *31*, 194–199. [CrossRef]
13. Mao, Z.; Garg, A.; Sinnott, S.B. Molecular Dynamics Simulations of the Filling and Decorating of Carbon Nanotubules. *Nanotechnology* **1999**, *10*, 273–277. [CrossRef]
14. Cao, Y.; Wang, W.; Ma, W. Analysis on the Diffusion and Mechanical Properties of Eucalyptus Dried via Supercritical Carbon Dioxide. *BioResources* **2022**, *17*, 4018. [CrossRef]
15. Pelaez-Samaniego, M.R.; Yadama, V.; Lowell, E.; Espinoza-Herrera, R. A Review of Wood Thermal Pretreatments to Improve Wood Composite Properties. *Wood Sci. Technol.* **2013**, *47*, 1285–1319. [CrossRef]
16. Atalla, R.H.; VanderHart, D.L. Native Cellulose: A Composite of Two Distinct Crystalline Forms. *Science* **1984**, *223*, 283–285. [CrossRef] [PubMed]
17. Wang, X.; Tang, C.; Wang, Q.; Li, X.; Hao, J. Selection of Optimal Polymerization Degree and Force Field in the Molecular Dynamics Simulation of Insulating Paper Cellulose. *Energies* **2017**, *10*, 1377. [CrossRef]
18. Paajanen, A.; Vaari, J. High-Temperature Decomposition of the Cellulose Molecule: A Stochastic Molecular Dynamics Study. *Cellulose* **2017**, *24*, 2713–2725. [CrossRef]
19. Theodorou, D.N.; Suter, U.W. Detailed Molecular Structure of a Vinyl Polymer Glass. *Macromolecules* **1985**, *18*, 1467–1478. [CrossRef]
20. Wang, W.; Wang, Y.; Li, X. Molecular Dynamics Study on Mechanical Properties of Cellulose with Air/Nitrogen Diffusion Behavior. *BioResources* **2018**, *13*, 7900–7910. [CrossRef]
21. Nosé, S. Constant Temperature Molecular Dynamics Methods. *Prog. Theor. Phys. Suppl.* **1991**, *103*, 1–46. [CrossRef]
22. Andrea, T.A.; Swope, W.C.; Andersen, H.C. The Role of Long Ranged Forces in Determining the Structure and Properties of Liquid Water. *J. Chem. Phys.* **1983**, *79*, 4576–4584. [CrossRef]
23. Berendsen, H.J.C.; Postma, J.P.M.; Van Gunsteren, W.F.; DiNola, A.; Haak, J.R. Molecular Dynamics with Coupling to an External Bath. *J. Chem. Phys.* **1984**, *81*, 3684–3690. [CrossRef]
24. Ewald, P.P. Die Berechnung optischer und elektrostatischer Gitterpotentiale. *Ann. Phys.* **1921**, *369*, 253287. [CrossRef]
25. Andersen, H.C. Molecular Dynamics Simulations at Constant Pressure and/or Temperature. *J. Chem. Phys.* **1980**, *72*, 2384–2393. [CrossRef]

26. Sun, H. Ab Initio Calculations and Force Field Development for Computer Simulation of Polysilanes. *Macromolecules* **1995**, *28*, 701712. [CrossRef]

27. Wang, W.; Ma, W.; Wu, M.; Sun, L. Effect of Water Molecules at Different Temperatures on Properties of Cellulose Based on Molecular Dynamics Simulation. *BioRes* **2021**, *17*, 269–280. [CrossRef]

28. Allen, M.P.; Tildesley, D.J.J. *Computer Simulation of Liquids*; Clarendon Press: Oxford, UK, 1987.

29. Wang, X.; Tu, D.; Chen, C.; Zhou, Q.; Huang, H.; Zheng, Z.; Zhu, Z. A Thermal Modification Technique Combining Bulk Densification and Heat Treatment for Poplar Wood with Low Moisture Content. *Constr. Build. Mater.* **2021**, *291*, 123395. [CrossRef]

30. Xu, B.; Chen, Z.; Ma, Q. Effect of High-Voltage Electric Field on Formaldehyde Diffusion within Building Materials. *Build. Environ.* **2016**, *95*, 372–380. [CrossRef]

31. Li, X.; Tang, C.; Wang, J.; Tian, W.; Hu, D. Analysis and Mechanism of Adsorption of Naphthenic Mineral Oil, Water, Formic Acid, Carbon Dioxide, and Methane on Meta-Aramid Insulation Paper. *J. Mater. Sci.* **2019**, *54*, 8556–8570. [CrossRef]

32. Nishiyama, Y.; Langan, P.; Chanzy, H. Crystal Structure and Hydrogen-Bonding System in Cellulose Iβ from Synchrotron X-Ray and Neutron Fiber Diffraction. *J. Am. Chem. Soc.* **2002**, *124*, 9074–9082. [CrossRef]

33. Wang, W.; Sun, L.; Wu, M.; Li, X.; Song, W. Molecular Dynamics Simulation of Bamboo Heat Treatment with Cellulose Based on Molecular Different Weight Fractions of Water. *BioRes* **2020**, *15*, 6766–6780. [CrossRef]

34. Kulasinski, K.; Keten, S.; Churakov, S.V.; Derome, D.; Carmeliet, J. A Comparative Molecular Dynamics Study of Crystalline, Paracrystalline and Amorphous States of Cellulose. *Cellulose* **2014**, *21*, 1103–1116. [CrossRef]

35. Bao, M.; Huang, X.; Jiang, M.; Yu, W.; Yu, Y. Effect of Thermo-Hydro-Mechanical Densification on Microstructure and Properties of Poplar Wood (*Populus Tomentosa*). *J. Wood Sci.* **2017**, *63*, 591–605. [CrossRef]

36. Du, D.; Tang, C.; Zhang, J.; Hu, D. Effects of Hydrogen Sulfide on the Mechanical and Thermal Properties of Cellulose Insulation Paper: A Molecular Dynamics Simulation. *Mater. Chem. Phys.* **2020**, *240*, 122153. [CrossRef]

Article

Effect of Bamboo Nodes on the Mechanical Properties of *Phyllostachys iridescens*

Xuehua Wang [1,*], Siyuan Yu [2], Shuotong Deng [1], Ru Xu [1], Qi Chen [3] and Pingping Xu [4,*]

[1] College of Furnishings and Industrial Design, Nanjing Forestry University, Nanjing 210037, China; dengshuotong@njfu.edu.cn (S.D.); 1016.xuru@njfu.edu.cn (R.X.)

[2] College of Materials Science and Engineering, Hunan University, Changsha 410082, China; yusiyuan@hnu.edu.cn

[3] College of Forestry, Sichuan Agricultural University, Chengdu 611130, China; qichen@sicau.edu.cn

[4] BASF (China) Co., Ltd., Shanghai 200137, China

[*] Correspondence: wangxuehua@njfu.edu.cn (X.W.); mark.a.xu@basf.com (P.X.)

Abstract: Bamboo is a significant natural resource, recognized for its rapid growth, lightweight composition, high strength, and excellent mechanical properties, making it increasingly valuable in the furniture and construction industries. A critical structural aspect of bamboo is its nodes, yet there has been limited research on their impact on bamboo's mechanical properties. This study investigates the mechanical properties of round bamboo tubes in three different states: internodes (S1), nodes with diaphragm removed (S2), and nodes with diaphragm (S3). The results show that the mechanical properties of S1 are a compressive strength (CS) of 29.72 MPa, a shear strength parallel to grain (SS_p) of 11.82 MPa, a radial stiffness (S_r) of 155.59 MPa, an impact toughness (IT) of 20.74 kJ/m^2, a modulus of rupture (MOR) of 16.45 MPa, a modulus of elasticity (MOE) of 408.53 MPa, a tensile modulus of rupture parallel to grain (MOR_T) of 189.62 MPa, and a tensile modulus of elasticity parallel to grain (MOE_T) of 431.05 MPa. Compared with S1, these above parameters change by CS +11%, SS_p 6%, S_r +100%, IT −29%, MOR +5%, MOE +63%, MOR_T −29%, and MOE_T −58% in S2 and CS +10%, SS_p 28%, S_r +250%, IT −31%, MOR +28%, MOE +92%, MOR_T −25%, and MOE_T −42% in S3. It demonstrates that the bamboo diaphragm and nodes significantly influence the mechanical properties of bamboo; they have a significant positive effect on the bending properties across the transverse grain, radial ring stiffness, and shear properties along the grain, but negatively impact the tensile properties along the grain.

Keywords: round bamboo; bamboo diaphragm; mechanical property; parallel and transverse to bamboo grain

Citation: Wang, X.; Yu, S.; Deng, S.; Xu, R.; Chen, Q.; Xu, P. Effect of Bamboo Nodes on the Mechanical Properties of *Phyllostachys iridescens*. *Forests* **2024**, *15*, 1740. https://doi.org/10.3390/f15101740

Academic Editor: Luis García Esteban

Received: 10 September 2024
Revised: 25 September 2024
Accepted: 30 September 2024
Published: 2 October 2024

1. Introduction

Bamboo resources are widely distributed across the globe, primarily concentrated in the tropical and subtropical regions of the Asia–Pacific, with the global bamboo planting area reaching 22 million hectares. Bamboo's remarkable characteristics—such as its rapid growth rate, high strength-to-weight ratio, and wide distribution—make it a versatile material extensively used in construction, home furniture, and daily necessities [1]. With appropriate modification treatments, the service life of bamboo products can exceed 30 years, making it an effective carbon sequestration material and a renewable resource with significant potential.

For many Asian and African countries, bamboo is a vital natural resource with substantial economic value. In Asia, it is extensively utilized in industries such as construction, furniture, and papermaking, contributing to economic development. In Africa, bamboo, as an emerging resource, is driving green economic growth, poverty alleviation, and environmental protection, demonstrating its significant development potential.

China is rich in diverse bamboo resources, boasting the largest bamboo forest area in the world [2]. The output value of the bamboo industry continues to grow, supported by numerous processing enterprises, and its products are widely employed in construction, furniture, papermaking, and other sectors. This has led to the formation of a complete industrial chain, making significant contributions to economic development.

Bamboo culms exhibit excellent bending toughness and serve as the primary material for bamboo processing and utilization. The bamboo culm is characterized by a thin, hollow wall, with its length segmented by nodes that appear every few centimeters to several tens of centimeters [3]. The cross-sectional area of the culm gradually decreases from the base upwards. This unique structure has evolved to help bamboo adapt to its growth environment, providing effective protection against external forces like wind and snow. The structural characteristics of these nodes vary across different bamboo species, with their form and frequency depending on the species and the height of the culm [4].

Inside the bamboo, there is a bamboo diaphragm corresponding to the position of the bamboo node. The presence of bamboo nodes significantly impacts both the processing efficiency and the product quality of bamboo [5]. For instance, bamboo nodes enhance the gluing properties of bamboo laminated timber but they reduce its flexural strength. During the bamboo flattening process, nodes can have a detrimental effect, exacerbating cracking during both the flattening and the dividing stages. Additionally, the importance of bamboo nodes in the intermediate growth and the material transport of bamboo should not be overlooked. The rapid growth of bamboo relies on the internode meristem, while the lateral transport of water and nutrients occurs within the bamboo nodes [6]. Consequently, bamboo nodes play a crucial role in maintaining the overall structural integrity of bamboo culm's hollow structure.

The differences in the microstructure of bamboo nodes and internodes are believed to influence the processing performance of bamboo. The vascular bundles and the parenchyma cells in the internode are mainly parallel to the bamboo's height in bamboo internode, while they are disordered in the node [7,8]. The effect of bamboo nodes on bamboo function and processing is influenced by the bamboo's microstructure, which differs between the nodes and internodes. In the internodes, the vascular bundles and parenchyma cells are primarily aligned parallel to the height of the bamboo, whereas they are more disordered within the nodes [9]. Additionally, the bamboo diaphragm inside the nodes forms a partition that effectively prevents splitting, damage, and deformation of the round bamboo [10].

Given the significance of the bamboo nodes, numerous studies have examined their influence on the properties of bamboo-based panels. For instance, Widjaja et al. explored the relationship between bamboo fibers and mechanical properties, finding that flexural strength, compressive strength, and tensile strength are closely related to the bamboo fiber [11]. Murphy et al. reported that the distinctive secondary wall structure of bamboo fibers, characterized by alternating width and thickening, significantly influences bamboo's mechanical properties [12]. Garcia et al. assessed the transverse elastic modulus, shear modulus, and Poisson's ratio of Guadua bamboo through flexural testing of ring specimens, providing data directly applicable to the numerical simulation of this bamboo species [13]. Despite these advancements, relatively few studies have examined the impact of bamboo nodes on these properties.

Currently, most studies about the influence of bamboo nodes on the mechanical properties of bamboo primarily use a bamboo strip as the test unit. However, the effects of the bamboo node and the diaphragm on mechanical properties of round bamboo are often not considered. This paper investigates the effects of the bamboo node and the diaphragm on the mechanical properties of bamboo, providing a scientific and effective basis for its practical application. These insights can enhance the utilization rate of bamboo resources and promote the healthy and sustainable development of China's bamboo industry.

2. Materials and Methods

2.1. Materials

Phyllostachys iridescens, commonly known as red bamboo, was selected as the research subject due to its frequent use in furniture manufacturing. The bamboo specimens were sourced from Anji, Zhejiang Province, and were between 4 and 6 years old. The bamboo sections selected for sample preparation were from 0.5 m to 2.3 m above the ground. They had an average diameter of 40.0 ± 1.0 mm, with a thickness of the bamboo wall of 4.0 ± 0.5 mm. The moisture content of the red bamboo was 12.0 ± 1.5%. Three types of bamboo were investigated in this study: bamboo internodes (S1), bamboo tubes with nodes but without diaphragms (S2), and bamboo tubes with nodes and diaphragms kept (S3) (Figure 1a).

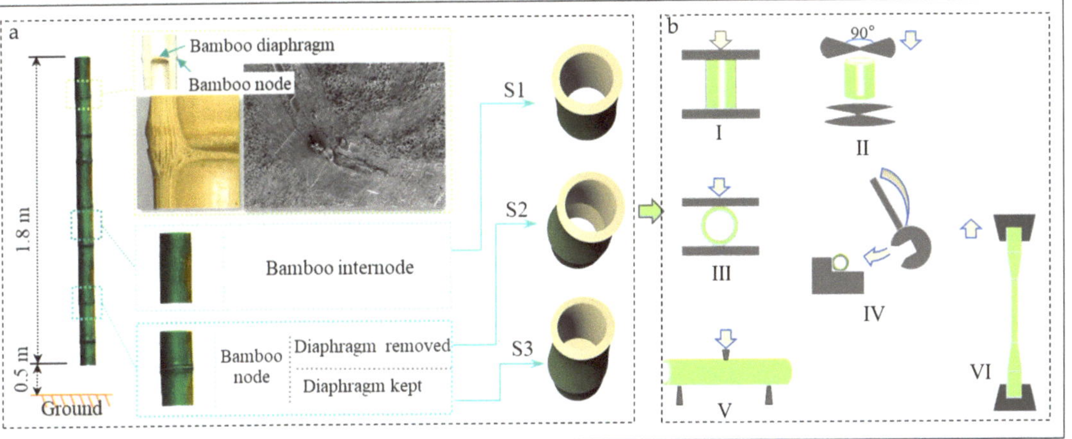

Figure 1. Diagram of the sampling and mechanical properties testing: (**a**) sample preparing, and (**b**) mechanical properties testing methods, I—compression strength, II—shear strength, III—radial stiffness, IV—impact toughness, V—bending properties, and VI—tensile properties.

2.2. Test Methods

The mechanical properties evaluated in this study include modulus of rupture (MOR), modulus of elasticity (MOE), impact toughness (IT), radial stiffness (Sr), shear strength parallel to grain (SSp), compressive strength parallel to grain (CS), tensile modulus of rupture parallel to grain (MOR_T), and tensile elastic modulus parallel to grain (MOE_T). Detailed sample information for each parameter is presented in Figure 1b. Six parallel samples were prepared for each test, and the arithmetic mean values were calculated, with the standard deviations represented as error bars.

The parameters MOR, MOE, SR, SSP, CS, MOR_T, and MOE_T were tested according to the LY/T2564-2015 standard, "Test Method for Physical and Mechanical Properties of Round Bamboo" [14]; an MMW-50 mechanical testing machine (Jinan Nair Testing Machine Co., Ltd., Jinan, China) was employed for testing these mechanical properties. IT was assessed following the GB/T 1927.17-2021 standard, "Test methods for physical and mechanical properties of small clear wood specimens-Part 17: Determination of impact bending strength" [15]; a JBS-300S pendulum impact testing machine (Beijing Time High Technology Ltd., Bejing, China) was used for the IT tests.

3. Results

3.1. Compressive Strength Parallel to Grain

At the beginning of the testing, the bamboo did not exhibit significant changes under loading. However, as the load gradually increased, slight bulging occurred in the bamboo tube near the loading end. In specimen S1, cracks developed along the longitudinal direction of the outer surface near the bulged area, extending towards the inner surface. For S2 and S3, the stiffening effect of the bamboo node reduced the extent of bulging compared to S1. Nevertheless, cracks still formed along the bamboo wall, extending from the wall to the center of the bamboo diaphragm in S3 (Figure 2a).

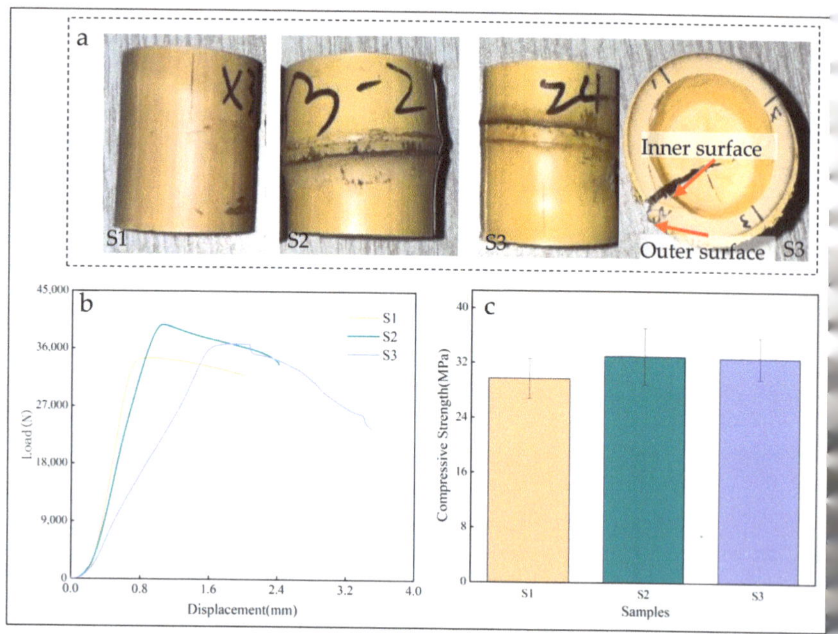

Figure 2. Compression property of different bamboo tubes: (**a**) the failure samples, (**b**) the load displacement curve, and (**c**) compression strength.

Figure 2b presents the load-displacement curves, which demonstrate similar behavior across all specimens. Initially, the load-displacement relationship exhibits a concave shape. Subsequently, the load increases linearly with the displacement, indicating the elastic deformation phase. When the load exceeds the proportional limit, the slope of the curve decreases. Upon reaching the maximum load, the load gradually declines, reflecting a reduction in bearing capacity. As the load continues to increase, the specimen eventually loses its bearing capacity, leading to complete failure. Notably, the slopes of the curves differ among the groups, with S1 showing the steepest slope, followed by S2 and then S3.

The compressive strengths of S1, S2, and S3 were 29.72 MPa, 32.98 MPa, and 32.79 MPa, respectively. Compared to S1, S2 and S3 increased by approximately 11% and 10%, respectively, although these differences were not statistically significant (Figure 2c). While the bamboo node enhances the compressive strength, the bamboo diaphragm had no significant influence. This observation is consistent with the findings of Hao [16] and Lin [17]. This suggests that the bamboo node, rather than the diaphragm, had a slight stiffening effect on the compressive strength. The modest positive effect may be attributed to the disordered arrangement of the cells, which contributes to a more stable structure compared to the single-direction arrangement in S1.

3.2. Shear Strength Parallel to the Grain

In shear resistance testing, no significant phenomena are observed initially. However, as the load increases, a sudden loud noise occurs, indicating that the sample has reached its failure load. After being removed from the test equipment, the specimen exhibits a crack along the grain at the shear location, with the bamboo pieces on either side of the cushion block misaligned. Additionally, the bamboo diaphragm in S3 shows localized tearing (Figure 3a).

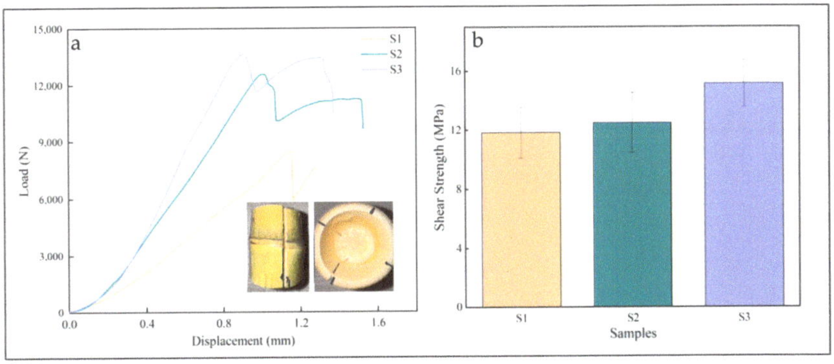

Figure 3. Shear property of different bamboo tubes: (**a**) the failure sample and load-displacement curve, (**b**) shear strength parallel to the grain.

Figure 3a shows the loading curves from the shear resistance tests, which exhibit similar trends between these three samples. Initially, the increase is gradual with a concave pattern. As the load increases, the load-displacement relationship becomes linear. The slope then decreases, indicating the transition to the elastoplastic stage. Upon reaching the ultimate load, the curve drops sharply, followed by subsequent stepped drops. This behavior indicates that the shear region does not fail entirely at once, allowing the specimen to sustain some load even after initial localized failure. Notably, the curves exhibit a sequential decrease in magnitude, with S3 showing the highest values, followed by S2 and S1.

Figure 3b presents the value of the shear strength parallel to the grain. The shear strength was 11.82 MPa for S1, 12.48 MPa for S2, and 15.15 MPa for S3. Compared to S1, the shear strength of S2 increased by approximately 6%, while S3 showed an increase of about 28%. These results indicate that the bamboo diaphragm enhances the shear strength of bamboo, whereas the bamboo node has minimal impact on its shear strength. This result aligns with the findings of Shao [18].

The single arrangement of fibers and parenchyma cells along the longitudinal direction made bamboo easy to crack; the positive effect of the bamboo node and the diaphragm on shear strength might be caused by the arrangement direction change in the bamboo node and the diaphragm. When S1 is subjected to shear force, the bamboo wall would be easy to crack due to the single arrangement of fibers and parenchyma cells [19]; the specimen loses its bearing capacity once the bamboo wall cracks. In contrast, for specimens S2 and S3, failure under shear force requires not only cracking the bamboo wall at the shear site, but also breaking the connection between the bamboo node, the diaphragm, and the bamboo wall [7]. This additional resistance provided by the bamboo diaphragm enhances the shear strength of the bamboo along the grain.

3.3. Radial Stiffness

During radial stiffness testing, the crack initiation and propagation patterns were similar across all groups, with initial cracks typically developing near the mid-height of the bamboo specimens (Figure 4a). Cracking began on the bamboo outer surface and progressively extended towards the bamboo inner surface as the load increased, eventually penetrating the entire bamboo wall and resulting in specimen failure. The highest number of cracks was observed in group S1. Specimens in group S3 exhibited significantly less deformation along the loading direction compared to the other two groups. Additionally, a vertical crack appeared at the center of the bamboo diaphragm, perpendicular to the direction of the loading plate, gradually extending outward until the specimen failed.

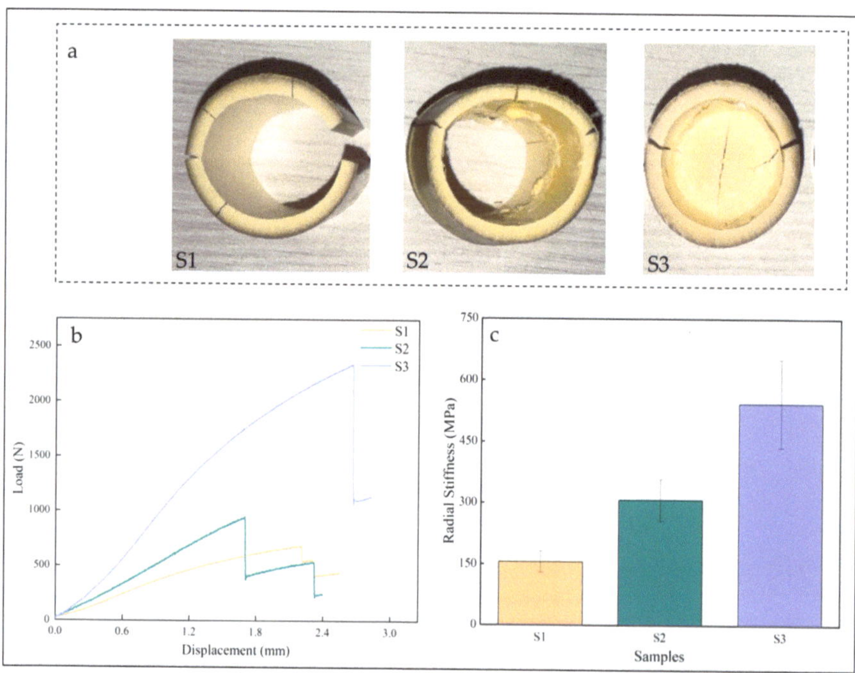

Figure 4. Radial stiffness of different bamboo tubes: (**a**) the failure samples, (**b**) the load-displacement curve, and (**c**) the radial stiffness.

In the load-displacement curves, the change trends were similar across all groups, but the slopes differed, with S3 showing the steepest slope, followed by S2 and then S1. The curves initially displayed a concave shape during the early loading stage, transitioning into a linear increase with displacement, which indicates elastic deformation where deformation is proportional to the load and is reversible. As the slope gradually decreased, the specimens entered the elastoplastic stage with an accelerated rate of displacement. Upon reaching the maximum load, the curves sharply declined; however, the specimens did not immediately lose their bearing capacity. The curves then underwent several cycles of rising and falling before the specimens ultimately failed (Figure 4b).

The radial stiffness is shown in Figure 2c. The average radial stiffness was 155.59 MPa for S1, 306.87 MPa for S2, and 542.75 MPa for S3. Compared to S1, the radial stiffness of S2 increased by approximately 100%, and the S3 showed an increase of about 250% (Figure 4c). These results indicate that both the bamboo node and the diaphragm significantly influence radial stiffness, positively enhancing it, which aligns with the findings of Yu [20].

The positive effects of the bamboo node and the diaphragm are likely due to their unique microstructure. In the bamboo internode, fiber bundles and parenchyma cells are predominantly arranged longitudinally, while in the bamboo node and diaphragm, the arrangement is more chaotic and disordered [21]. The fiber strength in the transverse direction is considerably lower than in the longitudinal direction [22]. During the radial stiffness test, fibers and parenchyma cells are subjected to vertical loading in S1 and partial horizontal loading in S2 and S3, effectively utilizing the high longitudinal strength of the fibers. Additionally, the bamboo diaphragm exerts a tightening effect on the bamboo tube limiting lateral deformation and enhancing the bamboo's radial stiffness [23].

3.4. Impact Toughness

Figure 5 presents the results of the impact toughness test. The average impact toughness of S1 was 20.74 kJ/m^2, S2 was 26.80 kJ/m^2, and S3 was 14.34 kJ/m^2. Compared to S1, the impact toughness of S2 increased by approximately 29%, whereas that of S3 decreased by about 31%. These results indicate that the bamboo node enhances the impact toughness of red bamboo, while the bamboo diaphragm reduces it.

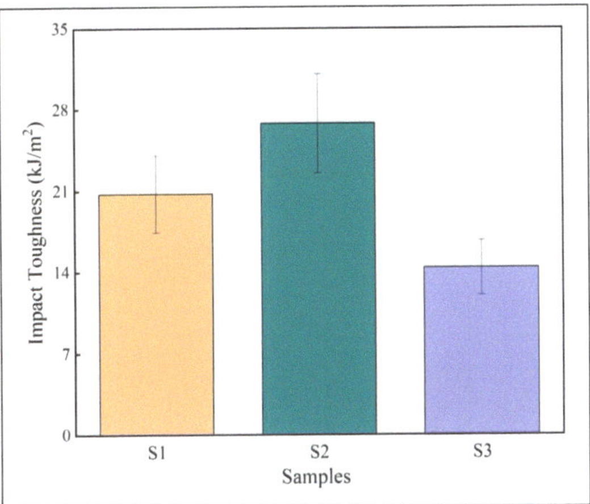

Figure 5. Impact toughness of different bamboo tubes.

Compared to S1, the IT in S2 increased, likely due to changes in the microstructure of the bamboo node. In bamboo internodes, fibers and parenchyma cells are arranged longitudinally, while in bamboo nodes, the cells display an irregular arrangement. In S2, the cells from the internode and the node formed a cross structure [24], which is more effective at withstanding impact loading than a single-direction arrangement [25,26], leading to the increased IT observed in S2.

However, the reduced IT in S3 can be attributed to the brittleness resulting from the high lignin content in the bamboo diaphragm. Bamboo primarily consists of three chemical components: cellulose, hemicellulose, and lignin. With higher lignin content, the bamboo exhibited increased brittleness [27]. The bamboo diaphragm has a higher lignin content compared to the bamboo internode [28], resulting in greater brittleness. Consequently, this increased brittleness of the bamboo diaphragm negatively affects its performance in impact toughness tests.

3.5. Bending Properties

During the initial stages of the bending tests, the bamboo specimens showed no significant changes. However, as the load increased, longitudinal cracks developed at the loading points across all groups. Notably, only the S1 specimens exhibited bending deformation. In S1, cracks formed along the grain direction at mid-height, initially appearing on the outer surface and then propagating towards the inner surface. After bending failure, S1 displayed penetrating cracks along the bamboo's length and through its wall, whereas no penetrating cracks were observed in S2 and S3 (Figure 6a).

Figure 6b shows the load-displacement curves for the bending test, which exhibit a similar trend across the three groups. In the initial loading stage, the load-displacement curves increase linearly. As the load increases, the deflection rate accelerates, causing the curves to decelerate. The deceleration is significantly more pronounced in S1 compared to S2 and S3, with S1 showing noticeable bending deformation. Upon reaching the ultimate load, the specimens fail instantaneously, leading to a sharp drop in the load-displacement curve accompanied by significant fluctuations near the ultimate load.

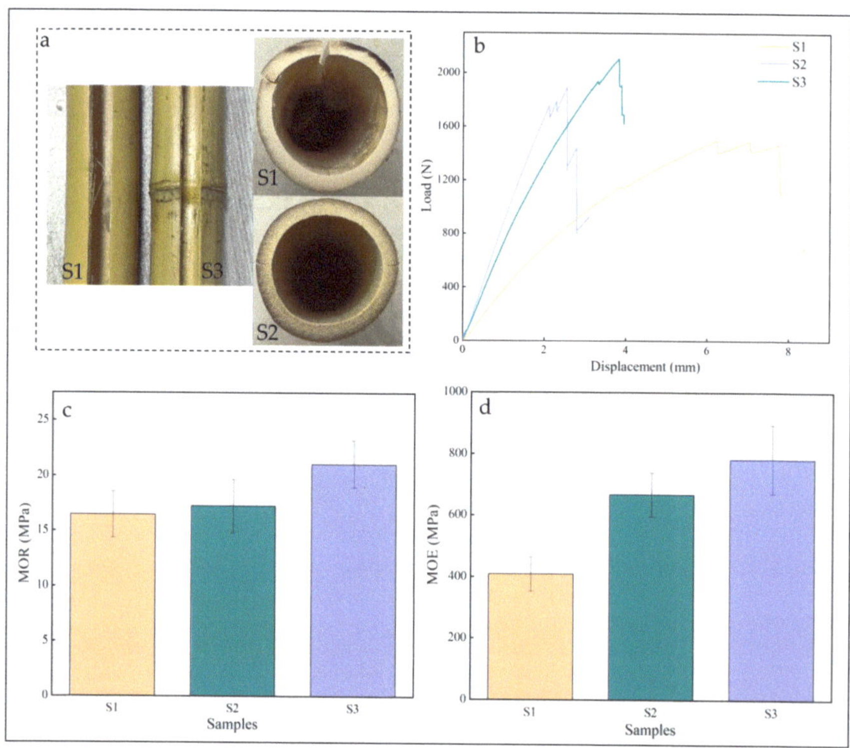

Figure 6. Bending properties of different bamboo tubes: (**a**) the failure samples, (**b**) the load displacement curve, (**c**) the modulus of rupture, and (**d**) the modulus of elasticity.

The MOR values for S1, S2, and S3 were 16.45 MPa, 17.23 MPa, and 21.05 MPa respectively, which is much lower than for bamboo strips, whose MOR values are about 150 MPa both in the bamboo node and internode species [29]. Compared to S1, S2 exhibited a 5% increase, while S3 showed an approximate 28% increase, which also shows a slightly positive effect of the bamboo node on MOR in bamboo strips [29]. Similarly, the MOE values for S1, S2, and S3 were 408.53 MPa, 667.48 MPa, and 783 MPa, respectively. Compared to S1, the MOE of S2 increased by 63% and S3 increased by 92% (Figure 6c,d). The results indicated that the presence of bamboo nodes had a slight positive effect on MOR but had a significant positive effect on MOE. In contrast, bamboo diaphragms had a notable positive effect on both MOR and MOE. These findings are consistent with the studies of Shao [18] and Yang [30].

The effects of bamboo nodes and diaphragms on the modulus of rupture (MOR) and the modulus of elasticity (MOE) are associated with structural changes introduced by these features within the bamboo tube. Bamboo's hollow tubular structure consists of cells arranged in a single longitudinal pattern at the internodes, with weak transverse connections between fiber cells [31], making the internode section susceptible to failure under bending stress. In contrast, the cell arrangement at the bamboo nodes and diaphragms is modified into a cross-linked structure that provides enhanced support for the bamboo tube. The bamboo diaphragm, in particular, reinforces the hollow tube, increasing its rigidity [32] and improving the MOE, especially in sections S2 and S3.

3.6. Tensile Properties Parallel to Grain

The tensile failure fracture of S1 occurred on the inner surface of the bamboo, resulting from the microstructural variation of the bamboo along the radial direction. The inner side is primarily composed of parenchyma cells, while the outer side is dominated by fiber cells. Since the tensile strength of parenchyma cells is significantly lower than that of fiber cells [33,34], the inner side is more susceptible to damage under tensile stress. In S2 and S3, the tensile fractures are concentrated at the bamboo nodes. Additionally, compared to S1, the fiber tear length at the tensile fractures of S2 and S3 is shorter (Figure 7a). This is attributed to the high presence of transversely arranged cells at the bamboo nodes [18], which have weak transverse bonding strength, making them more prone to tearing under tensile load.

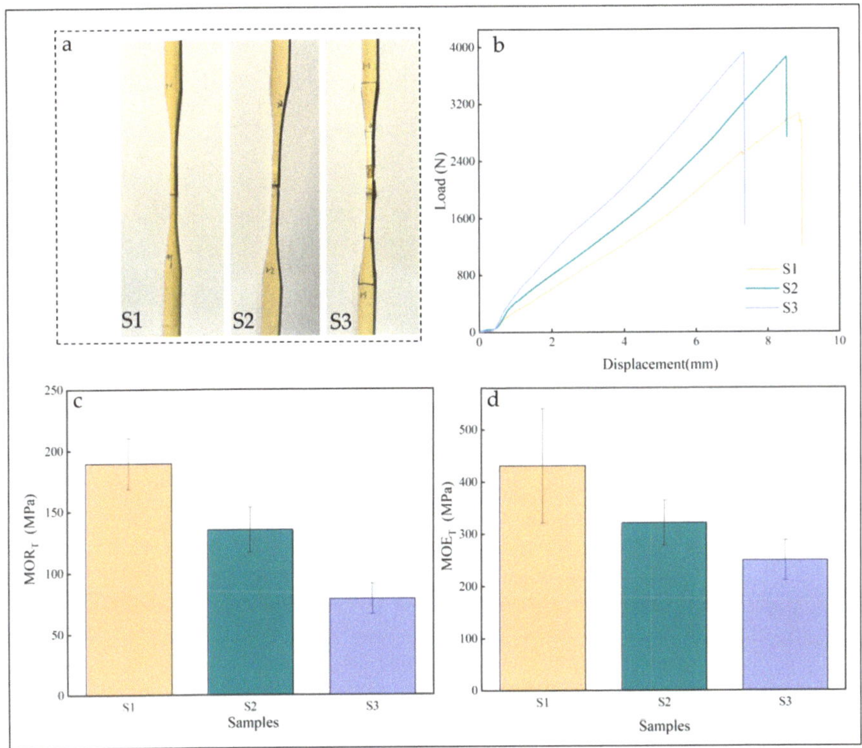

Figure 7. Tensile properties parallel to grain of different bamboo tubes: (**a**) the failure samples, (**b**) the load-displacement curve, (**c**) the modulus of rupture, and (**d**) the modulus of elasticity.

The load-displacement curves of the three specimens exhibit a similar trend. As the displacement increases, the load generally rises linearly, with the slope gradually increasing, and no distinct yield point was observed. Upon reaching the maximum load, the curve declines sharply, leading to a complete loss of bearing capacity, which indicates total failure of the specimen (Figure 7b). It is evident that the steepness for S3 is higher than that of S2, and S2 is higher than S1, suggesting that the brittleness of the bamboo node surpasses that of the bamboo internode, and the bamboo diaphragm strengthened its brittleness. The failure loads of S2 and S3 are comparable and both higher than those of S1, indicating that the bamboo nodes enhance tensile performance, while the bamboo diaphragm does not exhibit a significant positive effect.

The MOR$_T$ values for S1, S2, and S3 were 189.62 MPa, 135.16 MPa, and 78.76 MPa, respectively. Compared to S1, S2 exhibited a 29% decrease, while S3 showed an approximate 58% increase (Figure 7c). Similarly, the MOE$_T$ values for S1, S2, and S3 were 431.05 MPa, 321.02 MPa, and 249.13 MPa, respectively. Compared to S1, the MOE$_T$ of S2 decreased by 25% and S3 decreased by 42% (Figure 7d). The results indicated that both the bamboo node and the diaphragm had a significantly negative effect on MOR_T and MOE_T. The results are consistent with Qi's study in bamboo fiber-reinforced composite [35] and Wang's study in laminated bamboo lumber [36].

The negative impact of bamboo nodes and diaphragms on the modulus of rupture (MOR_T) and modulus of elasticity (MOE_T) in tension is closely related to the structural characteristics of the bamboo node section. In the bamboo internode, fiber cells and parenchyma cells are arranged longitudinally; whereas, in the bamboo node section, these cells exhibit a cross or disordered arrangement [21]. The polymer chains in fiber cells and parenchyma cells are primarily oriented along the longitudinal direction, providing the highest strength along this axis due to molecular alignment [37], resulting in the superior tensile properties of the bamboo internodes compared to the bamboo nodes and the diaphragms. The bamboo diaphragm exhibits the lowest MOR_T and MOE_T in tension likely due to the presence of weak-link areas that enlarge the tensile zone but do not positively contribute to its tensile performance.

4. Conclusions

The bamboo internode exhibits the highest tensile strength and modulus. Compared to bamboo internode, both the bamboo node and the diaphragm negatively impact tensile properties, reducing the tensile strength by 29% and 58% and the tensile modulus by 25% and 42%, respectively. While the bamboo node significantly enhances impact toughness by 29%, the bamboo diaphragm reduces it by 31%.

Compared to the bamboo internode, the bamboo node positively affects bending strength, bending modulus, radial stiffness, and shear strength, and these properties of the bamboo node can be further enhanced by the bamboo diaphragm. Among these parameters, radial stiffness showed the most significant increase, rising by 100% in the bamboo node and 250% in the bamboo diaphragm, respectively.

The bamboo node also contributes to an 11% increase in compressive strength; however, the bamboo diaphragm does not provide additional reinforcement, and it shows minimal changes compared to the bamboo node specimens.

Mechanical properties are critical reference parameters for the application of bamboo in construction. However, the influence of the bamboo's inherent characteristics on its mechanical properties remains relatively understudied, with limited theoretical foundations and incomplete testing methodologies available. This gap has led to a lack of comprehensive guidance for practical engineering applications, and relevant standards and norms are still underdeveloped. Although bamboo is increasingly recognized as an innovative material with broad applications in daily life, further research is needed to fully explore its potential in engineering contexts.

Author Contributions: Conceptualization, X.W.; formal analysis, S.Y.; data curation, S.D., R.X., and Q.C.; writing—original draft preparation, S.Y.; writing—review and editing, R.X.; project administration, X.W.; funding acquisition, X.W. and P.X. All authors have read and agreed to the published version of the manuscript.

Funding: This research was funded by the Ministry of Science and Technology of the People's Republic of China, the Natural Key R&D Program of China, grant No. 2023YFD2202101, and the National Natural Science Foundation of China, grant No. 32471976.

Data Availability Statement: During the preparation of this work, the authors used ChatGPT to polish this paper in order to make it more easily understood. After using this tool/service, the authors reviewed and edited the content as needed and take full responsibility for the content of the publication.

Conflicts of Interest: Pingping Xu is employed by BASF (China) Co., Ltd.; his employer's company was not involved in this study, and there is no relevance between this research and their company. The funders had no role in the design of this study; in the collection, analyses, or interpretation of data; in the writing of the manuscript; or in the decision to publish the results.

References

1. Li, H.; Zheng, X.; Guo, N.; Sheng, Y. *Modern Bamboo & Wood Buildings*; Architecture Publishing & Media Co. Ltd.: Beijing, China, 2020.
2. Yi, P.; Li, J.; Li, J.; Qin, N.; Tang, L.N. Application of the Modern Bamboo in Building Structure. *J. Chongqing Jianzhu Univ.* **2011**, *33* (Suppl. S2), 115–118.
3. Jia, S.; Wang, Y.; Wei, P.; Ma, X.; Wu, Y. Structure and Mechanical Properties of Bamboo Nodes: A Review of Current Research. *World Bamboo Ratt.* **2024**, *22*, 90–99.
4. Luan, Y.; Yang, Y.; Jiang, M.; Liu, H.; Ma, X.; Zhang, X.; Sun, F.; Fang, C. Unveiling the mechanisms of Moso bamboo's motor function and internal growth stress. *New Phytol.* **2024**, *243*, 2201–2213. [CrossRef]
5. Li, Z.; Chu, S.; Qin, L.; Lin, L. Influence of node on the bonding properties of glued laminated bamboo. *J. For. Eng.* **2022**, *7*, 80–85.
6. Xia, M.; Chen, A.; Yu, F. The Effect of the Node on the Development of Ground Tissue Cell Walls during the Rapid Elongation Growth of *Phyllostachys edulis* 'Pachyloen' Culm. *Acta Agric. Univ. Jiangxiensis* **2018**, *40*, 1178–1184.
7. Chen, Q.; He, Y.; Jiang, Y.; Qi, J.; Zhang, S.; Huang, X.; Chen, Y.; Xiao, H.; Jia, S.; Xie, J. Effect of bamboo nodes on crack generation of round bamboo and bamboo-based composites during drying. *Eur. J. Wood Prod.* **2023**, *81*, 1201–1210. [CrossRef]
8. Shi, J.; Li, Z.; Chen, H.; Wu, Z.; Ji, J.; Xia, C.; Zhong, T. Tunable bending characteristics of bamboo by regulating moisture content for bamboo curved component manufacturing. *Ind. Crops Prod.* **2024**, *221*, 119365. [CrossRef]
9. Chen, Q.; He, Y.; Lai, S.; Qi, J.; Zhang, S.; Jia, S.; Xiao, H.; Chen, Y.; Jiang, Y.; Fei, B.; et al. 3D characterization of vascular bundle in moso bamboo node and its effect on compressive properties. *Holzforschung* **2023**, *77*, 368–377. [CrossRef]
10. Xu, K. Experimental Study on the Mechanical Properties of Round Bamboo-Light Mortar Composite Wall. Master's Thesis, Xi'an University of Architecture and Technology, Xi'an, China, 2014.
11. Widjaja, E.; Riayad, Z. Anatomical. Properties of Some Bamboos Utilized in Indonesia. In *Chouinard Bamboo Research in Asia*; Lessard, G., Ed.; IDRC: Ottawa, ON, Canada, 1980; pp. 244–249.
12. Murphy, R.J.; Alvin, K.L. Variation in fiber wall structure in bamboo. *IAWA Bull.* **1992**, *13*, 403–410. [CrossRef]
13. Ghavami, K.; Garcia, J.J.; Torres, L.A. A Transversely isotropic law for the determination of the circumferential Young's Modulus of bamboo with diametric compression tests. *Lat. Am. Appl. Res.* **2007**, *37*, 255–260.
14. *LY/T2564-2015*; Test Method for Physical and Mechanical Properties of Round Bamboo. Standards of China: Beijing, China, 2015.
15. *GB/T 1927.17-2021*; Test Methods for Physical and Mechanical Properties of Small Clear Wood Specimens-Part 17: Determination of Impact Bending Strength. General Administration of Quality Supervision, Inspection and Quarantine of the People's Republic of China: Beijing, China, 2021.
16. Hao, J.; Qin, M.; Tian, L.; Liu, M.; Zhao, Q.L. Experimental research on the mechanical properties of *Phyllostachys pubescen* along the grain direction. *J. Xi'an Univ. Archit. Technol.* **2017**, *49*, 777–783.
17. Lin, F.; Chen, Y.; Dai, Z.; Liu, M.; Zhang, Q. Experimental Study on the Effect of Bamboo Nodes on the Pressure Resistance of Bamboo. *J. Bamboo Res.* **2019**, *38*, 72–77.
18. Shao, Z.; Huang, S.; Wu, F.; Liu, Y.M.; Arnaud, C. A Study on the Difference of Structure and Strength between Internodes and Nodes of Moso Bamboo. *J. Bamboo Res.* **2008**, *27*, 48–52.
19. Mouka, T.; Dimitrakopoulos, E.G.; Lorenzo, R. Effect of a longitudinal crack on the flexural performance of bamboo culms. *Acta Mech.* **2022**, *233*, 3777–3793. [CrossRef]
20. Yu, J.; Hao, J.; Tian, L. The study on the main influencing factors and mechanical properties of *Phyllostachys pubescens*. *J. Xi'an Univ. Archit. Technol.* **2018**, *50*, 30–36.
21. Han, S.; He, Y.; Ye, H.; Chen, F.; Liu, K.; Shi, Q.S.; Wang, G. Mechanical Behavior of Bamboo, and Its Biomimetic Composites and Structural Members: A Systematic Review. *J. Bionic. Eng.* **2024**, *21*, 56–73. [CrossRef]
22. Al-Rukaibawi, L.S.; Károlyi, G. Through-thickness distribution of bamboo tensile strength parallel to fibres. *SN Appl. Sci.* **2023**, *5*, 174. [CrossRef]
23. Li, S.; Liu, C.; Wang, Y.; Shang, L.; Liu, X.; Wang, S.; Yang, S. Three-dimensional visualization of the conducting tissue in a bamboo culm base. *Wood Sci. Technol.* **2024**, *58*, 1585–1603. [CrossRef]
24. Li, X.; Zhong, T.; Chen, H.; Li, J. Chemical composition and thermal stability of cells in different structures of *Phyllostachys edulis*. *J. Beijing For. Univ.* **2023**, *45*, 156–162.
25. Pi, X.; Tie, Y.; Hu, M. Anti-impact performance of plain woven composite patching structure based on multi-scale analysis. *J. Vib. Shock.* **2022**, *41*, 188–197.
26. Pruksawan, S.; Lim, J.W.R.; Lee, Y.L.; Lin, Z.; Chee, H.L.; Chong, Y.T.; Chi, H.; Wang, F. Enhancing hydrogel toughness by uniform cross-linking using modified polyhedral oligomeric silsesquioxane. *Commun. Mater.* **2023**, *4*, 75. [CrossRef]
27. Qin, W.; Yu, H.; Xu, M.; Zhuang, X.; Wang, H.; Ying, M.; Pan, X.; Liang, Y. Effects of microwave softening treatment on dynamic mechanical and chemical properties of bamboo. *J. Mater. Sci.* **2024**, *59*, 3488–3503. [CrossRef]
28. Mehramiz, S.; Oladi, R.; Efhamisisi, D.; Pourtahmasi, K. Natural durability of the Iranian domestic bamboo (*Phyllostachys vivax*) against fungal decay and its chemical protection with propiconazole. *Eur. J. Wood Prod.* **2021**, *79*, 453–464. [CrossRef]
29. Shi, J.; Li, Z.; Chen, H.; Wu, Z.; Ji, J.; Xia, C.; Wang, H.; Zhong, T. Optimizing processing strategies for eco-friendly bamboo curved components: Insights from bamboo internode and node differences. *Ind. Crops Prod.* **2024**, *216*, 118823. [CrossRef]
30. Yang, X.; Zhang, F.; Huang, Y.; Fei, B. Tensile and bending properties of radial slivers of Moso bamboo. *J. Beijing For. Univ.* **2022**, *44*, 140–147.

31. Liu, C.; Xizhi, W.; Li, X.; Liu, X. Structure and physical properties of high-density bamboo scrimber made from refined bamboo bundles. *Eur. J. Wood Prod.* **2024**, 1–12. [CrossRef]

32. Ma, Z.; Cai, C.; Yin, Y.; Dang, J.; Ming, W.; An, Q.; Chen, M.; Liu, G.; Li, J. Vibration characteristics and machining performance of carbon fiber reinforced shaft in poor rigidity machining tool system. *Int. J. Adv. Manuf. Technol.* **2024**, *134*, 2637–2652. [CrossRef]

33. Chan, C.H.; Wu, K.J.; Young, W.B. The effect of densification on bamboo fiber and bamboo fiber composites. *Cellulose* **2023**, *30*, 4575–4585. [CrossRef]

34. Su, Q.; Huang, A.; Chen, X.; Dai, C.; Fei, B.; Fang, C.H.; Ma, X.; Sun, F.; Zhang, X.; Liu, H. Anisotropic tensile performance of bamboo parenchyma tissue and its influencing factors. *Cellulose* **2023**, *30*, 9147–9160. [CrossRef]

35. Qi, J.; Xie, J.; Yu, W.; Chen, S. Effects of characteristic inhomogeneity of bamboo culm nodes on mechanical properties of bamboo fiber reinforced composite. *J. For. Res.* **2015**, *26*, 1057–1060. [CrossRef]

36. Wang, Z.; Li, H.; Yang, D.; Xiong, X.; Sayed, U.; Lorenzo, R.; Corbi, I.; Corbi, O.; Hong, C. Bamboo node effect on the tensile properties of side press-laminated bamboo lumber. *Wood Sci. Technol.* **2021**, *55*, 195–214. [CrossRef]

37. Xue, J.T.; Bai, Y.; Peng, L.; Huang, X.B.; Sun, Z.Y. Exploring the Interplay between Local Chain Structure and Stress Distribution in Polymer Networks. *Chin. J. Polym. Sci.* **2024**, *42*, 874–885. [CrossRef]

MDPI AG
Grosspeteranlage 5
4052 Basel
Switzerland
Tel.: +41 61 683 77 34

Forests Editorial Office
E-mail: forests@mdpi.com
www.mdpi.com/journal/forests